Parental Behaviour

For Wladek's wife Alice, and sons Tim and Andy

Wladyslaw (Wladek) Sluckin died during the final stages of preparation of this book. I lost a distinguished mentor and a dear friend.

Martin Herbert

Parental Behaviour

Edited by
Wladyslaw Sluckin
and Martin Herbert

Basil Blackwell

© Basil Blackwell Ltd 1986

First published 1986

Basil Blackwell Ltd
108 Cowley Road, Oxford OX4 1JF, UK

Basil Blackwell Inc.
432 Park Avenue South, Suite 1503,
New York, NY 10016, USA

All rights reserved. Except for the quotation of short passages for the purposes of criticism and review, no part of this publication may be reproduced, stored in a retrieval system, or transmitted, in any form or by any means, electronic, mechanical, photocopying, recording or otherwise, without the prior permission of the publisher.

Except in the United States of America, this book is sold subject to the condition that it shall not, by way of trade or otherwise, be lent, re-sold, hired out, or otherwise circulated without the publisher's prior consent in any form of binding or cover other than that in which it is published and without a similar condition including this condition being imposed on the subsequent purchaser.

British Library Cataloguing in Publication Data
Parental behaviour.
 1. Psychology, Comparative 2. Human behavior 3. Mammals — Behavior
 I. Sluckin, W. II. Herbert, Martin
 156   BF671
ISBN 0-631-13487-5

Library of Congress Cataloging in Publication Data
Parental behaviour.
 Includes index.
 1. Parenting. 2. Parental behavior in animals.
 3. Parent and child. I. Sluckin, W. II. Herbert, Martin.
 HQ755.8.P377 1986   155.6'46   86-6846
ISBN 0-631-13487-5

Typeset by DMB (Typesetting) Oxford
Printed in Great Britain by T. J. Press Ltd., Padstow

Contents

	Acknowledgement	vi
1	A Comparative View of Parental Behaviour W. Sluckin and M. Herbert	1
2	Parental Behaviour in Birds K. Immelmann and R. Sossinka	8
3	Parental Behaviour in Rodents J. C. Berryman	44
4	Parental Behaviour in Ungulates D. G. M. Wood-Gush, K. Carson, A. B. Lawrence and H. A. Moser	85
5	Parental Care in Carnivores O. A. E. Rasa	117
6	Parental Behaviour in Non-human Primates J. D. Higley and S. J. Suomi	152
7	Human Mother-to-Infant Bonds W. Sluckin	208
8	The Role of the Father in the Human Family C. Lewis	228
9	Substitute Parenting M. Shaw	259
10	Parental Responsiveness and Child Behaviour H. R. Schaffer and G. M. Collis	283
11	The Pathology of Human Parental Behaviour M. Herbert	316
12	Methods and Approaches to the Study of Parenting K. Browne	346
	Notes on Contributors	376
	Index	378

Acknowledgement

I would like to thank Margaret Frape for her invaluable assistance in preparing the manuscript.

Martin Herbert

1 A Comparative View of Parental Behaviour

W. Sluckin and M. Herbert

CARE-GIVING BEHAVIOUR AND
PARENTAL ATTACHMENT

Most birds and all mammals care in some way for their offspring. In birds this caring involves incubating eggs, feeding their immature young, keeping them warm and protecting them; but some bird species do none of this and many species exhibit only some of these modes of behaviour. In mammals the suckling of young by the females is the most prominent feature of care-giving. Parental behaviour in mammals also manifests itself, of course, in many other ways: in the provision of nourishment other than mother's milk, in the provision of other 'creature comforts' and protection from danger. Furthermore, parental behaviour may also involve 'the transfer of information' from parent to offspring; in the human species this is particularly marked and refers, broadly speaking, to the socialization of the young (see chapters 10 and 11).

We have set out in this volume to 'sample' parental behaviour in a selected range of animal species, and decided to deal only with certain aspects of parental behaviour in human beings. Thus, we have made no attempt to be encyclopaedic in our review. We have adopted this approach in order to highlight and discuss some general features of parenting and to draw, as far as possible, some broad conclusions concerning the nature of parental care.

In many species parental behaviour is the prerogative of the female; however, in many species males participate in care-giving, sometimes, even playing the leading role (see chapters 6 and 8). Generally where monogamy is the rule, as for instance in geese or foxes, parenting duties are divided more or less equally. In some species adult individuals tolerate no young other than their own; non-parental care of the young, however, is common among primate species, and the phrase 'aunting behaviour' is often used in the relevant literature to denote the mothering of infants which are not the caring female's own offspring.

In avian and mammalian species parental behaviour is influenced to some degree, and often very considerably, by the behaviour of the young

– the signals they give to the adults. Thus, parental functioning must be viewed in the context of the parent–offspring interaction. This is strikingly so in the human species (see chapters 10 and 11). While any species' anatomical structures and physiological functioning are relatively slow to adapt to changing environmental conditions, behaviour patterns, including parental behaviour, often change readily and rapidly in response to new environments.

Most patterns of parenting are rooted in genetics, but learning, however, can play an important role in the ontogeny of parental behaviour in many species. In our own, learning is of course most influential. Can the study of parental behaviour in non-human species help us understand the nature of human care-giving behaviour? Can cases of deficiency in parental functioning in human beings be better understood in the light of the relevant investigations of animal behaviour? Clearly, we shall be better able to answer such questions by the last chapter of this book rather than in the first. At any rate, although the study of care-giving and parental behaviour both in animals and in the human species is a sufficiently worthwhile aim in itself, we may at the outset express the hope that inter-species comparisons will prove factually informative and theoretically fruitful.

At this stage we must draw a distinction between, on the one hand, general *care-giving behaviour* directed towards the young and helpless of the animal's own species, that is, parental behaviour in the broadest sense, and, on the other hand, the provision of care specifically for the animal's own, or adopted, young, that is to say, *parental behaviour* in the restricted sense (see, for example, chapter 9). The former may involve varying degrees of learning, from virtually none to a great deal. The latter, however, that is specific parental behaviour, must clearly depend on the learning which is necessary to bring about the recognition that some particular young are different from all others, and that they deserve a treatment which is entirely special. The attention and care bestowed upon these particular infant individuals suggest that these are cases where parent-to-infant bonds of some kind have been established. Behaviour of this kind may be referred to as *parental attachment*; this must be distinguished from infant-to-parent attachment. The latter may well involve the process of imprinting; whereas in the case of parental attachment it would be inappropriate – as we shall see in chapters 6 and 7 – to invoke rapid olfactory or tactile imprinting as an explanation of the parent-to-infant bonds either in animals or in human beings.

The question of explaining parental behaviour and parental attachment may be tackled from a number of angles. It has been clear for a long time that 'why' questions about any kind of behaviour may refer to that behaviour's function or its phylogeny, or its causation, or its ontogeny (Tinbergen, 1951, 1963). The function of any particular behaviour refers to the usefulness of that behaviour to the organism – to its survival value. Phylogeny is concerned with the evolutionary origins of the behaviour in question. When we talk of the causation of behaviour, we generally have in mind the stimuli, or environmental cues, that precede and evoke the given behaviour pattern. Lastly, questions of ontogeny are questions concerning the development of the individual, that is, both maturation and learning. We shall now make some introductory observations about parental behaviour in relation to each of these different types of question.

THE FUNCTION OF PARENTAL BEHAVIOUR

In the study of any type of behaviour, one of the problems, perhaps the very first problem, that can present itself is that of the function or purpose of the particular behaviour. In the case of parental behaviour the answer is not far to seek: parental behaviour ensures the survival of the young and, thus, the preservation of the species. As noted earlier, not all bird species rely on parental behaviour for their survival, although all mammalian species do. The European cuckoo and American cowbird, for instance, lay their eggs in the nests of other birds and themselves show no parental behaviour of any kind; the survival of their young is ensured by the parental behaviour of the foster parents. Parental behaviour proper is also absent in megapodes, such as the Australian jungle fowl and brush turkeys; here the adults show care-giving behaviour towards their eggs, but no parental behaviour towards their young, who are able to fend for themselves driectly after hatching. Nevertheless, in most bird species parental behaviour is very conspicuous and chapter 2 is devoted to its detailed description and discussion.

In mammals maternal care is ubiquitous and it includes suckling the infants. The latter is, of course, a *sine qua non* of survival. But in addition, as we see in chapters 6 and 10, the chances of the young reaching maturity are greatly increased by other forms of care-giving, such as retrieving the young, carrying them, grooming them, playing with them, and so on. These other manifestations of care-giving are often displayed not only by mothers but also by virgin females. In many species adult males, too, show various forms of caring for the young.

How mothers and others take care of the young depends on the total adjustment of the species to its environment: how its members feed, what dangers threaten them, and so on. Chapter 5, for instance, highlights the relationship between the modes of parenting and the modes of hunting in the many different carnivorous species. Some species can only survive if the male feeds the lactating female and otherwise helps to rear the young, whilst some species, which hunt in packs, have developed a system of communal rearing of their young. In very many species, however, the mother alone takes care of her offspring (see chapters 3 and 4). Different types of parenting are rooted in different modes of life in general; that is, the mode of parenting is related to the particular ecological niche which is occupies by the given species. Broadly speaking, the consequence of each style of parenting is the maximization of the probability of survival of the young (see Gubernick and Klopfer, 1981). However, modes of behaviour that act against survival, such as cannibalism, also sometimes occur, but in circumstances of privation such modes of behaviour are not necessarily pathological. But maladaptations of parental behaviour do exist both in animals and in humans (see chapter 11). One of the tasks of comparative psychology is to try to explain the function of both normal and abnormal parental functioning.

THE PHYLOGENY OF PARENTAL BEHAVIOUR

It is clear that descendants of individuals that are deficient in care-giving behaviour have a poor chance of survival, and so this deficiency is not likely to perpetuate itself. Conversely pronounced care-giving behaviour, provided that it is not so extreme as to endanger the life of the care-giver, will promote the survival of offspring, and is consequently maintained generation by generation. Thus, on the face of it, the phylogeny of parental behaviour is readily explained in evolutionary terms.

However, much parental behaviour may be described as altruistic in that it benefits the young but entails some degree of sacrifice on the part of the care-giving individual. How, then, can such self-sacrificing tendencies be genetically transmitted? The problem is, of course, that altruistic adults are not the ones that are the most likely to continue to live and procreate; such individuals have less than an even chance of survival and hence of passing on their altruistic proclivities to future generations. To solve this puzzle, psychobiologists have advanced some ingenious theories (Dawkins, 1976). It has been said that the phylogeny of altruistic behaviour directed towards close blood relatives can be much more readily understood than can altruistic behaviour in cases in which relatives are not involved (Hinde, 1974).

In general terms, theories of altruism in animals try to show how acts of self-sacrifice, which as such are disadvantageous to the individual, can at the same time confer certain hidden selective advantages. There is plenty of scope for theorizing in this area. For instance, it might be suggested that a suitable model for the inheritance of altruism is that of the human sickle-cell gene, which bestows on its carrier an organic weakness but also provides her/him with resistance to malaria. Perhaps the genes which underlie altruistic tendencies help survival in some ways but hinder it in other ways; but on balance help more than hinder. There is some scope for testing theorizing of this kind in relation to animal behaviour, although this is not an easy matter. As for human behaviour, it is very doubtful whether psychobiology can materially contribute to our understanding of the complexities of human parental activities (Mussen, 1982).

THE CAUSATION OF PARENTAL BEHAVIOUR

Parental behaviour is directed towards certain suitable (and sometimes unsuitable) objects, and it is elicited by certain specific stimuli. What, then, are the characteristics of the stimuli that evoke or 'release' parental behaviour? This is, of course, the type of question to which ethologists have traditionally addressed themselves.

Most mammalian infants share certain features, such as a proportionately large head, relatively thick extremities, and so on, which are thought to be cues to adult care-giving behaviour, at least in human beings (for example, Eibl-Eibesfeldt, 1970); evidence for this view is strongly suggestive rather than firmly established experimentally.

General care-giving behaviour may be brought about in different species by a wide variety of olfactory, visual, or auditory stimuli. Selective parental behaviour, directed towards the adult's own or adopted offspring is elicited by sensory cues which are, of course, narrowly specific. Thus, crying in human babies is a very potent releaser of maternal behaviour and, as such, has been extensively studied. Furthermore, there is good evidence that mothers can rapidly learn to discriminate between the crying of their own and other infants (Wiesenfeld and Malatesta, 1982) and this also appears to be the case in some infra-human primates. At any rate, it can be said that non-specific maternal behaviour is readily elicited by neonates in the adults of most primate species (Swartz and Rosenblum, 1981).

THE ONTOGENY OF PARENTAL BEHAVIOUR

This is an area where the nativist-environmentalist debate has considerably influenced both the direction of empirical studies and the reported findings. There have been, for example, reports of profound environmental effects in the development of maternal behaviour in rats. However, it has been cogently argued that the relevant experimental results are open to more than one interpretation (Hinde, 1970). On the whole, in infra-primate species the emergence of parental behaviour is a matter of maturation. On the other hand, in human beings in particular, there appears to be a large element of social learning in the development of parental functioning.

The emergence of maternal and paternal behaviour in mammals shows striking variations from species to species. Some mammals build nests for their offspring while others do not. Among those that do, females of some species build before parturition, while females of other species build after parturition; in some species males help with nest building and in others they do not (Moltz, 1971). The interaction of parents and infants develops as a function of changing physiological conditions of the adults and also depends on the changing stimulus characteristics of the young.

THE COMPARATIVE PERSPECTIVE

Starting with the incubation process in birds and with parturition in mammals, species differences in terms of parental behaviour are very striking indeed. As mentioned earlier, there are enormous variations in the degree of help provided by the male. In some bird species the males, and not the females, incubate the eggs. In many species, males participate in the incubation of eggs and in the feeding of the nestlings. The urge to feed their own (and other) young appears to be very strong in most birds and is genetically programmed (Challinor, 1983). This tendency may also be seen in many carnivorous mammals. In some mammalian species, assistance with parturition is provided by conspecifics, both female and male. The help given by some human fathers at and after birth may have some genetic basis, although cultural influences, as is shown in chapter 8, are clearly the dominant factor in determining the role of the father in different human parenting patterns.

Despite the vast inter-species variations, many features of parental behaviour are characteristic of more than a single species. Shared features of behaviour may derive from common ancestry or from parallel, but independent, evolutionary developments. Comparative studies of parental behaviour can bring out clearly the existing inter-specific similarities and differences. Where there is a large element of learning in parental behaviour, the comparative approach can help to indicate which features of behaviour have genetic roots and which have not. As Hinde (1970) points out, the differences and the similarities in behaviour revealed by comparative studies are both of interest and of significance. Comparative studies of diverse species, including our own, can help shed light on problems of function, phylogeny, causation and ontogeny of parental behaviour.

There are dangers, however, in expecting too much from the comparative approach. Extrapolating general principles from observations of behaviour of one particular species can be greatly misleading. Popular ethology has often loosely applied animal behaviour concepts such as territoriality, or imprinting, to the explanation of human behaviour.

Hinde (1982) points out how tempting, and how unjustifiable, it would be to assume that (for instance) contact comfort plays the same role in human mother–infant interaction as it does in the lives of non-human primates. As Lehrman (1974) has argued, virtually any conclusion can be reached about human functioning by comparing selected aspects of human and animal behaviour. At the same time, the lack of the comparative perspective can lead – and has led in the past – to erroneous interpretations of experimental results. Thus, biologically rooted behaviour can be too readily assumed to be wholly socially determined. Comparative studies of behaviour in general, and of parental behaviour in particular, can, by broadening one's horizons, be helpful in enabling the research worker to abstract general principles and to avoid overgeneralizations. We shall see how this works in practice in the succeeding chapters.

REFERENCES

Challinor, D. 1983: Nature's midwives and nursemaids. *New Scientist*, 99, 641.
Dawkins, R. 1976: *The Selfish Gene*. Oxford: Oxford University Press.
Eibl-Eibesfeldt, I. 1970: *Ethology*. New York: Holt, Rinehart and Winston.
Gubernick, D. J. and Klopfer, P. H. (eds) 1981: *Parental Care in Mammals*. New York: Plenum Press.
Hinde, R.A. 1970: *Animal Behaviour*. Tokyo: McGraw-Hill Kogakusha.
Hinde, R. A. 1974: *Biological Bases of Human Social Behaviour*. New York: McGraw-Hill.
Hinde, R. A. 1982: The uses and limitations of studies of non-human primates for the understanding of human social development. In L. W. Hoffman, R. Gandelman and H. R. Schiffman (eds), *Parenting, its Causes and Consequences*, Hillsdale, New Jersey: Erlbaum.
Lehrman, D. S. 1974: Can psychiatrists use ethology? In N. F. White (ed.), *Ethology and Psychiatry*, Toronto: Toronto University Press.
Moltz, H. 1971: The ontogeny of maternal behavior in some selected mammalian species. In H. Moltz (ed.), *The Ontogeny of Vertebrate Behavior*, New York: Academic Press.
Mussen, P. 1982: Parenting, prosocial behaviour and political attitudes. In L. W. Hoffman, R. Gandelman and H. R. Shiffman (eds), *Parenting, its Causes and Consequences*, Hillsdale, New Jersey: Erlbaum.
Swartz, K. B. and Rosenblum, L. A. 1981: The social context of parental behaviour. In D. J. Gubernick and P. H. Klopfer (eds), *Parental Care in Mammals*, New York: Plenum Press.
Tinbergen, N. 1951: *The Study of Instinct*. Oxford: Oxford University Press.
—— 1963: On aims and methods of ethology. *Zeitschrift fur Tierpsychologie*, 20, 410–433.
Wiesenfeld, A. R. and Malatesta, C. Z. 1982: Infant distress: variables affecting responses of caregivers and others. In L. W. Hoffman, R. Gandelman and H. R. Shiffman (eds), *Parenting, its Causes and Consequences*, Hillsdale, New Jersey: Erlbaum.

2 Parental Behaviour in Birds

K. Immelmann and R. Sossinka

An evaluation of parental care, as of other behaviour systems, is possible only if the specific adaptive situation is considered of the particular taxonomic group of animals. This chapter, therefore, will try to elucidate the characteristic features of parental care in birds and to correlate them to the specific morphological and physiological attributes and ecological needs of this class of vertebrates. As the rest of the book is devoted to non-human and human mammals, special attention will be given to a comparison between the two homoiothermic groups of animals, that is, birds and mammals. The aim of this chapter is to demonstrate that the mammalian system of parental care is only one of many possibilities realized in the animal world and to show that even within the two fairly closely related and comparatively advanced groups of vertebrates very different solutions to the necessity of raising offspring can be found. These in turn do have profound effects on the development of the offspring and even on their adult life.

1 PARENTAL CARE IN BIRDS AND MAMMALS - A COMPARISON

Except for their feathers, the capacity of flight is the main characteristic of the vertebrate class of birds. The ability to fly, which probably evolved from bipedal running and jumping with balancing wing movements (Caple et al., 1983), has led to a multitude of special adaptations which can be found in almost all aspects of morphology and physiology. The main selection pressure on birds – except for those species which have become flightless again – lies on a reduction in body weight.

With regard to parental care, the two most important consequences of this selection pressure are a total lack of viviparity and, with few exceptions, the absence of the use of parental body substances as a food source for the offspring. Such constraints not only lead to differences during embryonic development between birds and mammals but also cause a multitude of consequences for postnatal developmental processes and even for aspects of adult behaviour.

The most obvious consequence refers to the distribution of 'parental duties'. In mammals, the female – by means of the developing embryo until birth and of lactation thereafter – 'automatically' has the greater share in such duties and in many species does raise the offspring all on her own. Contrastingly, in birds, except for egg production, all duties, theoretically, can be performed by both sexes. Parental care therefore is more evenly distributed in birds than it is in mammals. This in turn has led to differences in the prevailing mating systems: in most species of mammals where the male has few or no parental duties the male is 'free' to engage in other behaviours, for example in attracting and courting more than one female. Therefore, different types of polygyny are widespread in mammals, whereas monogamy is comparatively rare. However, in birds polygyny is the exception, and monogamy is by far the most widespread mating system; more than 90 per cent of all species of birds have been found to be monogamous. But polygamy and occasionally polyandry also do occur (Emlen and Oring, 1977).

With regard to parental care itself, lack of viviparity and lactation in birds have the following consequences:

a Birds have to provide shelter for their eggs. Most species construct more or less sturdy nests or breed in holes. Only some ground and rock nesters place the eggs on the bare ground or, as do penguins and guillemots, on their feet. In mammals, on the other hand, it is mainly the small altricial species which construct nests or use burrows, whereas in the majority of precocial species no nest or equivalent is used at all.

b Birds have to provide external heat for the eggs and, for a certain period of time after hatching, also to the young. Although the nest may help to maintain high temperatures, the actual source of heat – except for the megapodes (see section 10) – is provided by one parent or by both the parents, which, through a special differentiation of the ventral skin (the so-called brood patch), transfer the necessary heat to the eggs or young. This in turn requires the almost constant presence of one parent on the nest. Mammalian embryos, in contrast, are automatically provided with heat and the female is free to engage in other activities.

c The avian embryo, due to its development within the egg and apart from the mother, unlike placental mammals, has no access to the maternal circulatory system. As a consequence, hormonal influences by the mother on the developing embryo, so pronounced in mammals (Erhardt and Hever-Bahlburg, 1981) are not possible in birds. On the other hand, the avian embryo can be more directly exposed to sensory stimulation, and cases of acoustic communication from egg to egg and between embryos and parents, and even of prenatal learning, have been described (see section 6c).

d Birds, with very few exceptions (see section 5c), have to collect food, carry it to the nest and present it to their offspring. This requires efficient methods of food collection and special behaviour adaptations to food transfer (begging behaviour of young, as well as mechanisms of transfer, for example regurgitation, in the adults). In mammals, the female has little to do other than adopt her nursing position or, in primates, assist the suckling young by supporting it with her hand. It is only in the case of older offspring of some precocial mammals, mainly carnivores, that additional food is actively carried to the young.

Apart from attributes which are directly correlated with parental care there is one other difference between birds and mammals which has indirect consequences in exerting an influence on the duration of parental care: the most remarkable characteristic of the ontogenetic development of birds is its rapidity; birds may reach their adult size and weight within about one per cent of their total life expectancy, whereas some mammals need 30 per cent or more, as is the case in most of the higher primates (Frazer, 1977; Rasa, chapter 5, this book). Such speed in general development has commonly been understood to be an adaptation to the need for a quick acquisition of the ideal ratio between body weight and wing surface (Mayr, 1963).

As a consequence of their rapid development, birds usually disperse from their parents quite early and so parental duties are restricted to a fairly brief period of time. Among mammals there does not seem to be as strong a selection pressure for early dispersal as in birds. Mammalian infants, therefore, stay with their parents (especially with the mother) much longer than do avian offspring, and in many species far beyond nutritional weaning. Probably such prolonged contact provides the offspring with additional opportunities to learn further survival techniques about, for instance, food sources or social relationships. In birds, by contrast, the time available for learning is usually much shorter and it is certainly not by chance, therefore, that the much discussed phenomenon of imprinting is especially widespread in avian species (Immelmann and Suomi, 1981; Mason, 1979) (see section 6c).

From this listing of differences between birds and mammals it follows that parental care in birds must be characterized by a number of complex special adaptations, which will be described below.

2 NEST-BUILDING

As outlined in the introduction to this chapter, the main function of the nest is to facilitate the transfer of heat into the egg and thus provide a homiothermic environment. The energy derives from the parents, except in those few cases where external heat sources are used (Megapodidae, see section 10). A further aim of constructing a nest is to provide shelter from predators, from unfavourable climatic conditions and from physical dangers. In addition, the act of nest building itself has a role; being part of the courtship behaviour in some species, it allows one potential partner to estimate the value of the other. Furthermore, the building activity itself has some influence on the release of hormones and thus synchronizes a breeding pair as well as autocatalyses the succession of reproductive activities within a single parental bird (Hinde, 1970; Sossinka et al., 1980).

Bearing in mind the main tasks of a nest (as a heat transfer-box and a shelter) the immense variety of details of nest-building behaviour in birds boils down to some general strategies. Given a certain climate and equipment, a bird has not very much choice as to what type of nest should be constructed and which technique should be used. But there are of course several other factors influencing nest building, for example the particular mating system (which often is a result of nutritional conditions) and the phylogenetic history of the species. It should be noted, however, with regard to phylogeny, that in many families or sub-families, species resemble one another in their technique of nest construction, but certain species within a family may be adapted to very different habitats and will have therefore developed quite different modes of nesting behaviour. For example, whereas most gulls mould a simple depression in sandy or stony ground and gather a few twigs or grasses only for 'upholstery', the cliff-nesting kittiwake (*Rissa tridactyla*) builds a perfect platform out of vegetation and forms a barrier preventing the egg from falling off the ledge.

Nest-building birds manage nearly all imaginable crafts: they select, gather and transport materials, they mould, pile up, weave and knot them, they model and glue, they insulate, mend and repair the nest. Materials used can be small stones, twigs, grass, strings of leaves, plant wool, mud, saliva, down, spider webs or cactus-spines. Despite the birds having gained the air space to live in, (some species spend most of their time in the air, feeding, sleeping and even copulating without touching the ground), they all need a substrate to deposit their eggs for incubation, high up in a tree, on the ground, or underground several feet deep inside a hole.

The duration of nest-building activities vary between species, from no time-investment to highly time-consuming elaborate constructions. While in some species only a decision when[1] and where to place the egg is required, other species need extensive instruction for the construction

[1] In this chapter, the problem of timing of reproduction will not be mentioned as it is reviewed elsewhere.

itself, as well as for the orientation of the nest, (for example, of the entrance, of the optimal site in relation to rain-shielding leaves and in relation to objects providing shade or shelter from wind). No nests are built by the Guillemot (*Uria aalge*) and some of its relatives since they nest on small ledges of the cliff and deposit their conical eggs on the bare rock. Very poor nests are constructed by some sand-dwelling species, for example the male ostrich (*Struthio camelis*) moulds a shallow depression in the sand to aggregate the great number of eggs, produced by several females. The ringed plover (*Charadrius hiaticula*) and related species have a little mould in wasteland or coastal areas. Protection against predators is provided by the excellent cryptic coloration of the eggs. Many sandpipers, terns and gulls make similar moulds, which are generally upholstered with pieces of shell or plant material. A much better upholstery is provided by ducks, which put a layer of down plucked from their own breasts inside the mould. The nests are particularly solid among the cold-adapted sub-arctic species, for example the common eider (*Somateria mollissima*), having far better thermal insulation (which is why eider nests are harvested by humans).

Those species living in marshlands, the black-headed gull (*Larus ridibundus*) or stilt (*Haemantopus haemantopus*) for example, make heaps out of plant material and thus raise the nest platform in case the water level rises. Piles of reeds are used by grebes and are constructed so that they can float if necessary. Heaps of branches and twigs are piled up by many birds of prey as well as herons and storks. Spacious, but not very carefully composed, they are equipped with a final layer of soft material, grass or even down, to take the eggs. Since, from these species, many individuals return to the nest every year and continue to add to the construction, such eyries sometimes reach enormous dimensions; some nests 2.5 metres in diameter and 2 metres high have been recorded.

To make a better shelter for the young, many bird species construct a circular wall and thus build cup-shaped nests. The size of the nest in relation to the parent bird varies; some nests are like tiny thimbles, invisible beneath the incubating bird, as in the crested swift (*Hemiprocne longipennes*); some nests are like small cups, not big enough to accommodate the head and tail of the parent bird, as in the paradise flycatcher (*Tchitrea paradisi*) or in humming birds; some nests are like bowls, which hide the incubating bird completely. An even more perfect protection is given by a roofed nest. The European wren (*Troglodytes troglodytes*) builds a nearly spherical enclosure with a side entrance; so does the willow warbler (*Phylloscopus trochilus*). The tailor bird (*Orthotomus* sp.) sews one or two big leaves together, perforating the edges and connecting them by pulling through some plant fibres, which forms a deep sack into which it puts grass and coconut fibres. Several species of weaver birds knot and weave strings of reed or grass and construct a hanging globe, often provided with a long entrance tunnel. It is often suspended from the very end of a branch, preferably a branch hanging over a water surface. The same pendulous position of the nest is found in several more species of birds (vireos, orioles, oropendolas, broadbills and sunbirds) since the hanging position reduces the risk of predation by tree snakes or small climbing mammals. Like weaver birds, the penduline tit (*Remiz pendulinus*) weaves hanging spherical nests with an entrance tunnel. These are upholstered and insulated with the 'wool' of the fruits of willow and popular trees. Such nests are so stable and warm that they are used in some parts of the world as slippers for children. Finally, oven birds (genus *Funarius*) model out of mud a globular nest with an entrance chamber.

A similar amount of time is spent nest building by those species which dig a nest in the ground or make holes in wood. They, too, invest many days in intensive building and they achieve nests which afford even better protection against predators. The bank swallow, several species of parrot, the kingfisher (*Alcedo atthis*), and the European bee-eater (*Merops apiaster*) all dig tunnels, which can be up to 2 metres deep, into steep muddy banks. Best-known for making holes in tree trunks are the woodpeckers. Quite a number of species nest in holes: most parrots, some ducks, many owls and many passerine birds, such as tits. Some species produce such holes themselves, while other species look for natural holes and merely alter the shape of them. In fact there is much competition for nesting holes, not only within species but also between species. Apart from the avoidance of predators, this competition may be the reason for some species laying 'bricks' at the entrance of their holes; the nuthatch (*Sitta europea*) as well as some hornbills close the entrance up, leaving a small window which allows the passing of food, but not the entry of the male which stays outside, or the exit of the female which stays inside the nest. Several species of hole-nesters provide some plant material for padding, but some others, for instance several woodpeckers or parrots, only deposit a few twigs in the nest. Thus, strictly speaking, nest construction is greatly reduced in these species. Nest building is also reduced for some species living in extreme environments. The palm swift (*Cypsiurus parvus*) hides its 'nest' in the lower dead leaves of palm trees, where it merely sticks an agglutination of feathers and cotton. This untidy little cup would not safely keep the eggs, especially in windy conditions, which is why the palm swift glues its two eggs into the nest with saliva. For the fairy tern (*Gypis alba*) not even a reduced nest is found. Not incubating in the moulded ground like other terns, this bird puts its single egg on a bare branch and has to balance it between its feet during incubation. The most extreme example of this method is found in the emperor penguin (*Aptenodytes forsteri*), which lives in the Antarctic region and nests on the ice. Instead of building a nest it places the egg on its feet and shelters it in the ventral fold of its skin. Likewise, no nest building is found in brood parasites like many cowbirds and the European cuckoo (*Cuculus canorus*). In general, the simpler nests are found among those species where the young leave the nest immediately or soon after hatching (precocial species); the more complicated and elaborate nests are found among species where the young live for several weeks within the nest (altricial species).

According to local conditions and varying selection pressures, one finds variations in the nesting behaviour. Colonial breeding and the degree of synchronization is influenced by the predation risks within some species. Some populations prefer the neighbourhood of stinging wasps or ants and fix their nest quite close to the hymenopter's nest. Often a passerine nest has been found in the close proximity to a raptor's eyrie. And recently woodland birds have been found to try to escape egg robbery by corvids or cuckoos by building their nests close to human settlements, sometimes just under the roof of a cottage or house. Another example of differently tuned response to varying environmental conditions is seen in the degree of insulation in nest construction at different altitudes (humming birds for example) or latitudes (e.g. the chaffinch, *Fringilla coelebs*) (Palmgreen and Palmgreen, 1939).

The time consumed by the nest building varies considerably; extremes range from a few minutes to more than four weeks. But how is the work divided within the pair? Is it mostly done by the male or the female? Do they share the jobs equally, or is each sex specialized in its own skill? There are probably more than 10,000 living species of birds, and as might be expected, different arrangements occur in different cases. There is quite a diversity in mating systems and family structures: promiscuity is characteristic of the leck systems, for instance the ruff (*Philomachus pugnax*) or in most species of grouse. Polygyny may be either facultative, as in the wren or marsh harrier (*Circus aeruginosus*) or more regular as amongst the great reed warbler (*Acrocephalus arundinaceus*) or the marsh wren (*Telmatodytes palustris*). Polyandry, occurs either as a consecutive event, amongst some sandpipers, such as the spotted sandpiper (*Actitis macularia*) for example, or, more rarely, as simultaneous, for example in the Tasmanian waterhen (*Tribonyx mortieri*). A more or less strict monogamy exists either as a temporal one, lasting for one breeding period, for instance among the white stork (*Ciconia guttata*) or as a lifelong one, among the zebra finch (*Taeniopygia castanotis*) for example (for review see *Lack*, 1968; *Emlen* and *Oring*, 1977; *Oring*, 1982). Accordingly, in some species it is the male who builds several nests and the female who selects one of them (for example the weaver birds and European wrens), or it is the female alone (for example mallards and grouse) which builds the nest.

In many species both sexes are involved equally, in several species there is a more or less pronounced specialization, whereby the male looks for potential nest sites and shows them to the female and the latter selects one of them. Subsequently the male carries most of the rough material whereas the female builds in the soft upholstery. In bower birds it looks as if an over-specialization has taken place; the male builds a hut-like construction out of twigs, which it decorates with colourful ornaments, but subsequently this bower is used only to attract a female and to facilitate copulation. Later on, the female builds by herself a simple nest to incubate the eggs. In some species several pairs join together and build a common nest construction, as in the case in the green parakeet (*Myiopsitta monachus*) and the sociable weaver (*Philetairus socius*). In the latter species, hundreds of pairs may contribute to a giant nest construction, where, under a common roof, each pair has its own nest chamber and entrance. Sometimes an old nest becomes so heavy that eventually the tree on which it is fixed breaks down.

The movements and behaviour patterns of nest building are generally species specific and more or less stereotyped. For example, carrying of nest material in some species is done with the feet (by birds of prey for example), or twigs are taken singly in the bill, or bundles are transported with the bill. In the genus *Agapornis*, the lovebirds, some species carry the bundles of leaves in the bill (for example *A. personata*, *A. fisheri*); others stick them between the feathers of the whole rump (for example *A. cana*) or of the back part of the rump only (*A. roseicollis*) (Dilger, 1960).

3 INCUBATION

Birds are the incubators of their eggs. Until the first experiments with artificial incubators had been carried out, it was not known how complicated the process of incubating avian eggs really is. There are four main factors which have to be regulated: temperature, humidity, ventilation (gaseous environment) and mechanical treatment (that is, the orientation of eggs to gravitation). Thus, a bird actively has to regulate processes which in mammals proceed 'on their own'; that is to say, the active behaviour of the parental bird influences factors which inside the mammalian womb are governed by vegetative 'automatic' processes. That may be one of the reasons why in most bird species investigated it was found that the hatching rate is lower in inexperienced pairs than in those breeding for the second or third time.

Whereas temperature regulation and mechanical treatment of the eggs are well understood, very little is known about the influence and regulation of partial pressure of gases such as oxygen and carbon dioxide, and relatively little is known about the regulation of humidity. Oxygen, being present in abundance, does not seem to play a major part, because variations in oxygen pressure up to more than 25 per cent hardly influence the development of the embryo, as proved by comparing nests of lowland and mountain populations (Carey et al., 1982). Even within the bird-built incubator of the megapode (*Alectura lathami*) which is composed of decaying leaves and sand (see section 10), where fermenting vegetable matter not only produces heat but also relatively high carbon dioxide levels and low oxygen levels (Baltin, 1969), development within the egg continues. More critical still seems to be the loss of water by evaporation when the relative humidity outside the egg is not suitably adjusted. In such cases there is a steady weight loss day by day. Normally the weight of the egg remains nearly constant in most bird species, at least up to the time when lung ventilation begins (Dilger, 1973). In the species where developmental time may be extended because of irregular incubation, water vapour conductance of the egg shell is reduced (Vleck and Kenagy, 1980).

As far as the temperature is concerned, the optimal value is between 34°C and 38°C, depending on the species. In the domestic fowl, no egg will hatch if the temperature is permanently either below 35°C or above 40.5°C (Lundy, 1969). Tolerance of overheating is low in most bird species, but a little increase for a short time is tolerated by desert-adapted birds. Embryonic tolerance of chilling is found more often, but in most species it decreases during incubation (Baldwin and Kendeigh, 1932). In storm petrels, humming birds and estrildid finches, even prolonged reduction of temperature will not kill the embryo but will merely retard its development (Vleck and Kenagy, 1980).

Another need of the developing egg – especially in the first half of the incubation phase – is a repeated turning round. Without turning, in most species, a reduced hatching rate will result due to premature adhesions of the extraembryonic membranes and aberrant position of the embryo (Robertson, 1961). Furthermore, by tilting and shifting the eggs, each egg is brought in the position it 'prefers' because of weight-asymmetries (Lind, 1961). This non-random position is important, as it predetermines the most favourable position for the hatching process (Dilger, 1973).

How does a parental bird manage to ascertain all these needs? As far as gaseous exchange is concerned, it has to prevent an encrustation of the permeable egg shell by mud or faeces. In very dry environments, the parent has to cover the egg and thus restrict evaporation and it may even moisten its breast feathers before returning to the nest. Also there is a lot of pushing and pulling of the eggs. Each time a bird arrives at the nest to start incubation, before sitting down it works with its bill between the eggs. In birds sitting for many hours on the eggs, a repeated lifting, standing up, and turning of the eggs can be seen. In large clutches, besides the turning, an exchange of egg arrangement is effected: eggs from the periphery are moved into the centre and vice versa. In species like ducks and tits, which sometimes lay 12 or more eggs, there is a gradual decrease in temperature towards the periphery despite the fact that all eggs are covered by the incubating bird.

The heating transfer proceeds by contact with the underside of the incubating bird. In many species the skin of the breast gets featherless, either because of the bird actively plucking its feathers, as is the case in most ducks and geese, or by a hormonally induced moult-like loss of feathers in that area. Different taxonomic groups develop one, two or three brood-patches. During incubation, these areas become well vascularized and the amount of circulating blood – and thereby emitted heat – can be regulated. In the gannet (*Sula bassana*), the intertarsal skin of the feet has this function. The birds actually measure the temperature of the environment of the egg and, according to these values, administer heat as required. This can be seen in the correlation between (a) the intervals and length of sitting periods during incubation and (b) the external air temperature or the surface temperature of the eggs; the higher the temperature, the shorter and less frequent are the incubation periods (Kluyver, 1950; Drent, 1970, 1972). In experiments with artificially cooled or heated eggs, it *can* be demonstrated that very cold eggs are heated intensely by the shivering of body muscles which increases heat production, and by feathers fluffed for better insulation. In the case of overheated eggs, the bird covers the clutch with its brood-patch and tries to conduct the heat from the eggs by actively cooling its body by gular flutter and evaporative cooling (Franks, 1967; Calder and King, 1963). In warm but not excessively hot environments, sometimes a shading of the eggs alone is sufficient to ensure their optimal temperature (McLean, 1967).

During incubation the incubating parent provides some shelter from predation and physical damage. Ground-nesting species retrieve the egg that has somehow slipped out of the nest by a typical pulling movement with the underside of the bill (the same movement is also used to turn and shift eggs inside the nest). Such rolling movements in geese became famous through the well-known ethological investigations by Lorenz and Tinbergen (1938); they suggested the concepts of 'Erbkoodination' (fixed action pattern) and 'Taxis'. In some species (some ducks and geese), the parent covers the eggs when leaving them for a short time, both for thermal isolation and camouflage. In several species, the incubating bird itself is optimally camouflaged: a breeding nightjar (*Caprimulgus europaeus*) looks like a dead branch and will not move, even if a predator passes by very closely.

As with nest-building, the duties according to sex-role are distributed quite differently in different species. There are several species in which the male does the incubation all by himself, for example in most populations of the African ostrich (*Struthio camelis*), and some species of sandpipers. In the phalaropes and in the turnix quails, the males, which incubate exclusively, have a duller coloration than the females. The females are, of course, duller-looking in those bird species, where females incubate alone. The latter is the case for polygynous birds, but also for some monogamous types, as in some *anseriform* species or in some song birds (for example, song sparrow, *Melospiza melodia*), where the female sits on the nest, but the male stays nearby and defends the territory. In cases of accidental death of the female, males are sometimes able to take over her job.

Approximately equal division of incubation duties, and the occurrence of brood-patches in both sexes, is quite common in birds, but individual variability occurs. In most species there is a relatively continuous period spent sitting on the eggs, since the rewarming of eggs which have lost heat consumes more energy than keeping them continuously warm (Vleck, 1981); also, the continual presence of the parent reduces the risk of egg predation. Special ceremonies can take place when one mate is taking over the 'incubation service' from the other: woodpeckers knock at the tree trunk to announce replacement, and overtaking cormorants present a piece of nest material to the sitting bird.

The length of incubation duty can vary considerably. For tits it is about 30 minutes, whereas for doves it lasts several hours. The ring dove (*Streptopelia risoria*) has a firmly fixed schedule according to which the male sits for 6 hours in the middle of the day and the female incubates continuously for the remaining 18 hours. For some petrels one sitting takes two to three days; and the male emperor penguin keeps the egg on top of his feet, close to the warm belly, for nearly two months until the female has finished her long journey over the ice to the open sea, and has regained her weight loss resulting from egg production; when she comes back she takes over the incubation for a period of several weeks.

Amongst the groove-billed anis (*Crotophagus sulcirostris*) communal nesting takes place with up to four monogamously paired females laying eggs in a single nest. The last-laying female takes over most of the incubation and may cause the loss of some earlier laid eggs by pushing them out of the nest (Vehrencamp, 1978). Thus her greater effort is balanced by the larger benefit she derives from propagating her progeny.

When and why do birds start incubation and thereby alter completely the rhythm of their daily activity? The onset of the incubation period is sometimes difficult to identify since in the later stages of nest building there are increasingly frequent phases when the bird sits inside the nest. In some species one or two hours of sitting in the nest (but without fluffing the breast feathers) may occur before egg laying. In other species, even after two-thirds of the eggs are laid, no prolonged incubating can be seen; it only starts after the last egg has been produced. As a result, in the clutches of eggs of this type, nestlings will hatch at approximately the same time, otherwise hatching mirrors the sequence of laying. Among large birds, eagles for example, which lay the two eggs at two days' interval, there is a pronounced age difference in the nestlings, which under natural conditions nearly always leads to a sibling killing. The incubation period is highly dependent on the size of the eggs (which is related to the size of the bird) and varies from 11 days for some small passerines up to 80 days for the kiwi (*Apteryx australis*) and large albatrosses. For some species, for example petrels, hummingbirds and some estrildid finches, there are high intra-specific variations in incubation time; this can increase up to 50 per cent in cases of prolonged cooling resulting from nutritional inadequacy in the parents. The end of the incubation phase is mainly determined by releasers emitted by the hatchlings, by the empty egg shells, and by the vocalization of the hatchlings still inside the egg. The beginning of incubation is brought about by external stimuli and the hormonal condition of the bird.

All the different phases of the reproductive cycle in birds are accompanied by distinct hormonal patterns. For several passerine birds during territory establishment testosterone and dihydrotestosterone predominate, to be succeeded by an increasing oestrogen level during the peak of the courtship activity. A phase during which oestrogen is predominant is correlated with nest building activity, and is succeeded by increased progesterone levels at the onset of incubation. High progesterone levels during incubation and increasing prolactin levels during the feeding phases terminates the cycle; in the case of double-brooded species the cycle will start again with a rise in gonadotrophic hormones (Elsner, 1960; Hinde and Steel, 1978; Lehrman, 1965; Feder et al., 1977; Sossinka et al., 1980; Wingfield and Farner, 1980). Considering the correlations between high hormone levels and the predominance of certain behaviour patterns, it is not easy to establish which is the cause of which. Oestrogen increases nest building in canaries (*Seninus canaria*, var. *domestica*), but the rise in oestrogen level is caused by the presence of a courting male and of nesting material. Furthermore, oestrogen causes the development and sensitization of the brood-patch, the tactile stimulus reception of which is responsible for progesterone and prolactin production (Hinde, 1970). In the ring dove (*Streptopelia risoria*) the presence of a mate brings the birds into a condition of readiness to incubate, and this effect is greatly enhanced by the presence of nesting material; but the onset of incubation can also be triggered by the administration of progesterone (Lehrman, 1958a, 1958b). The presence or absence of broodiness in different breeds of the domestic fowl is associated with hormone levels which control the thresholds of response (Sharp et al., 1979). Thus, the long-term shift in the dominance of certain behaviour patterns, incubation for instance is the result of interaction between external stimuli, hormone systems with mutual feedback, and behavioural inputs and outputs (Hinde, 1970).

There are some species of birds which differ in their parental behaviour completely: the brood-parasites. Several species are only partially parasitic and usually build a nest of their own and care for their young, but they do sometimes lay their eggs in other nests. There are several species of ducks which do this, and the redhead (*Aythya americana*) lays its eggs in other nests more often than not. The black-headed duck (*Heteronetta atricapilla*) and nearly 80 other bird species are completely parasitic, for example the European cuckoos, most American cowbirds and the widow-birds of Africa. The young as well as the parents exhibit a great number of special adaptations – some very 'costly' – to achieve this 'relief' from duties (for review see Friedmann 1929, 1955; Lack, 1968; Nicolai, 1964; Payne, 1973; Gärtner, 1981). Instead of nest building and incubating, the parental behaviour of these species consists of a number of peculiarities. Females are generally able to accomplish an extremely fast oviposition with nearly no labour, lasting less than 10 seconds for the great spotted cuckoo (*Clamator glandularius*) for example (von Frisch, 1969). Immediately before egg laying, the European cuckoo (*Cuculus canorus*) picks up an egg of the host with its bill and then eats it (Löhrl, 1979). The female European cuckoo produces many more eggs than do females of other species of comparable size. This is due to a much lower hatching probability because nests are often abandoned by some hosts. Furthermore, parasitic females must be able to visit the hosts' nests unseen and at an early stage of their egg laying, which makes several re-visits and checks necessary. Also, they have to find the nest of the particular species to which they are adapted (that is, where there is some similarity of egg size and coloration between host and parasite). This adaptation may possibly be due to early imprinting on the host species or on the type of nest or habitat of the host. Even the complicated mouth-markings of the nestlings of the host are mimicked by the offspring of the parasitic widow-bird and thus a high host-specificity is required, so the widow-male learns the song of the host and makes it a part of its own song. The female, having learned the song of its foster-father, will select only those males in which the song resembles the host's song, and thus guarantee a maximization of genes in the offspring adapted to the host species (Nicolai, 1974).

4 HATCHING

Hatching is the birth of a bird. It is not as dramatic an event, however, as the birth of a mammal. In mammals, the oxygen requirements of the developing embryo are provided by the maternal ciculatory system, and so true lung respiration does not start until immediately after birth. In contrast, birds start breathing several days before hatching, as soon as the beak has reached the air space, hence 'birth' does not alter the method of oxygen acquisition. Furthermore, in mammals, where the embryo is exposed at most only to some weak acoustic influences, the sensory environment changes drastically after birth. For birds, however, quite massive acoustic and perhaps even some diffuse visual stimulation takes place before hatching (Gottlieb, 1976; Bateson, 1982).

Another difference concerns the distribution of activities between the mother and the offspring. In mammals, the mother, who presses out the embryo with her muscular contraction, plays the active part. In birds, the embryo itself does the active work by emerging from inside the egg and the problem here is that the egg-shell must be broken. As an adaptation to this biological problem, almost all birds at the time of hatching are equipped with an egg-tooth, a horny shield often with a tiny sharp projection on their upper and lower mandibles. With the help of the egg-tooth the young bird, by turning its body, cuts out a skull-cap through which it is able to emerge from the egg (Clark, 1961). The only exception to this method of hatching is by the megapodes which break the egg-shell with their extremely strong feet. A second structure which helps the chicks to open the shell from inside is the hatching muscle, a paired muscle which extends broadly from the upper hind-neck to the back of the head, which reaches its maximum size just before hatching (Fischer, 1958).

Such anatomical adaptations made by the embryo to enable it to hatch, together with special patterns of prehatching and hatching behaviour (Oppenheim, 1972), demonstrate the active part played by the offspring during this stage of ontogenetic development. The parents, however, may also share in this, although rather a minor one. This consists of acoustic contact with the prehatching embryo and of removing the egg-shells shortly after hatching.

For precocial birds, which leave the nest shortly after hatching, it is essential that hatching is strongly synchronized in order to allow the whole group of siblings to set out, guided by the mother, at the same time. In mallards (*Anas platyrhynchos*), for example, a clutch (11–13 eggs) hatches within 3–8 hours (Björvall, 1967). Amongst captive bobwhites (*Colinus virginianus*), synchronous hatching has been observed even if the start of incubation was more than 48 hours later for some eggs than for others (Johnson, 1969). Such synchrony is achieved in two ways: by acoustic contact between the embryos while still in the nest and by acoustic contact between the unhatched embryo and its mother.

Social interactions between embryos have been demonstrated in bobwhite and European quails. In these species, the embryos produce a variety of sounds and vibrations. Apparently the most important signal is a rhythmic ticking, which is associated with the chick's respiratory movements and usually begins roughly 24 hours before hatching (Driver 1967). By encouraging some chicks to work faster and others to slow down, such mutual interactions promote synchronous hatching (Vince, 1969).

In addition to vocal contact between siblings, acoustic interactions have also been observed between the mother and the embryos. The incubating female mallard responds to sounds from her eggs, and the hatchlings respond to their mother. It is to be assumed that such contact also contributes to achieving synchronous hatching either directly or by stimulating the offspring to call to each other (Hess, 1972).

The second mode of parental care at the time of hatching is the removal of the egg-shell. This is most obvious in song-birds which usually either eat the egg-shells or take them out of the nest as soon as the nestling has wriggled out. The shell is dropped in flight at some distance from the nest. In precocial birds, egg-shell removal is less pronounced, yet many species practise it. Gallinaceous birds and ducks, however, tend to leave the shells in the nest.

The biological function of shell removal is obvious: the egg-shell being white on the outside in some species (especially hole-nesters) and on the inside in all species of birds, may attract predators to the nest. Its disappearance, therefore, contributes to the protection of the hatchlings. Only those species, like ducks, which leave the nest shortly after hatching, do not require a protective device of this kind.

5 PARENTAL CARE AFTER HATCHING

a Taxonomy of developmental stages at hatching

The amount and kind of parental care after hatching depend on the developmental stage at which hatching occurs and major differences within the class of birds are to be observed here. In some species the young hatch at a comparatively late stage and are equipped with full motor and sensory capacities at the time of hatching; they are covered in down, have their eyes open and leave the nest within the first day or two (precocial or nidifugous birds). In others the hatchlings are still naked, their eyes are closed, and they are unable to locomote (altricial or nidicolous birds). Between these two extreme cases, there are several intermediate situations. Nice (1961) has proposed the following classification which subsequently has been modified slightly by several authors but is still widely used:

Precocial 1   The young are completely independent from their parents from the beginning (megapodes, see section 10).
Precocial 2   The young follow their parents but find their own food (e.g. ducks, shorebirds).

Precocial 3	The young follow their parents but are shown food by the parents (most gallinaceous birds).
Precocial 4	The young follow their parents and are fed by them, although the food is not carried to them (grebes, rails).
Semi-precocial	The young are able to leave the nest within a day or two, but they remain near it because they are still dependent on their parents for warmth and because, in contrast to Precocial 4, food is carried to them by the parents (gulls, terns).
Semi-altricial	The young are covered with down but are unable to leave the nest. Their eyes are open (herons, hawks) or closed (owls).
Altricial	The young are not covered with down. Their eyes are closed (parrots, passerines).

The same distinction between precocial and altricial species can be made in mammals. However, there is an interesting difference in birds, the precocial situation is found in the more 'primitive' groups, including the ratites (ostrich-like birds) which are commonly placed at the very base of the 'phylogenetic tree' of recent birds. The most advanced species, on the other hand, like passerines and parrots, are strictly altricial. In mammals, the opposite is the case: the less advanced species, like insectivores and rodents, are altricial, and it is the most highly evolved species like carnivores and ungulates which are precocial (with the primates providing an interesting, special and somewhat intermediate situation). Therefore, in birds precocity is a primitive character, but in mammals an advanced one.

Parental care after hatching includes brooding, feeding, protection and defence, nest-sanitation and, in precocial species, guiding behaviour.

b Brooding

Brooding is essential for the regulation of temperature. In precocial species it is required only for several hours or up to 1–2 days because due to their covering of down the young are soon able to thermoregulate. For the naked nestlings of altricial species, on the other hand, brooding is necessary for many days or (in slow-developing, larger species) even for weeks (up to 20 days in the *Capercaillie*). The 'division of labour' of brooding is the same as that of incubation: if the female alone incubates she is also the only sex to brood; if both parents incubate in turn, the same is the case for brooding. Thermoregulation by the parents may also include shading, especially in hot regions; but generally it involves shading the young in exposed situations.

c Feeding

Feeding varies greatly between species as far as the nature of the food is concerned, its preparation, the frequency of feeding, the method of transfer of food to the young, and the participation of parents. In contrast to the similarity of participation of the sexes in incubation and brooding, the situation can be different as regards feeding. In many passerines, for example, only the female takes care of incubating the eggs and brooding the young, whereas both parents carry food to the nestlings and fledglings.

There are two principal methods of feeding in birds: directly from the bill or by regurgitation. Which method is used depends largely on the kind and size of the food and on the distance it has to be carried. Regurgitation is mainly used if small items are being fed for example grass and other seeds in some fringillid and all estrildid finches, or small marine organisms, as in the case of petrels and shearwaters, which are gathered and swallowed at great distances from the nest. A method of food transfer which is, in a way, the opposite to regurgitation is found in pelicans and cormorants, the young of which dig with their beak deeply into the throat of their parents to take out partly digested fish.

Feeding directly from the bill is the common method if large prey items have to be carried, for example, in those passerines which feed the young on insects, insect-larvae or earth-worms, as well as in raptors and owls. In the latter two groups the parents usually pull off small parts from the prey item in order to present them to their offspring. In some species of raptors, a certain division of labour occurs; the male catches the prey and carries it to the nest where the female is responsible for its division and transmission to the young.

Three groups of birds have a special method of feeding which bears a slight similarity to the feeding method of mammals insofar as it includes the transfer of substances from the parent's own body: pigeons regurgitate to their young a curdled substance which is commonly called 'crop-milk' or 'pigeon's milk'. It is rich in proteins, fat and ash, but does not contain any sugar being produced from fatty cells which become detached from the crop's epithelium (by a process which is somewhat similar to that which yields milk in the mammary gland of mammals). (This analogy can even be carried one step further: in both pigeons and mammals, the secretion of 'milk' is stimulated by the hormone prolactin). Flamingos feed their young with a liquid secreted by cells of the upper digestive tract (from the pharynx to the glandular stomach). This secretion is very nourishing and contains 15 per cent fat and 8–9 per cent proteins. Due to the admixture of blood and its contents of carotenoids, its colour is bright red (Land, 1963). Finally, albatrosses, petrels and shearwaters secrete oil (the so-called 'stomach oil') in their proventriculus (a part of the stomach) for the nourishment of their young (Rich and Kenyon, 1962). A small amount of own-body substances (for example, enzymes and globulines) is provided by the parents in those species which feed their young by regurgitation (parrots and many passerines for instance). Despite a certain similarity of these methods to the lactation of mammals there is one profound difference: in all mammals the production of milk is strictly confined to females whereas in all species of birds which feed their young with own-body substances the latter are produced by both parents.

The rate at which nestlings are fed varies considerably from species to species. In general, those species which carry food in their bills (with the exception of raptors and owls which catch large prey items) feed their offspring more frequently than those which provide food by regurgitation. Other factors which determine the frequency of feeding include the time of the day, the size of the brood, the size of the food, the age and behaviour of the nestlings and the age and experience of the parents. Many birds feed most actively during early morning hours. As a rule, feeding frequency is higher for large as compared to small broods, but there is a ceiling effect (Royana, 1966; Henderson, 1975; Tinbergen, 1981). The rate of feeding is inversely related to the size of the food and directly related to the age of the nestlings. Great tits (*Parus major*), for example, have been observed to make up to 900 trips to the nest per day (Kluyver, 1961; van Balen, 1973).

The begging behaviour of the nestlings, especially the begging calls in hole-nesting species and the strength and speed of begging movements in open nesters, also exert an influence. The rate at which the parents bring food may be determined by the begging behaviour of the hungriest of the nestlings (von Haartman, 1953). Finally, the age and experience of the parents may also be an important factor: in kittiwakes (*Rissa tridactyla*), for example, nestlings of broods of two young gained weight more rapidly if they were cared for by experienced parents rather than parents breeding for the first time (but not when there was only a single young in the nest) (Coulson and White, 1958). Similar data are available for other species. Lehrman and Wortis (1967) were able to demonstrate that in such cases experience seems to be even more important than age. In ring doves (*Streptopelia risoria*) they found that birds which had already nested once were more efficient in hatching eggs and rearing young than those of the same age which were nesting for the first time. Interestingly, pairs in which one partner of either sex was experienced and the other was not were as successful as pairs of two experienced partners; obviously, the experienced partner compensated for the other's deficiencies. The transfer of food from the parents to the nestlings requires efficient and neatly tuned ways of communication. Usually, signals from the parent birds elicit begging behaviour in the nestlings which, in turn, causes the adults to feed the young. Stimuli from the parents may include visual, acoustic or tactile signals. Their relative importance depends on environmental conditions and on the sensory capacities of the nestlings. Tactile stimuli provided automatically (for example, if the adult bird lands on the rim of the nest) and they are of great importance in altricial species shortly after hatching when the eyes of the nestling are still closed. Vocalizations are most relevant in hole-nesting species (Vince, 1974). Lastly, visual stimuli provide the most important cues for most precocial birds and for older nestlings of altricial species.

The immediate reactions of newly hatched young to visual cues provided by the parents have been the subject of one of the first experimental studies of early ethology. Hungry chicks of the herring gull (*Larus argentatus*) peck at the red spot on the yellow bill of the adult bird and thus trigger the regurgitation of food. The cues on the gull's head most effective in releasing the pecking response in the chick are determined with aid of cardboard models on which various combinations of colours of the spot, bill and head were painted. The results of the experiments show that the necessary cues are located on the bill, while the shape of the head, its size, and its colour have no effect. On the bill, the red spot plays an important role. It is effective by its colour as well as by the degree of contrast with the background of the bill. Thus, if models with medium-grey bills and with spots ranging from white to black through various shades of grey are presented, the chick's pecking is released more frequently by the 'whiter' or the 'blacker' spot, that is, by those that contrast more strongly with the background. On the other hand, a model of the natural yellow bill with a red spot is more effective than one with a black spot, although the latter offers more contrast (Tinbergen and Perdeck, 1950). Similar reactions to colours and combinations of colours have since also been found in many species of gulls and terns and in other species of birds.

Nestlings, in turn, emit stimuli which cause the parents to feed them. The most widespread signals are begging calls (which are especially noisy in many hole-nesters, such as woodpeckers, which are well protected from predators inside their hole) and brightly coloured mouth-markings which become visible when the beak is wide open for receiving food. In the estrildid finches, for example, the mouth-markings consist of various spots and lines on the palate and tongue. In some species there are additional luminous gape-spots at the base of the bill. In the dim light inside the domed nest, which is characteristic of estrildid finches, such 'candles' probably help to lead the parents to the mouth of the nestlings (Immelman, 1982). The effect of the colour markings may be enhanced by specific head and tongue movements. Particularly bright and colourful markings on the roof of the mouth, on the tongue and at the corners of the bill, are also found among woodpeckers, kingfishers and, above all, in parasitic birds, for example the European cuckoo. For the foster parents, they may provide a 'super-normal stimulus' which compensates for 'deficiencies' with respect to other characteristics ('wrong' size, number of nestlings and 'wrong' calls and movements).

d Water support

In most species of birds, the food given to the nestlings also contains the necessary moisture. There are a few exceptions, however, mainly concerning birds which breed in extremely hot arid environments. A particularly interesting and well-studied example is provided by the sandgrouse (*Pteroclidae*), a family of birds which inhabit the arid areas of Africa and Asia. The precocial young of the sandgrouse feed on dry seeds and need some extra water supply. In these species, the male flies to a water hole in the morning, soaks his ventral feathers in a special way, flies back to the brood and presents his abdomen by standing in an upright position with his feathers fluffed out. The young take the wet feathers in their beaks and remove the water by 'stripping' movements.

The ventral feathers of the male sandgrouse are characterized by structural peculiarities which are responsible for their unusual water-retaining capacity; their barbules have no hooks or grooves, instead they are flattened at their base and are coiled into helices along both sides of the barbs. In addition, the helices of neighbouring barbules intertwine so that the whole network of coils is very resistant to mechanical damage. Due to these adaptations in feather structure, 25–40 ml of water can be held in the belly feathers of a male sandgrouse and can be carried over distances of up to 20 miles. This quantity, obviously, is sufficient to satisfy the need for water in the young sandgrouse, which live in an extremely dry environment, feed on an extremely dry kind of food and do not benefit from the additional moisture which is provided by the parents of those species of seed-eating birds which feed their young by regurgitation (Cade and Maclean, 1967). The phenomenon of carrying water to the young also occurs in other species of birds, but obviously has not (yet) led to structural adaptations in the morphology of the feathers as complicated as in the sandgrouse. On very hot days, little ringed plovers (*Charadrius dubius*), for example, carry water to their chicks on the feathers of their underparts. The water has a cooling effect on the young, but they may possibly also drink from the wet feathers of the adult birds. In contrast to the sandgrouse, here both parents are involved in providing the water supply for the young (Gatter, 1971). A similar behaviour has been described in the case of the Egyptian plover (*Pluvianus aegyptius*) (Howell, 1978).

e Nest sanitation

Birds of many species remove the faeces of their offspring from the nest. This behaviour is most efficiently performed by the most advanced groups of birds, especially the passerines. In most species of this order, the excrement of the nestlings is enclosed in a fairly tough, gelatinous sac, which facilitates its removal by the parent. Frequently, the nestlings defaecate immediately after being fed, and the parents, having fed the youngsters, 'wait' for this to occur. The faeces are taken directly from the anus, which in many species is surrounded by a circle of bright white feathers and made even more conspicuous by the nestlings' lifting their abdomen shortly before defaecation. Probably, this provides the necessary stimulus for the parent bird to take the excremental sac.

As with the removal of egg-shells, nest sanitation may be a device to protect the nest by taking away the conspicuous white faeces. In addition it may also prevent the occurrence of diseases and parasites and may stop the susceptible down feathers of the nestlings from sticking together.

6 PARENTAL CARE AFTER FLEDGING

In the vast majority of bird species, the parents continue to care for the young after the latter have left the nest. The kind and duration of such post-fledging parental care, however, differ considerably between species. In the species denoted as Precocial 2 it consists only of protecting and leading the young to favourable places (to the water, for example, for most species of ducks) – so-called guiding behaviour. In birds other than Precocial 1 and 2 the parents also show or bring food to the fledglings. The way the food is transferred to the young and the participation of the parents in feeding is still largely the same as before fledging, although in some species of passerine birds, the male may have the main share in feeding older nestlings while the female starts to lay and incubate another clutch.

a Brooding and transporting the young

During the first days after fledging, especially in those species which leave the nest very early, some brooding may still be provided at times during the day, but mainly it takes place at night. A special type of care is found in some estrildid finches. During the first days after fledging the young are led back to the nest by the parents for the night, and sometimes also during the day to be fed inside the nest (Immelmann, 1962). One other aspect of parental care after fledging which, however, is not very common, is the transport of the young. It occurs regularly only in swans, loons and grebes, where the young ride on the back of the parents. The occasional transport by parents has been observed in several other species, with various methods being used: under the wing –sungrebe (*Heliornis fulica*), African jacana (*Actophilarnis africanus*);

Parental Behaviour in Birds 29

between the legs – woodcock (*Scolopax rusticala*), Java whistling duck (*Dendrocygna javanica*); with the bill – several rails, Montagu's harrier (*Ciraus pygargus*) (review in Skutch, 1976).

b Formation of crèches

An interesting phenomenon in the parental care of fledged young is the formation of crèches. In its most specialized form, as it occurs for example in the common eider (*Somateria mollissima*) and the shelduck (*Tadorna tadorna*), a crèche is an association of several mothers with their young, although sometimes other adult females which have lost their own young also join the group. The advantage for the mothers lies in the fact that not all of them have to stay with the young continuously and hence they have more 'free time' for foraging to make up for the loss in weight which occurred during egg production and incubation (Williams, 1974; Gorman and Milne, 1972). Probably crèche formation of this kind can be seen as a case of reciprocal altruism.

In a wider sense crèches are also found among a number of altricial birds. They occur at a time when the young have already left the nest but are still fed by their parents and hence remain in the vicinity of the former nesting place. Here, too, only a few adults are necessary to 'guard' the young releasing others to take off for foraging. They return to the crèche but feed selectively only their own young. Crèches of this type have been described in penguins, flamingos, pelicans, terns, cockatoos and jays (Skutch, 1976; Balda and Balda, 1978).

c Individual recognition

The most essential difference between the situations before and after fledging concerns the recognition of the young. As long as the offspring are in the nest it is sufficient for the parents to remember the location of the nest. After fledging, in contrast, they have to recognize their young individually and perhaps also under varying environmental conditions. Individual recognition of young is especially important in colonial breeders where many fledglings may stay in close proximity to each other and the risk of a mistake is high. The efficiency of mutual recognition can be increased further, especially in those species in which the young actively approach the parents to be fed, if the offspring are also able to recognize their own parents.

The evidence available indicates that both abilities are indeed to be found in birds and that there is a neat correlation between the time of fledging and the onset of individual recognition. Such correlation points to the strength of the selection pressure favouring the availability of mechanisms of recognition at the very age at which they are needed for the first time.

In the extremely altricial birds, for example ducks and gallinaceous birds, recognition of the mother is learnt during the first hours or days of life through a process called filial imprinting. This term refers to a very brief and early kind of learning which, like other imprinting processes, is characterized by two main criteria: the restriction to a sensitive phase and a high degree of stability of the results of learning (Hess, 1973; Immelmann and Suomi, 1981). It is these two characteristics which are also most important with regard to parental care: the early onset of the sensitive phase ensures that the young starts to learn to recognise its mother immediately after hatching, whereas the early end of the sensitive phase and the subsequent stability of social preferences acquired during imprinting are able to prevent new and perhaps 'wrong' social signals being learnt after the bird has left the nest.

Semi-precocial species such as gulls are not in such a hurry. Even in this case, however, close temporal correlations have been found: young blackbilled gulls (*Larus bulleri*), for example, are able to discriminate the 'mew' calls of their parents from those of other adults at an age of three or four days. This is roughly the age at which they abandon the nest and begin to intermingle with members of other families (Evans, 1970). Laughing gulls (*L. atricilla*), on the other hand, do not begin to restrict their responses solely to the calls of their parents until they are at least one week old. In this species the pairs breed considerably farther apart from each other than do black-headed gulls and so their chicks are usually not faced with situations demanding individual recognition by the parents until the second week of age (Beer, 1970).

A really dramatic example of the adaptive significance of individual recognition between offspring and parents is provided by the guillemot (*Uria aalge*). In this species, which nests on narrow ledges on rocky coasts and islands, the adults respond, much as does the mallard female, to the calls from the egg shortly before hatching. The chick thus learns to recognise its parents and can eventually distinguish them from neighbouring birds. Thus by the time the chick hatches it is already able to recognise its parents individually. Such extremely early learning has again to be understood as an adaptation to the breeding biology of the species: guillemots breed in extremely dense colonies in which the adults almost touch each other as they incubate their only egg without a nest. Under these conditions the danger of a mistake exists from the very moment the chick hatches, and individual recognition between parents and offspring is already vital at this early age (Tschanz, 1968).

Similar correlations between the occurrence of individual recognition and the breeding biology of the species occur also with regard to individual recognition of the young by their parents. A classical example is again provided by gulls: ground-nesting species learn to recognise their young individually within a few days after hatching, and if one attempts to exchange the young after that age they are no longer accepted. By contrast, the parental rock-nesting kittiwakes never learn to recognize their young individually and they will accept strange young as old as four weeks. The biological explanation for such failure is easy: kittiwakes nest on even the smallest ledges on cliffs, just large enough to offer space for their two young. As an adaptation to this nesting strategy, young kittiwakes do not leave their nest until they fledge. Hence the parents never need to search for and identify their young at another location (Cullen, 1957).

Further examples of temporal correlations are also to be found in the true altricial species: among the bank swallow (*Riparia riparia*), there is no chick discrimination up to day 15 after hatching, but by day 17, when the young are mobile and capable of wandering from their own nest-tunnel to another, the parents reject strange young and feed only their own, even in a strange burrow (Hoogland and Sherman, 1976). Galahs (*Cacatua roseicapilla*) reject strange young even later, at an age of about 40 days; but in this species fledging (during the seventh week of life) also occurs considerably later than in swallows. In the zebra finch, parents know their offspring at the age of fledging (about day 20); but the young need a few more days to avoid begging from non-parents (which frequently results in getting pecked).

7 HELPERS AT THE NEST

One aspect of parental care in birds which has been known for a long time (see Skutch, 1961) but which has not been adequately explained until recently is the fact that nestlings and fledglings in many species are cared for not only by their own parents but also by other individuals. As a rule, 'helpers' are relatives of the parents, perhaps offspring from a preceding brood (and as such siblings of the nestlings they feed). In some cases, however, unrelated helpers have also been observed (Brown, 1978; Reyer, 1980).

Helpers have been noted in more than 150 species of birds. The possible benefits of helping for both helper and helped have been discussed a great deal in the literature and opinions do still differ. At first sight, helping seems to be a clear case of altruism, especially as it has been shown in many species that there is some positive correlation between the number of non-breeding helpers and the breeding success of the parents; this may be due to the fact that helpers reduce the number of feedings by the parents and thus lighten the load of the parents. On the other hand, helpers may also gain personal advantages from helping since they are able to gain experience of parental care which may be an advantage once they start their own first brood. They may also achieve, by means of helping, the reward of being tolerated beyond the usual time of weaning and dispersal within the familiar parental territory or the natal colony. Also, the helper may have a chance of succession in case of death of one of the parent birds. Apart from such individual advantages, however, those helpers which are closely related to the parents may also gain indirect advantages. They are able to increase their 'inclusive fitness' by promoting the survival of individuals which carry certain percentages of the helpers' genes (for detailed discussions, see Brown, 1978).

Altogether, the phenomenon of helping is so varied between species, and even between populations within a species, that the controversy over whether or not helping can be classified as true altruism has not been settled yet and many more data are necessary in order to arrive at a taxonomy and a group of explanations which cover all aspects and all known cases of this behaviour. The data available, however, have already led to a revision of the traditional assumption of the exclusive role of the parents in the raising of their offspring.

In mammals, helping seems to be less widespread than in birds. A case of 'helping' has been described, however, in several species of primates where other females, the so-called 'allomothers', elder sisters perhaps, participate in carrying or grooming young infants (Hrdy, 1976; Wolters, 1978). Assistance to the offspring of other individuals has also been found in dolphins, canids, viverrids and, among the lower vertebrates in some amphibians and fish. These cases, however, have not as yet been as well-studied as the helping behaviour in birds, and the term 'helpers' seems to be less commonly used in species other than birds (Hrdy, 1976; Taborsky and Limberger, 1981).

8 ACTIVE PROTECTION OF NEST AND YOUNG

During the whole period of incubation and pre- and post-fledging care of the young, the parents attempt to protect both the nest and eggs as well as the nestlings and fledglings. This can be done (mainly by larger-sized species) by threat and actual attack but there are other, sometimes very specialized, means of protection.

The common eider, for example, in which only the cryptically coloured female incubates, lacks an attack response towards predators. Instead, when a predator is approaching, the female defaecates on her eggs as she flies off. The fluid ejected at this time is quite different from normal faecal material and has a repelling effect on various potential predators, such as ferrets and rats (Swennen, 1968).

A widespread way of protecting the young is the so-called distraction display. This is defined as 'the elaborate stereotyped activities performed by a parent bird that tend to concentrate the attention of potential predators on it and away from the nest or young' (Armstrong, 1964). This consists of a sequence of complex behaviour patterns during which conspicuous plumage markings may be exposed, making the bird distinctly visible. There are two types of distraction displays, the 'injury feigning' and the 'rodent run'. During the first type, the parent imitates an injured bird which is unable to fly, whereas during the second type it runs like a mouse and thus resembles another prey item. The distracting parent always stays just outside the reach of the predator and is thus able to lure him away from the nest or the young (Simmons, 1955). Distraction displays occur in both precocial and altricial species. Within the first group, parental distraction is most marked at the time of hatching, in the latter it is most frequent at the time of fledging (Armstrong, 1956).

9 PARENTAL CARE AND LEARNING

The general occurrence of very intensive parental care in most birds and mammals has led to the reliable presence of the parent(s) during extended parts of the ontogenetic development of the offspring. Such availability of a 'model' and of experienced individuals from whom the young may learn has another evolutionary consequence: it allows the pre-programmed portions of the mechanisms for species recognition or the recognition of natural resources, which are predominant in many invertebrates and lower vertebrates, to decrease and to be supplemented by learning; this enables the young to participate in the many advantages of 'open programs', as discussed by Mayr (1974).

The full extent of this aspect of parental care in birds and mammals has been realised only fairly recently. The evidence available indicates that in addition to the learning processes involved in mutual recognition of parents and offspring, the parents may also exert an influence on learning processes which are of importance beyond the period of parent-offspring relationships. In mammals, an extended duration of social contact between parent(s) and young may serve to establish, for example, food traditions and may enable the offspring to learn the correct kind of food to obtain and the best methods of killing and/or preparing the prey for consumption (Ewer, 1968).

In birds, a similar influence of the parents has been found. The mother red junglefowl, for example, adopts a particular stance, pointing to available food for the chicks. When the young are old enough to wander away on their own, the mother still indicates food sources, this time by exaggerated and apparently stereotyped perching movements (Stokes, 1971). The consistency with which fish are presented to newly hatched chicks of the common tern (*Sterna hirunda*) may enhance the development of the chicks' ability to handle and manipulate food items (Godfeld, 1980).

One area of learning, however, which is virtually unknown in mammals but represents one of the most essential processes in many species of birds is the acquisition of vocalizations. In birds, particularly in the oscines, many characteristics of their vocalizations, as expressed not only in the song but also in several types of calls, need to be acquired entirely or at least in part. There seems to be in birds a phylogenetic trend from a rather definite to a very weak genetic determination in this respect. In mammals, by contrast, even in the most advanced forms (for example, non-human primates), an amazingly large number of purely innate vocalizations has been found. Unlike in birds, deprivation experiments in mammals do not result in permanent alterations to the development of the physical structures of vocalizations; and calls of juveniles have been found to be identical or similar to those of the adults. In mammals learning seems to be involved in the social context in which vocalizations are used, while acoustic features of the vocalizations remain unaltered (Delius, 1983).

For birds much information is available about the factors influencing song learning; and this also is the area in which parental care, associated with the song-tutoring by the father, can be of importance. In the passerines, the song has to be acquired, to a greater or lesser extent, early in life. In many species, there is a rather brief sensitive phase during which the song has to be learnt. After this period no changes in song structure are any longer possible – a fact which has frequently led to a comparison of song learning with imprinting and to the introduction of the term 'song imprinting' for those species in which the sensitive phase and a subsequent stability of the song are very pronounced (Thorpe, 1959).

If the song has to be learnt, mechanisms are required which ensure that it is the correct one, that is the conspecific type of song, which is acquired. Two main mechanisms have been described: a tendency to learn vocalizations of a certain quality and tonal structure, and a tendency to learn selectively the song of the father. In both cases, the father is involved as a tutor for the song-development of the young male; and it has been suggested that an increase in song output which can be observed in some species towards the end of the breeding season when territorial defence is unimportant may serve mainly the purpose of providing the necessary information for the male offspring.

A particular involvement of parental care may be assumed in those cases in which there is a selective acquisition of the father's song. This has first been demonstrated in the bullfinch (*Pyrrhula pyrrhula*) (Nicolai, 1959) and has since been described in the domesticated Bengalese finch (*Louchura striata*) (Dietrich, 1980) and the zebra finch (Immelmann, 1969; Böhner, 1983). In the zebra finch, young males develop a song which is very similar to the song of their father even if, under experimental conditions, another male is singing in the immediate vicinity and more frequently than the father (Böhner, 1983).

Mechanisms like these certainly favour song traditions which may be of importance, for example the development and maintenance of song dialects, the possible biological function of which has been discussed very actively during recent years (Krebs and Kroodsma, 1980; Baker, 1982).

It is along these lines that social interactions between parents and offspring in birds, as in mammals and occasionally perhaps also in 'lower' animals, have reached a level which is additional to the 'original' functions of parental care which is focused on the sheer survival of the young. Together with other early learning processes, like the above-mentioned filial imprinting, but also sexual imprinting, habitat and locality imprinting and other imprinting-like processes, song learning indicates that learning in birds takes place more quickly and is restricted more markedly to early and rather brief sensitive phases than is the case in mammals. Such speed in learning, which may be a consequence of the general rapidity of the ontogenetic development of birds, certainly requires a high intensity and high quality of parental care in the broadest sense.

10 THE MEGAPODES - A 'CONCORDE FALLACY'?

There is one group of birds to which almost none of the descriptions of parental care given so far applies, and which certainly deserves special mention: the megapodes (*Megapodiidae*), a family of the gallinaceous birds (*Galliformes*), the 12 species of which are distributed over the Indo-malaysian and Polynesian islands and the Australian continent. In this most remarkable group of birds, the parents do not incubate the eggs and do not directly care for their young. Nevertheless they show, as far as the amount of time and energy involved is concerned, the most extended parental care of all species of birds. Such seeming contradiction calls for an explanation.

The most essential feature of megapode biology is their ability to make use, for the development of their eggs, of the heat provided by sunshine, fermentation of decaying vegetation, hot springs and hot volcanic soils. The simplest method these animals use – which may be the phylogenetic origin of the habit – is to drop their eggs close to hot springs or on sun-warmed beaches. Most species, however, which live in less favourable environments, cannot rely on such 'natural incubators' but have to construct special mounds in which to place their eggs (because of this habit, the whole family is also known as 'mound builders'). The mound consists of different amounts and kinds of vegetation and sand or other soils, and their composition may vary from species to species, between different geographic areas within one species, and even in one and the same mound between different times of the year (Frith, 1956; Clark, 1964; Diamond, 1983).

Those species of megapodes which live in areas with rather pronounced seasonal variations in climatic conditions, like the Australian brush turkey (*Alectura lathami*) and, above all, the mallee fowl (*Leipoa ocellata*), have developed sensory capacities and very complex behaviour patterns to detect and regulate the temperature of the mound. In the mallee fowl, for example, the annual schedule is as follows. During the winter months the mound is opened and filled with organic matter. During spring all the heat reaching the eggs is derived from fermentation, and –if the rate of fermentation is such that more heat is created than necessary –the male opens the mound at or before dawn to allow sufficient heat to escape. In summer the rate of heat provided by fermentation gradually decreases and solar heat becomes increasingly important. Early in summer, therefore, the mound is open during the day to expose the egg chamber to the sun's rays; later in summer, in contrast, when solar heat increases, the mound is opened early in the morning for cooling but it is restored to its full height towards evening for insulation purposes. In autumn, when there is no further heat available from organic matter and little from the sun, the mound is opened late in the morning, when the sun is shining directly on to it, and is rebuilt again with thoroughly heated soil in mid-afternoon (Frith, 1956).

In the mallee fowl, which constructs mounds up to a size of 15 feet across and 2.5 feet in height, work at the mound takes several hours per day and the mound is attended for up to eleven months of the year – an amount of parental investment which is far above any other species of bird. The phylogenetic development of the mound building behaviour probably has to be seen as follows. In the beginning there were those species which lived in areas with an equally warm climate throughout the year and with other sources of heat, such as hot soils and springs. In these conditions, through dropping the eggs at suitable places, such behaviour really can result in saving time and energy which would otherwise be used for incubating eggs. As a result, the female has a degree of 'freedom' which is similar to that of a pregnant female mammal. This fact might have provided a selective advantage promoting a decreasing and vanishing motivation to incubate the eggs. However, with deteriorating climatic conditions or, more probably, with the extension the range of habitation into less favourable areas, some care had to be taken to keep the eggs at a constant temperature of about 92°F by providing some additional heat when ambient temperatures are low, and some cooling when they are high. Such selection pressure should favour the evolution of capacities to control and regulate temperature by constructing more and more complex mounds consisting of organic matter and soil. Following this line of thought it is not surprising that the most complicated mounds and the greatest amount of time spent at the mound is to be observed in the Australian brush turkey and, above all, in the mallee fowl which is to be found in habitats where the largest daily and annual fluctuations in temperature occur. Under these conditions a behavioural trait of lack of incubation, which initially had been an improvement by allowing the female more 'freedom' has turned into an extra large burden for both the parents or, as in many species of megapodes in which the male does all the work at the mound, particularly for the male. In sociobiological terminology, a phylogenetic development like this, which allows no 'return' but requires larger investments, is called the 'concorde fallacy'.

With regard to parental care, the megapodes are remarkable also from another point of view. They represent the only group of birds – in fact the only group of warm-blooded animals – which does not care at all for their offspring after hatching. Instead, the offspring, once they have worked their way to the surface of the mound, immediately disperse from the mound without obvious social contact with their parents.

At the time of hatching, young megapodes are equipped with a full set of flight-feathers and so they are probably able to reach a roosting tree on the very first day of their life. Furthermore, unlike the offspring of other gallinaceous birds, they are also able to forage by themselves with no assistance from either parent and are thus the most precocial young of all birds.

There is another difference between these and other gallinaceous birds. In the majority of species the offspring do have a strong following reaction and while following their mother or parents they acquire, through filial imprinting, the necessary knowledge of social signals. Young megapodes, however, do not possess a following reaction at all. They either must have some kind of an innate knowledge of the species – specific characteristics – or, as some recent observations vaguely indicate, there must be some (probably acoustic) contact between parents and offspring at or before hatching, and/or the emerging offspring must be able to learn the characteristics of their parents by mere exposure to them as they walk away from the mound, but without the usual social contact (Immelmann and Böhner, 1984).

With all their peculiarities, the megapodes provide an example of an entirely 'alternative' strategy of breeding biology and parental care. They demonstrate – and this is the reason why they have been treated at some length at the end of this chapter – that very different solutions to the problem of raising offspring may be successful in avian species.

11 CONCLUSIONS

The aim of the preceding survey was not only to review some of the essential features of parental care in birds but also to demonstrate that birds, as far as parental care is concerned, are quite different from mammals. Of course, one has to keep in mind that differences between the bird families and species within each class are very pronounced. Furthermore, there are great individual differences even within a population, the analysis of which had to be omitted completely in this survey. If, despite this heterogeneity, a cautious attempt was made to see the 'system bird' in comparison to the 'system mammal', it was to gain some insight into the adaptiveness of evolution, and to see recent structures as a compromise between historical dispositions and developmental plasticity. On the one hand, there are striking examples of similarity between non-related taxonomic groups of birds as well as between some avian and mammalian species. Adaptations to avoid predators manifested in nest construction, or the role of the sexes according to parental duties, in some cases exhibit an almost completely identical picture in certain birds and mammals. On the other hand, besides these few analogies which are results of a convergent development, there are innumerable differences between the two classes of vertebrates.

Therefore if one discusses parental care in mammals (and especially in non-human primates and man) one should be aware that there are other solutions to the problems posed by parent–offspring relationships which may be as highly evolved as the mammalian way. One of many possible conclusions and thought-provoking findings is the fact that, as a rule, the father seems to be much more important in birds than he is supposed to be in many mammalian species or – to express it another way – that the avian mother really is enjoying equal rights!

REFERENCES

Armstrong, E. A. 1956: Distraction display and the human predator. *Ibis*, 98, 641–54.
—— 1964: Distraction display. In A. L. Nethersole-Thompson (ed.), *A New Dictionary of Birds*, London: Nelson.
Baku, J. H. v. 1973: A comparative study of the breeding ecology of the great tit *Parus major* in different habitats. *Ardea*, 61, 1–93.
Balda, R. P. and Balda, J. H. 1978: The care of young piñon jays (*Gymnorphina cyanocephalus*) and their integration into the flock. *Journal für Ornithologie*, 119, 146–71.
Baldwin, S. P. and Kendeigh, S. C. 1932: *Physiology of the Temperature of Birds*. Scientific Publications of the Cleveland Museum of Natural History.
Baltin, S. 1969: Zur Biologie und Ethologie des Talegalahuhns (*Alectura lathami* Gray) unter besonderer Berücksichtigung des Verhaltens während der Brutperiode. *Zeitschrift für Tierpsychologie*, 26, 524–72.
Bateson, P. 1983: The interpretation of sensitive periods. In W. A. Oliver and M. Zappela (eds), *The behaviour of Human Infants*, New York, Plenum Press, 57–70.
Beer, C. G. 1970: On the responses of laughing gull chicks to the calls of adults. *Animal Behaviour*, 18, 652–77.

Bjarvall, A. 1967: The critical period and the interval between hatching and exodus in mallard ducklings. *Behaviour*, 28, 141–8.

Böhner, J. 1983: Song learning in the zebra finch (*Taeniopygia guttata*): selectivity in the choice of a tutor and accuracy of song copies. *Animal Behaviour*, 31, 231–7.

Brown, J. L. 1978: Avian communal breeding systems. *Annual Review of Ecology and Systematics*, 9, 123–55.

Cade, T. J. and Maclean, G. L. 1967: Transport of water by adult sandgrouse to their young. *Condor*, 69, 323–43.

Calder, W. A. and King, J. R. 1963: Evaporative cooling in the Zebra Finch. *Experientia*, 19, 603–6.

Caple, G., Balda, R. P. and Willis, W. R. 1983: The physics of leaping animals and the evolution of preflight. *American Naturalist*, 121, 455–76.

Carey, C., Thomson, E. L., Vleck, D. M. and James, F. C. 1982: Avian reproduction over an altitudinal gradient: incubation-period, hatching mass and embryonic oxygen consumption. *Auk*, 99, 710–18.

Clark, G. A. 1961: Occurrence and timing of egg teeth in birds. *Wilson Bulletin*, 73, 268–78.

—— 1964: Life histories and the evolution of megapodes. *Living Bird*, 3, 149–67.

Coulson, J. C. and White, E. 1958: The effect of age on the breeding biology of the kittiwake *Rissa tridactyla*. *Ibis*, 100, 40–51.

Cullen, E. 1957: Adaptations in the kittiwake to cliff-nesting. *Ibis*, 99, 275–302.

Diamond, J. 1983: The reproductive biology of mound-building birds. *Nature*, 301, 288–9.

Dietrich, K. 1980: Vorbildwahl in der Gesangsentwicklung beim Japanischen Mövchen (*Lonchura striata* var. *domestica*, Estrildidae). *Zeitschrift für Tierpsychologie*, 52, 57–76.

Dilger, W. C. 1960: The comparative ethology of the African parrot genus *Agapornis*. *Zeitschrift für Tierpsychologie*, 17, 649–85.

Drent, R. H. 1970. Functional aspects of incubation in the Herring Gull. *Behaviour*, Supplement. 17, 1–132.

—— 1972: Adaptive aspects of the physiology of incubation. In *Proceedings of the XV International Ornithological Congress*, Leiden: E. J. Brill, 255–6.

Ehrhardt, A. A. and Meyer-Bahlburg, H. F. L. 1981: Effects of prenatal sex hormones on gender-related behaviour. *Science*, 211, 1312–18.

Eisner, E. 1960: The relationship of hormones to the reproductive behaviour of birds, referring especially to parental behaviour: a review. *Animal Behaviour*, 8, 155–79.

Emlen, S. T. and Oring, L. W. 1977: Ecology, sexual selection, and the evolution of mating systems. *Science*, 197, 215–23.

Evans, R. M. 1970: Parental recognition and the 'mew call' in black-billed gulls (*Larus bulleri*). *Auk*, 87, 503–13.

Ewer, R. F. 1968: *Ethology of Mammals*. London: Logos Press.

Feder, H. H., Storey, A., Goodwin, D., Reboulleau, C. and Silver, R. 1977: Testosterone and 5α-dihydrotestosterone levels in peripheral plasma of male and female Ring Doves (*Streptopelia risoria*) during the reproductive cycle. *Biology of Reproduction*, 16, 666–77.

Fisher, H. J. 1958: The 'hatching muscle' in the chick. *Auk*, 85, 391–9.

Franks, E. C. 1967: The responses of incubating Ringed Turtle Doves (*Streptopelia risoria*) to manipulated egg temperatures. *Condor*, 69, 268–76.

Frazer, J. F. D. 1977: Growth of young vertebrates in the egg or uterus. *Journal of Zoology*, 183, 189–201.

Friedmann, H. 1929: *The cowbirds*. Springfield: Thomas.

—— 1955: The honey-guides. *U.S. Nat. Mus. Bull*, 208.

Frisch, O. and H. von 1967: Beobachtungen zur Brutbiologie und Jugendentwicklung des Häher Kuckucks (*Clamador glanderius*). *Zeitschrift für Tierpsychologie*, 24, 129–36.

Frisch, K. von 1974: *Tiere als Baumeister*. Frankfurt, Berlin, Wien: Ulstein Verlag.

Frith, H. J. 1959: Breeding habits in the family *Megapodiidae*. *Ibis*, 98, 620–40.

Gärtner, K. 1982: Zur Ablehnung von Eiern und Jungen des Kuckucks (*Cuculus canorus*) durch die Wirtsvögel. *Vogelwelt*, 103, 201–24.

Gatter, W. 1971: Wassertransport beim Flußregenpfeifer (*Charadrius dubius*). *Vogelwelt*, 92, 100–3.

Godfeld, M. 1980: Learning to eat by young common terms: consistency of presentation as an early cue. *Proceedings Colonial Waterbird Group*, 3, 108–18.

Gorman, M. L. and Milne, H. 1972: Creche behaviour in the Common Eider (*Somateria m. mollissuna* L.). *Ornis Scandinavica*, 3, 21–5.

Gottlieb, G. 1976: Conceptions of prenatal development: behavioural embryology. *Psychological Review*, 83, 215–34.

Haartman, L. von 1953: Was reizt den Trauerfliegenschnäpper, *Muscicapa hypolenca*, zu füttern. *Vogelwarte*, 16, 157–64.

Henderson, B. A. 1975: Role of the chick's begging behaviour in the regulation of parental feeding behaviour of *Larus glaucescens*. *Condor*, 77, 488–92.

Hendrichs, H. 1978: Die soziale Organisation von Säugetierpopulationen. *Säugetierkundliche Mitteilungen*, 26, 81–116.

Hess, E. H. 1972: 'Imprinting, in a natural laboratory. *Scientific American*, 227, 24–31.

—— 1973: *Imprinting*. New York: Van Nostrand.

Hinde, R. A. 1965: Interaction of internal and external factors in integration of Canary reproduction. In F. A. Beach (ed.), *Sex and Behavior*, New York: Wiley, 381–415.

—— 1966: *Animal behaviour*. London: McGraw-Hill.

Hinde, R. A. and Steel, E. 1978: The influence of day length and male vocalizations on the estrogene-dependent behaviour of female canaries and budgerigars. *Advances in the Study of Behaviour*, 7, 39–73.

Hoogland, J. L. and Sherman, P. W. 1976: Advantages and disadvantages of bank swallow (*Riparia riparia*) coloniality. *Ecological Monographs*, 46, 33–58.

Howell, T. R. 1979: Breeding biology of the Egyptian plover *Pluvianus aegyptius*. *University of California Publications in Zoology*, 113, 1–76.

Hrdy, S. B. 1976: Care and exploitation of nonhuman primate infants by conspecifics other than the mother. *Advances in the Study of Behaviour*, 6, 101–58.

Immelmann, K. 1962: Beiträge zu einer vergleichenden Biologie australischer Prachtfinken. *Zoologische Jahrbücher Systematik*, 90, 1–196.

—— 1969: Über den Einfluß frühkindlicher Erfahrungen auf die geschlechtliche Objektfixierung bei Estrildiden. *Zeitschrift für Tierpsychologie*, 26, 677–91.

—— 1973: Role of the environment in reproduction as a source of predictive information. In D. S. Farner, (ed.), *Breeding biology of birds*, Washington, National Academy of Science, 121-47.
—— 1982: *Australian Finches*. (3rd edn). Sydney: Angus & Robertson.
—— 1984: The natural history of bird learning. In P. Marler and H. S. Terrace (eds), *The Biology of Learning*, Berlin: Springer.
Immelmann, K. and Böhner, J. 1984: Beobachtungen am Thermometerhuhn in Australien. *Journal für Ornithologie*, 125, 141-55.
Immelmann, K. and Suomi, S. J. 1981: Sensitive phases in development. In K. Immelmann, G. W. Barlow, L. Petrinovich, M. Main (eds), *Behavioral Development*, New York: Cambridge University Press, 395-431.
Johnson, R. A. 1969: Hatching behavior of the bobwhite. *Wilson Bulletin*, 81, 79-86.
Kluijver, H. N. 1950: Daily routines of the Great Tit, *Parus m.major* L.. *Ardea*, 38, 99-135.
Kluyver, H. N. 1961: Food consumption in relation to habitat in breeding chickadees. *Auk*, 78, 532-50.
Krebs, J. and Kroodsma, D. 1980: Repertoires and geographical variation in bird song. *Advances in the Study of Behaviour*, 11, 143-77.
Lack, D. 1968: *Ecological adaptations for breeding in birds*. London: Metheun & Co.
Lang, E. M. 1963: Flamingoes raise their young on a liquid containing blood. *Experientia*, 19, 532-3.
Lehrman, D. S. 1958a: Induction of broodiness by participation in courtship and nest-building in the Ring Dove (*Streptopelia risoria*). *Journal of Comparative and Physiological Psychology*, 51, 32-6.
—— 1958b: Effects of female sex hormones on incubation behavior in the Ring Dove (*Streptopelia risoria*). *Journal of Comparative and Physiological Psychology*, 51, 142-5.
—— 1965: Interaction between internal and external environments in the regulation of the reproductive cycle of the Ring Dove. In F. A. Beach, (ed), *Sex and Behavior*, New York: Wiley, 355-80.
Lehrman, D. S. and Wortis, R. P. 1967: Breeding experience and breeding efficiency in the Ring Dove. *Animal Behaviour*, 15, 223-8.
Lind, H. 1961: Studies on the behaviour of the Blacktailed Godwit (*Limosa limosa*). *Meddelelse fra Naturfredningsrådets reservatudvalg. Nr. 66*, Munksgaard, Copenhagen.
Löhrl, H. 1979: Untersuchungen am Kuckuck, *Cuculus canorus* (Biologie, Ethologie und Morphologie). *Journal für Ornithologie*, 120, 139-73.
Lundy, H. 1969: A review of the effects of temperature, humidity, turning and gaseous environment in the incubator on the hatchability of the hen's egg. In T. C. Carter and B. M. Freeman (eds), *The fertility and hatchability of the hen's egg*, Edinburgh: Oliver & Body, 143-176.
Mason, W. 1979: Social ontogeny. In J. G. Vandenberg and P. Marler, (eds), *Social Behaviour and Communication*, New York: Plenum, 1-28.
Mayr, E. 1963: The role of ornithological research in biology. *Proceedings XIII Ornithological Congress*, Ithaca 1962, 27-38.
—— 1974: Teleological and telenomic, a new analysis. *Boston Studies in the Philosophy of Science*, XIV, 91-117.

McLean, G. L. 1967: The breeding biology and the behaviour of the Doublebanded Courser *Rhinoptilus africanus* (Temminck). Ibis, 109, 556–69.

Murton, R. K. and Westwood, N. J. 1977: *Avian breeding cycles.* Oxford: Clarendon Press.

Lorenz, K. and Tinbergen N. 1938: Taxis und Instinkthandlung in der Eirollbewegung der Graugans. *Zeitschrift für Tierpsychologie.* 2, 1–29.

Nice, M. M. 1962: Development of Behavior in *Precocial Birds.* New York. Transactions of the Linnean Society.

Nicolai, J. 1959: Familientradition in der Gesangsentwicklung des Gimpels (*Pyrrhula pyrrhula* L.) *Journal für Ornithologie*, 100, 39–46.

—— 1964: Der Brutparasitismus der Viduinae als ethologisches Problem. *Zeitschrift für Tierpsychologie*, 21, 129–204.

—— 1973: Das Lernprogramm in der Gesangsausbildung der Strohwitwe Te traenura fischeri Reichenow. *Zeitschrift für Tierpsychologie*, 32, 113–38.

Oppenheim, R. W. 1972: Prehatching and hatching behavior in birds: a comparative study of altricial and precocial species. *Animal Behaviour*, 20, 644–55.

Orcutt, A. B. 1974: Sounds produced by hatching Japanese quail (*Coturnix coturnix japonica*) as potential aids to synchronous hatching. *Behaviour*, 50, 173–84.

Oring, L. W. 1982: Avian mating systems. In D. S. Farner, J. R. King and K. C. Parkes (eds), *Avian Biology*, vol. VI, New York: Academic Press, 1–92.

Palmgren, M. L. and Palmgren, P. 1939: Über die Wärmeisolierungskapazität verschiedener Kleinvogelnester. *Ornis. Fennica*, 16, 1–6.

Payne, R. B. 1973: Individual laying histories and the clutch size and numbers of eggs of parasitic cuckoos. *Condor*, 75, 414–38.

Reyer, H. -U. 1980: Flexible helper structure as an ecological adaptation in the Pied Kingfisher (*Ceryle rudis rudis* L.). *Behavioural Ecology and Sociobiology*, 6, 219–27.

Rice, D. W. and Kenyon, K. W. 1982: Breeding cycles and behavior of Laysan and black-footed albatrosses. *Auk*, 79, 517–67.

Robertson, I. S. 1961: The influence of turning on the hatch ability of hen's eggs. *Journal of Agron. Science*, 57, 49–69.

Royana, T. C. 1966: Factors governing feeding rate, food requirement and brood size of nestling great tits, *Parus major. Ibis*, 108, 313–47.

Sharp, P. J., Scanes, C. G., Williams, J. B., Harvey, S. and Chadwick, A. 1979: Variations on concentrations of prolactin, luteinizing hormone, growth and progesterone in the plasma of broody bantams (*Gallus domesticus*). *Journal of Endocrinology*, 80, 51–8.

Simmons, K. E. L. 1955: The nature of predator-reactions of waders towards humans, with a special reference to the role of the aggressive escape-, and brooding drives. *Behaviour*, 8, 130–73.

Skutch, A. F. 1961: Helpers among birds. *Condor*, 63, 198–226.

—— 1976: *Parent Birds and their Young.* Austin: Texas University Press.

Sluckin, W. 1973: *Imprinting and Early Learning.* (2nd edn). Chicago: Aldine.

Sossinka, R., Pröve, E. and Immelmann, K. 1980: Hormonal mechanisms in avian behaviour. In A. Epple and M. H. Stetson (eds), *Avian Endocrinology*, New York: Academic Press, 533–47.

Stokes, A. W. 1971: Parental and courtship feeding in the red jungle fowl. *Auk*, 88, 21–9.

Swennen, C. 1968: Nest protection of eiderducks and shovelers by means of faeces. *Ardea*, 56, 248–58.

Taborsky, M. and Limberger, D. 1981: Helpers in fish. *Behavioural Ecology and Sociobiology*, 8, 143–5.

Thorpe, W. H. 1959: Learning. *Ibis*, 101, 337–53.

Tinbergen, J. M. 1981: Foraging decisions in starlings (*Sturnus vulgaris* L.) *Ardea*, 69, 1–67.

Tinbergen, N., Broekhuysen, G. J., Feekes, F., Houghton, J. C. W., Kruuk, H. and Szulc, E. 1962: Egg shell removal by the black-headed gull *Larus ridibundus*, L.; a behaviour component of camouflage. *Behaviour*, 19, 74–117.

Tinbergen, N. and Perdeck, A. C. 1950: On the stimulus situation releasing the begging response in the newly hatched herring gull chick (*Larus argentatus*). *Behaviour*, 3, 1–39.

Tschanz, B. 1968: Trottellummen. *Zeitschrift für Tierpsychologie, Beiheft*, 4, 1–103.

Vehrencamp,. S. L. 1978: The adaptive significance of communal nesting in Groove-billed Anis (*Crotophaga sulcirostris*). *Behavioural Ecology and Sociobiology*, 4, 1–34.

Vince, M. A. 1969: How quail embryos communicate. *Ibis*, 111, 441.

—— 1974: Development of the avian embryo. In B. M. Freeman, and M. A. Vince, (eds), *Development of the Avian Embryo, a Behavioural and Physiological Study*, London: Chapman & Hall.

Vleck, C. M. and Kenagy, G. J. 1980: Embryonic metabolism of the forc-tailed storm petrel. *Physiological Zoology*, 53, 32–42.

Vleck, C. M. 1981: Energetic cost of incubation in the Zebra Finch. *Condor*, 83, 229–37.

Williams, C. 1974: Crecking behaviour of the Shelduck *Tadorna tadorna* L.. *Ornis Scandinavica*, 5, 131–43.

Wingfield, J. C. and Farner, D. S. 1978: The endocrinology of a breeding population of the white crowned sparrow (*Zonotrichia leucophrys pergetensis*). Physiological Zoology, 51, 188–205.

Winter, P., Handley, P., Ploog, D. and Schott, D. 1973: Ontogeny of squirrel monkey calls under normal conditions and under acoustic isolation. *Behaviour*, 47, 230–9.

Wolters, H. J. 1978: Some aspects of role taking behaviour in captive family groups of the cotton-top tamarin *Saguinus oedipus oedipus*. In H. Rothe, H. J. Wolters, J. P. Hearn, (eds), *Biology and Behaviour of Marmosets*, Göttingen: H. Rothe.

3 Parental Behaviour in Rodents

J. C. Berryman

INTRODUCTION

It is estimated that there are 1,680 rodent species and yet of these very few have been the subject of parental behaviour research. Of those rodents studied it is the female's behaviour which has been the focus of research interest, and indeed any caring behaviour shown by males in relation to the young is often described as *'maternal'*. Whatever the label attached to this behaviour in the male, his role in parental behaviour has been little researched and thus parental behaviour occupies only a small space in the majority of reviews on parental behaviour, a pattern reflected in Elwood's (1983a) comprehensive text, and the current chapter.

Definitions of maternal behaviour vary. Rheingold (1963) described the term as 'somewhat troublesome' (p. 5) and used it to refer to behaviour shown by the female, as well as any other members of the species which have commerce with the young. Richards (1967), on the other hand, used the phrase to cover behaviour shown by females from the time of conception until the weaning of the young. He did not restrict its use to behaviour related solely to mother–young interactions, and his review also included animals other than pregnant or lactating females. Noirot (1972a) used maternal behaviour to cover nest-building, activities during parturition, cleaning and nursing, but she also discussed it in relation to non-maternal females and males (Beniest-Noirot, 1958).

It is clear then that the term maternal behaviour has been used in relation to animals of both sexes and is thus not specifically related to the behaviour of pregnant or lactating mothers. As a result of this, paternal

---

I should like to thank Dr U. Weidmann and Dr J. Tattersall for their help in translating various papers, and Ms Lesley Hand for typing the manuscript. In writing this chapter I have come to feel that some of the experiments which I have described involved unnecessary pain, or death, to the animal subjects. I would like to draw the reader's attention to the ethical problems which arise when carrying out this type of research and would suggest that the future use of some of the procedures reported in rodent parental behaviour research should be seriously questioned by all of us working in this field.

behaviour is an even less precise term and it is also less often defined (for example, Elwood, 1983b). In order to avoid confusion when referring to the many papers on maternal behaviour, I shall use a broad general definition of maternal behaviour which includes behaviour of non-maternal females as well as that of mothers, and activities that are directly and indirectly (nest-building for instance) concerned with care of the young. The term paternal behaviour will be used to cover the behaviour of males towards their own or conspecific's young, and both terms will cover behaviour which may not be described as 'caring'.

Research on rodents does not cover the three major rodent suborders (Simpson, 1945) with equal depth, nevertheless this chapter will include some comment on the research available on the animals in these three groups. It should be noted however that the taxonomic classification of rodents is still a subject of debate (Simpson, 1974); and Wood (1965) has refined this earlier suborder classification to give a superfamilial classification (discussed in Eisenberg, 1981). Nevertheless I propose to use the three suborder labels since these are widely known, and are also used by Eisenberg (1981) as a label for related families. Thus I shall refer to the myomorphs: the *Myomorpha* (mouse-like rodents); the sciuromorphs: the *Sciuromorpha* and related families (the squirrel-like rodents); and the hystricomorphs: the old and new world hystricomorphs or *Hystricomorpha* (the porcupine-like rodents). The final group excludes those rodents which Eisenberg (1981) described as the 'Families of Uncertain Affinities' and these are the gundis and African mole-rats (brief mention will be made of a species of the latter, (*Heterocephalus glaber*), a most atypical rodent.

A wide variety of adaptations to different habitats are found amongst rodents, and this together with a range of types of social organization are relevant to any discussion of the nature of parental behaviour in these animals. Members of the myomorphs show terrestrial, arboreal and semi-aquatic adaptations. Many myomorphs are colonial, but solitary species are also found, for example *Psammomys obesus* (Daly and Daly, 1975). In general myomorphs are altrical with the genus *Acomys* (spiny mice) the notable exception.

Amongst the sciuromorphs adaptations to aquatic, terrestrial and arboreal living are also seen. The beaver (genus *Castor*) exemplifies the former group, and in the latter 'flying' adaptations are also recorded. A variety of types of social organization are found, including mated pairs (the beaver), and even within a genus (for example, *Marmota*) great variations have been documented (Barash, 1974a). The young of these rodents are altrical.

The hystricomorphs are also found in terrestrial, arborial and aquatic habitats. Their social organization includes mated pairs, as in the mara (*Dolichotis patagonum*), colonial living is reported in various cavy species, and more solitary living in *Proechimys* and *Coendou* genera Precocial young are typical in hystricomorphs.

Amongst the rodents which have been less easy to classify there is a species which has extreme adaptations to fossorial life and a form of social organization more typical of social insects than mammals, this animal is the African mole rat (*Heterocephalus glaber*), recently studied by Jarvis (1981).

Elwood (1983b) has divided the types of social organization amongst rodents into five major categories: monogamous pairs with young pre- and/or post-weaning, polygynous groups of various types, two forms of gregarious groupings with *more* or *less* contact between the sexes, and solitary animals in which contact after weaning is at mating only. Of these, male parental behaviour is potentially possible in all but the last category. It has been suggested that the breeding system of mammals, involving pregnancy in the female and infant dependency on the mother's milk, is conducive to mate desertion by the male, because the opportunity to desert the young arises for the male first that is, after copulation (Dawkins and Carlisle, 1976; Elwood, 1983b). Paternal care may also be inhibited by the fact that males are vunerable to cuckoldry (Trivers, 1972), and indeed, amongst rodents multiple paternity may occur within a litter, as is documented for the deermouse (discussed in Daly and Wilson, 1978).

Elwood (1983b) suggested that paternal care arose as a secondry adaptation in rodents. He pointed out that as many rodents have a post-partum oestrus the male may benefit by waiting with a female until after her parturition, in order to mate again with her, and having waited he may also improve his reproductive success by investing in his offspring. Other factors such as difficult environments may also be a determinant of parental care in males, if the survival of the young is jeopardized by mate desertion. Nevertheless it should be noted that conspecifics other than the father may also be involved in the care or protection of the young.

The nature of the parental care shown within a species is influenced by the maturity of the young at birth. Rodent young may be altrical and thus highly immature at birth, or quite advanced as in the precocial species. Parental care need not necessarily be greater in the first group, as is sometimes thought. Indeed many of the precocial species are very susceptible to predation and parental behaviour may be necessary for the protection of young from predators. This aspect of parental care cannot easily be studied in the laboratory and in view of the absence of predators in this context this may have led to a misconception about the nature of parental behaviour in the precocial species.

This chapter is a selective review of parental behaviour in rodents and, where possible, examples are taken from each of the rodent groups outlined earlier. Particular attention is paid to maternal behaviour in pregnancy, at parturition, during the suckling period, and in communication between mothers and infants, with some comment on the behaviour of non-maternal females in relation to the young. The behaviour of the male in relation to the young and his effects on infant development are also discussed. Neither aggression nor cannibalism are covered here, and a full account of these can be found in Ostermeyer (1983). For a detailed study of rodent parental behaviour the reader is directed to Elwood (1983a).

MATERNAL BEHAVIOUR PRIOR TO THE BIRTH OF YOUNG

Maternal behaviour is not confined to the postpartum period when young are present, it can be elicited in non-maternal animals, and in the pregnant animal changes in behaviour which are related to the imminent arrival of young are observed long before parturition (Rosenblatt and Siergel, 1983). Some of these elements of maternal behaviour have been correlated with the hormonal changes of pregnancy and these will be considered later.

Amongst myomorph rodents a considerable volume of research on rats is available. Maternal behaviour in these rodents is not generally evident in virgin females and first behavioural signs of imminent motherhood are observed during pregnancy when the rat's self-licking or grooming pattern changes (Birch, 1956; Roth and Rosenblatt, 1967), with the female concentrating her licking on her posterior regions, the nipples and the anogenital area. It is estimated that licking stimulation accounts for nearly half of the development of secretory tissue and milk synthesis that occurs in pregnancy (Roth and Rosenblatt, 1968). Evidence in support of this comes from Herrenkohl and Campbell's (1976) research which showed that gland development is greatly impaired if rats are put in collars to prevent licking, whilst collared rats which are given daily tactile stimulation in the nipple region (using sable brushes) show full gland development.

Maternal behaviour can be induced in pregnant and virgin female rats by a procedure called 'concaveation' (Weisner and Sheard, 1933) or 'sensitization' (Noirot, 1972a), in which different groups of pups are placed daily, for 24 hours, in the female's cage. Female rats so treated and then tested on day 16 of pregnancy (the duration of which is 22 days), showed shorter retrieval latencies (3 days relative to approximately 7 days) than virgins, or rats at an earlier stage in pregnancy. A number of studies have explored retrieval, and other forms of parental behaviour in non-pregnant animals (Weisner and Sheard, 1933; Cosnier, 1963; Cosnier & Couturier, 1966; Rosenblatt, 1967; Fleming and Rosenblatt, 1974), and research has shown that sensitized virgin females can be induced to display the full cycle of maternal behaviour if given fresh pups daily, appropriately advancing in age (Reisbick et al., 1975). Rats do not usually eat the placentas of newborn pups presented to them during pregnancy, but a small percentage of females have been found to do so (Kristal et al., 1981).

Some strains of mice, unlike rats, are much more responsive to pups both during and prior to pregnancy. Beniest-Noirot (1958) has demonstrated that laboratory mice display maternal behaviour on first exposure to pups (males also show similar behaviour). However work by Jakubowski and Terkel (1982) revealed that, in contrast to laboratory mice, naïve wild house mice cannibalize young. However, after experience of young, through social rearing with parents and their offspring, these authors found that wild male mice will exhibit paternal behaviour whilst the majority of wild virgin females continue to show infanticide. These authors suggested that research on the domestic mouse may not reliably represent the species *Mus musculus*, and the artificial selection exerted in developing various strains of laboratory mice has acted against the trait of infanticide. This point should be borne in mind when considering the behaviour of all domesticated species in relation to their wild counterparts.

In hamsters, nest-building occurs early in pregnancy (Richards, 1969), but few other signs of maternal behaviour are evident. Sensitization can be effective in pregnant female hamsters (Buntin et al., unpublished work cited in Rosenblatt and Siegal, 1983) although the incidence of retrieval was found to be extremely low until day 15 of pregnancy (typically 16 days in length). Virgin females attack and cannibalize young (Richards, 1966), but as in rats Buntin et al. showed that maternal behaviour can be elicited once females have been sensitized. The Mongolian gerbil (*Meriones unguiculatus*) is like the hamster in exhibiting pup cannibalism during pregnancy (Elwood, 1977), but in very late pregnancy maternal responsiveness is shown.

In sciuromorphs and hystricomorphs little systematic research is available on early signs of parental behaviour in females. Changes in behaviour during pregnancy include decreased activity reported in the guinea pig (Louttit, 1927) and the green acouchi (Kleiman 1972), and in the latter the characteristic purr call, used by mothers with their young, is heard in the week before parturition. No change in the grooming pattern, characteristic of the rat, is seen in green acouchi (Kleiman, 1972). Nonpregnant female guinea pigs may be unresponsive to infants with which they are housed (Berryman and Fullerton, 1976) but some forms of parental care, licking, for example, are shown by such females in the green acouchi (Kleiman, 1972).

MATERNAL BEHAVIOUR IN THE PREPARTUM PERIOD AND PARTURITION

In the period immediately prepartum a variety of hormonal changes take place which initiate parturition and lactation, and in a number of species (of the myomorph type) maternal behaviour is fully developed at this time. In rats nest building and retrieving are shown 34 and 28 hours (respectively) prior to parturition (Rosenblatt and Siegel, 1975). In domestic mice this test is, of course, inappropriate since maternal behaviour can be readily elicited by presentation of pups, but in the strains where cannibalism is shown, no prepartum onset is reported (Jakubowski and Terkel, 1982). In hamsters full maternal behaviour (retrieving crouching and suppression of cannibalism) is shown by about three-quarters of nulliparous females 2½ hours before parturition, with earlier maternal responsiveness shown by primiparous females (Siegel and Greenwald, 1975), and Mongolian gerbil females (Elwood, 1977) show a variety of parental responses when they are near term such as licking, retrieving and nest-building, but behaviour such as assuming the nursing posture is observed only rarely.

The mother's first experience of her young comes at parturition and yet, with the exception of the rat, parturition has been described in detail for relatively few rodents. It is already clear that parturition itself is not necessary for maternal behaviour to be shown by certain strains of mice and rats; indeed rat mothers of pups delivered by Caesarian section treat their own or other pups with appropriate maternal care within 24 hours, despite the mother's inability to lactate normally (Weisner and Sheard 1933; Labriola, 1953; Moltz et al., 1966). Nevertheless Rosenblatt and Siegel (1983, 34) have suggested that 'the heightened arousal which characterizes the parturient female and the attractive properties of the newborn' make parturition an optimal period for facilitating the interaction between mother and young.

Rat behaviour at parturition has been described by Weisner and Sheard (1933) who recorded that rats lie quietly during parturition but as soon as a foetus passes into the vagina most animals assume the 'head between heels position' (22) to lick the foetus at delivery. Rosenblatt and Lehrman (1963) described four phases during the birth of an individual rat pup, and Dollinger et al., (1980) in their study of rat parturition recorded that after the first pup birth there is a high incidence of pup-oriented behaviour (licking), then a phase when the female is self-orientated.

Amongst other rodents the details of parturition have been described for a number of species including the spiny mouse (Dieterlin, 1962) where 'obstetrical aid' to parturient females is provided by other females; the guinea pig (Avery, 1925; Kunkel and Kunkel, 1964; Naatgeboren, 1962, 1963); and the green acouchi (Kleiman, 1972). In the latter species the mother crouches during delivery with her hindquarters raised and the infant is dropped behind her. This posture in the green acouchi is said by Kleiman, to be typical of the mothers of precocial young, but guinea pigs at parturition have been observed to curve their heads around sideways, across the front leg, in order to lick the emerging foetus (Kunkel and Kunkel, 1964).

HORMONAL INFLUENCES ON MATERNAL BEHAVIOUR

Turning to the hormonal basis of maternal behaviour, it is the rat, once again on which the bulk of work has been done, although some work on the mouse, hamster and gerbil is also available.

The rat shows little or no maternal responsiveness towards pups prior to giving birth (except after prolonged sensitization in experimental studies) and thus it is the hormonal changes towards the end of pregnancy and in lactation, which may be influential in eliciting maternal behaviour. The end of pregnancy in the rat is marked by a decline in the circulating levels of progesterone and a rise in blood levels of oestradiol and prolactin. An oestrus cycle follows parturition. First evidence that hormones played a role in the initiation of maternal behaviour came from a study by Terkel and Rosenblatt (1968). These researchers showed that if a non-pregnant female rat was injected with the blood plasma of a postpartum female (within 6 hours of parturition) maternal behaviour was stimulated and females tested with pups retrieved them on average within two days. More recently Prilusky (1981) has demonstrated the possibility of the induction of maternal behaviour in non-pregnant females by treatment of these females with extracts of brain obtained from lactating mothers. However neither of these studies revealed the crucial factors in maternal behaviour elicitation.

Hysterectomy in the rat produces a rapid decline in circulating progesterone (Rosenblatt and Siegel, 1983) and studies in which pregnant females were hysterectomized on day 8, or at intervals later into pregnancy (Lott and Rosenblatt, 1969; Rosenblatt and Siegal, 1975), showed that, as pregnancy advanced, the onset of maternal behaviour was more rapid after hysterectomy relative to that in females which remained pregnant. A hysterectomy may be carried out in order to simulate the normal changes which occur at the end of pregnancy, but this of course does not indicate precisely whether the decline in progesterone is the decisive factor. There is indirect evidence that a rise in ovarian oestrogen secretions also occurs after hysterectomy (Bridges et al., 1978; Rosenblatt et al., 1979) since lordosis behaviour and ovulation occur about 48 hours after this surgery (direct measurements of oestrogens have not been made). Oestrogen may therefore also be an important factor in the onset of maternal behaviour after hysterectomy. Hysterectomy in late pregnancy is also associated with an increase in prolactin levels in the two days after surgery (Rosenblatt and Seigel, 1983. Thus it is possible that a number of hormones may be influential in stimulating maternal behaviour.

Ovariectomy and hysterectomy in pregnant female rats produce even longer latencies for maternal behaviour than does hysterectomy alone, implicating the ovarian hormone in the elicitation of maternal behaviour. Oestradiol benzoate injections in ovariectomized and hysterectomized pregnant females were found to induce a rapid onset of maternal behaviour (Siegel and Rosenblatt, 1975). This result suggested that the rise in oestrogen at the end of pregnancy, as early as 72 hours before parturition (Rosenblatt and Siegel 1975; Siegel and Rosenblatt, 1978), is responsible for the onset of maternal behaviour at parturition, and that the absence of progesterone during the period of oestrogen stimulation is a necessary condition for this behaviour to occur. However Bridges et al., (1977) have shown that even in the absence of the ovaries, after hysterectomy-ovariectomy in late pregnancy, maternal behaviour was still shown briefly about 24 hours after surgery. It is suggested that (Rosenblatt and Siegel, 1983) this short-lived increase in the instances of maternal behaviour may be due to the withdrawal of progesterone, which may have the effect of unmasking the stimulating effect of oestrogen which is also present, but at a low level, throughout pregnancy.

If the normal decline in progesterone is prevented by giving females progesterone in the last four days of pregnancy, initiation of maternal behaviour is prevented in about half of the nulliparous females so treated (Moltz et al., 1969). Moltz and Weiner (1966) showed that ovariectomy has the same effect, and Rosenblatt and Siegel (1983) have suggested that this is probably because this procedure eliminates the terminal rise in oestrogen. However, later ovariectomy on, or after, day 20 does not have this effect (Terkel, 1970, unpublished work cited in Rosenblatt and Siegel, 1983).

Studies exploring the role of prolactin secretions at the end of pregnancy in stimulating maternal behaviour (Numan et al., 1977; Rodriquez-Sierra and Rosenblatt, 1977) indicate that it has no effect on oestrogen-induced maternal behaviour.

In summary, the major hormonal stimulus to the *onset* of maternal behaviour in rats (Rosenblatt and Siegel, 1981) is oestrogen, provided a prior decline in progesterone has occurred.

Turning to other rodents, the mechanisms controlling maternal behaviour appear to be surprisingly different from those described in the rat. In certain strains of mice nearly all aspects of maternal behaviour can be elicited by exposure to appropriately aged young (Noirot, 1972a), and this applies to *both* males and females. However nest-building has been explored in relation to hormonal correlates. Koller (1952, 1956) observed that there is a sharp increase in nest-building in pregnant mice around the time that there is a rise in progesterone levels (around the 4th or 5th day of pregnancy), and he has shown that progesterone increases nest-building activity in both intact and ovariectomized females, to levels similar to those of pregnant females. Lisk et al., (1969) have explored the relationships between progesterone and oestrogen and have evidence to suggest that nest-building occurs when oestrogen is low and progesterone high. There is some evidence that prolactin may also increase nest-building (Voci and Carlson, 1973). Indirect evidence suggesting that hormones influence the maternal behaviour of mice is reviewed by Rosenblatt and Siegel (1981).

The hamster is unlike either the mouse or rat, showing a different hormonal profile at the end of pregnancy. Progesterone and oestrogen both decline, whilst prolactin is at a relatively low but unchanging level (Rosenblatt and Siegel, 1983). The administration of oestradiol, and/or progesterone in late pregnancy does not, surprisingly, affect the quality or onset time of hamster maternal behaviour (Siegel and Greenwald, 1975). Aggression to intruders, which is a notable feature of hamster behaviour during pregnancy and the post-partum period, appears to be controlled by prolactin (Wise and Pryor, 1977), neither postpartum hysterectomy reduce aggression, or, indeed, maternal behaviour.

Research on hormonal influences on other rodents is relatively sparse or inconclusive – some work on the gerbil is available (see Rosenblatt and Siegel, 1981) but amongst other types of rodents (the hystricomorphs, sciuromorphs and related families) indirect evidence is available. In the guinea pig there is suggestive evidence that hormonal changes may be important (Seward and Seward, 1940) and this work will be discussed later.

A number of studies have shown that maternal behaviour in non-pregnant animals can be stimulated by hormone treatment regimes and these studies have been discussed by Rosenblatt and Siegel (1983).

MATERNAL BEHAVIOUR POSTPARTUM

Priestnall (1983) has suggested that the hormonal changes occurring at parturition provide 'a kind of momentum in maternal behaviour' (69) which carries the animal through the transition period between the hormonal regulation of behaviour to the regulation of behaviour through pup stimulation. However Rosenblatt and Siegel (1983) believe that, in the rat, the *transition* between the hormonal regulation of the *onset* of maternal behaviour, and the *maintenance* of maternal behaviour post-partum, is dependent on contact with the pups immediately postpartum. Evidence for this comes from studies in which rat mothers are deprived of their pups; the immediate removal of pups produces a rapid decline in maternal behaviour, whereas if the females are allowed only a brief period of exposure to pups immediately postpartum they respond rapidly to test pups given to them 25 days later, unlike females which are totally deprived postpartum until a similar test at 25 days (Bridges, 1975, 1977).

Rosenblatt and Siegel (1983) have suggested that *maintenance* of maternal behaviour in the rat is 'almost entirely dependent upon pup stimulation' (57) and the role of hormones in the postpartum period is chiefly concerned with lactation and milk let-down, as well as stimulation of sexual behaviour shortly after parturition. These authors have reported a variety of studies in which the removal of either the ovaries, adrenal glands, or pituitary gland, does not interfere with the maintenance of maternal behaviour.

In certain mouse strains pup stimulation is not a necessary factor in the maintenance of maternal behaviour because, as has already been reported, virgin animals will show maternal care on first exposure to the young. However, research on those mice which are non-maternal or cannibalistic to pups, has not been carried out (Rosenblatt and Siegel, 1983). In hamsters pup-stimulation appears to be necessary for about 48 hours postpartum to maintain maternal behaviour for two weeks, but whether the postpartum maternal behaviour is hormonally based is yet to be established (Rosenblatt and Siegel, 1983). Very little research of this nature is available on other rodent species.

A number of myomorph rodents have a postpartum oestrus. This occurs in the rat, mouse and gerbil, with a single oestrous cycle following parturition almost immediately. In the hamster the first oestrous cycle is delayed for about 3 weeks (Rosenblatt and Siegel, 1983). The postpartum oestrus is followed by a prolonged dioestrous phase during which there is a high level of progesterone output, but output of gonadotrophins is low. Pregnancies resulting from postpartum matings appear to be common: this was found to be the case in wild rat populations (Davis and Hall, 1951). The effect of this postpartum oestrus and mating on the mother's interaction with her pups does not appear to be great (Gilbert et al., 1980), and it has been noted that the time of mating tends to occur when there is a natural ebb in maternal behaviour (Ader and Grota, 1970).

Priestnall (1983) has pointed out that during lactation a variety of factors pertaining to both the mother (her hormonal state and her experience) and her pups (their changing size and appearance) all change concurrently. This makes it difficult to establish, without considerable interference, which factors are crucial in determining changes in maternal behaviour.

Rosenblatt and Lehrman (1963) studied changes in maternal responsiveness in the rat by measuring nursing attempts, retrieval and nest-building during a four-week postpartum period (weaning of rat pups typically occurs at about 3 to 3½ weeks (Weisner and Sheard, 1933; Richards, 1967)). The first two measures involved daily testing during which the mother's own pups were replaced by foster pups of a standard age, the nest-building was measured by removing each female's nest and litter and then giving each female access to nest-building material for a limited period before returning the female's litter and observing nest-building. In normal (that is, undisturbed) conditions both retrieving and nest-building are relatively minor elements in postpartum maternal behaviour, so that the artificial nature of this type of test should be considered. Nevertheless Rosenblatt and Lehrman (1963) found during this study that changes in these three components of maternal behaviour resembled those in undisturbed mothers. All three components were present at high levels in most animals during the first 14 days of lactation, with nest-building declining quite rapidly during the third week of lactation, and the other measures showing a similar decline a day or two later. These authors believe that the decline in maternal behaviour is due, in part, to the mother's condition, since foster pups were of a standard age – but evidence that pups also play a role comes from studies in which pups are removed at various stages in lactation. The immediate removal of pups at birth effectively prevents the cycle of maternal behaviour occurring, and removal at 9 days postpartum shortened lactation (Rosenblatt and Lehrman, 1963), whereas if an older litter is replaced by a younger litter (under 10 days of age) maternal responsiveness can be extended beyond the usual period (Weisner and Sheard, 1933). Any disturbance to the mothers caused by disruptive testing procedures has been examined by Grota and Ader (1969), who found that undisturbed mothers showed a more gradual decline in maternal behaviour than was indicated by Rosenblatt and Lehrman's study (1963).

In mice, the weaning of young occurs at about the same time as it does in rats (Richards, 1967), and Noirot (1964 a, b, c, 1965, 1966) has carried out detailed studies of maternal responsiveness in virgin, primiparous and multiparous mice, measuring nursing, retrieving, licking and nest-building. The mice were presented with a standard stimulus pup for the first three measures and all four responses declined, but at different rates, over lactation. Priestnall (1972), using a slightly less disruptive procedure than Noirot, found that licking and nest-building declined more sharply than nursing and the time spent in the nest. The changes observed over the postpartum period for nursing, licking and nest-building were similar to those observed by Noirot, but the incidence of the last two was lower in Priestnall's work, probably because of his less disruptive testing procedure.

Research on gerbils has been carried out by Elwood (1975) who observed pairs of gerbils living together during the first 24 days of lactation. Nest attendance was found to be high throughout lactation, dropping off only slightly towards the end. Other behavioural patterns, such as licking and nest-building, were shown by both parents (although with a higher incidence by females during the earlier part of lactation). Elwood found that retrieval was rarely observed in undisturbed conditions, but work by Waring and Perper (1979) on retrieval, which involved the removal of pups from the nest, showed that this behaviour was less likely in the first

2 days than in the succeeding 2 days postpartum. It would appear that retrieval improves with practice.

The hamster's behaviour is rather different from that of the rodents described above. This rodent has no immediate postpartum oestrus and females are much more aggressive than mice, rats and gerbils. Females presented with foster-pups at various stages during the female's lactation period are most likely to accept pups at 7-14 days of age (Rowell, 1960). Females presented with foster litters of a standard age (that is, not growing older over lactation), ceased to lactate at the *same* time as females rearing their own young. Rowell (1960) suggested that the mother is not affected physiologically by her litter. Hamster mothers are aggressive during weaning, which occurs at about 3 to 3½ weeks (Richards, 1967), since pups may be attacked at this time and so the female builds herself a new and separate nest towards the end of the suckling period.

Amongst the rodents so far described, litter size is highly variable. Research on mice indicates that size can be affected by a variety of factors: parity of the mother (Biggirs et al., 1962); age of the mother (Rugh and Wohlfromm, 1967); and strain (Festing, 1968). Individuals in smaller litters tend to have a higher birth weight and this increased size may be maintained during the suckling period. The number of young within the litter must influence the amount of stimulation received by the mother during suckling and through a variety of other sensory stimuli. Studies on mice (Sietz, 1954, 1958; Priestnall, 1972) indicate that, with smaller litters, females show more maternal responsiveness (Sietz produced a total maternal behaviour score) and spend more time in the nest. Less time in the nest is also spent by gerbil mothers of larger litters (Elwood and Broom, 1978).

The effect of nest temperature on the mothers time in the nest will be discussed later.

Little work is available on sciuromorph rodents although Barash (1974b) recorded some observations of maternal behaviour in *Marmota monax* (the woodchuck). He was unable to distinguish nursing clearly, but mother-infant contact was seen to rise to a peak during the first week of lactation and then decline towards weaning, which occurs at approximately 38 days. The aggressive responses of the mother towards her young increase towards weaning, when the dispersal of the young occurs. Woodchucks are the most solitary of the marmot species and in some species (for example, *M. flaviventris*) dispersal of the young may not occur until the following year.

In hystricomorphs the young are typically precocial and thus the kinds of parental behaviour displayed are rather different from those described above. Nest-building occurs less frequently although some provision for the young may be made: mara (*Dolichotis patagonum*) make a burrow for the young but the mother never uses it herself; young are born at the entrance and they occupy the burrow (often with other offspring) for the first six months of life (MacNamara, 1981). Retrieval of young is rare or absent in many members of this group (Kleiman, 1974) with the exceptions of *Octodon* and *Octodontomys* which have less mature offspring.

Nursing may, in many cases, occur in an open exposed area and Kleiman (1974) reported that a number of species nurse young other than their own: *Cavia, Microcavia, Galea* and *Octodon* show this behaviour.

MacDonald (1981) reported indiscriminate suckling by capybara (*Hydrocherous hydrochaeris*), and indeed a female may be seen with very large numbers of young clustered around her (one instance of 18 was recorded, when the average litter size is four). Whether these crèches are beneficial to young is not known, although Kleiman (1974) has suggested that promiscuous nursing may be an anti-predator strategy to provide milk at very short notice.

Suckling bouts vary greatly in length in hystricomorph species (Kleiman, 1974); amongst those genera which have nests (*Octodon, Octodontomys* and *Proechimys*) females may spend periods of half an hour or so huddled over their young during which time there are brief periods of sucking, interspersed with quiescent periods with the young still attached to the nipple. In other genera the caviids and dasyproctids (agoutis), the young of which suckle in the open, suckling may last for a few minutes only. This short nursing bout is likely to be influenced by the vulnerability to predation of these animals. Kleiman (1974) noted that the establishment of teat ownership may be important to reduce fighting and hence make the most efficient use of the time available for nursing in species which suckle in exposed conditions. *Myoprocta* young have a 'teat order' and *Pediolagus* young alternate teats on the side which they are sitting, but they do not change sides within a nursing bout.

Although many hystricomorphs can eat solids within a day or two of birth, they often suckle for quite long periods. Weaning age may be at 8, 12 or even 16 weeks in some genera (see table 1). Thus the pattern of maternal care is quite unlike that seen in the myomorph rodents described earlier where weaning ceases relatively early. Late weaning may be an adaptation to maintain the bond between mother and young during the period when the young are highly vulnerable to predation (Kleiman, 1974). Thus precocity at birth does not indicate necessarily that parental care is less important; indeed it may be an adaptation which, combined with parental care, is vital for infant survival.

Weaning need not mark the end of the young's dependence on mother or parents. Kleiman (1974) suggested that in hystricomorphs a distinction should be made between: independence from mother's milk, the end of lactation, the end of suckling, and the termination of the mother–infant bond.

Amongst hystricomorphs female parental behaviour has been studied most notably in the green acouchi (Kleiman, 1972) and the guinea pig

(Seward and Seward, 1940; Kunkel and Kunkel, 1964; Berryman, 1974, 1981; Fullerton, et al., 1974), and it is the latter which will be considered in detail here.

Guinea pigs show a postpartum oestrus (Young, 1969), but the effect of postpartum matings on the female's care of the young has not been studied. In guinea pigs the lactation period is usually about three weeks (Seward and Seward, 1940), but it may be a week or two longer in multiparous animals (Nagasawa et al., 1960). Seward and Seward (1940) measured the 'strength' of maternal drive by recording the latency of mothers to cross a barrier or 'hurdle' between herself and her young, and they found that latencies increased with days since parturition, but were independent of the amount of milk present in the mammary glands. These authors also controlled factors such as habituation to the test situation, and to the age of stimulus infants, concluding that maternal behaviour was primarily under hormonal control, since females were not more maternal in attempting to reach infants younger than their own.

Observational work by Fullerton et al., (1974) showed that whilst suckling occurred during up to 40 per cent of the daily observations (by time sampling) recorded, remaining in contact with the mother or other lactating females present was the other major activity observed (see figure 1). Huddling together may well be an important factor in keeping young guinea pigs warm and the very high fat content of guinea pig's milk (Read, 1912) may also help to offset heat loss.

Apart from providing milk and warmth, the mother also fulfils the essential sanitary needs of her offspring. Guinea pigs, like other rodents, cannot urinate or defaecate without the stimulus of licking from the mother and this activity is normally observed during the first 5 days or so after birth (own observations). In the laboratory guinea pigs may be successfully weaned at 5 days, but the absence of predators, and adequate food and warmth, has in the part misled observers into thinking that parental care in these animals is not very pronounced (Kunkel and Kunkel 1964).

Litter size is a major factor in determining the length of the gestation period (Goy et al., 1957), the size of the young (Ishii, 1920) and the nursing behaviour of the mother. Mothers of larger litters show more nursing behaviour (Stern and Bronner, 1970), but with smaller litters there is a marked drop in nursing time by the end of the fourth week, giving support to the Sewards' (1940) view that hormonal control may be the major factor in determining maternal behaviour. Suckling does not appear to be extended by the presence of more, typically smaller, young (Ishii, 1920), although studies attempting to extend lactation by the replacement of older infants by younger foster infants do not appear to have been reported.

Finally the maternal behaviour of a most unique rodent species (which falls outside the three types so far considered) must be mentioned.

Table 1 Variations in nursing patterns in hystricomorph rodents

Genus	Average duration of nursing bouts (min)	Nursing position	Nursing site	Teat position	Promiscuous nursing	Weaning age (weeks)	Average litter size	Reference
Hystrix	–	sitting	–	lateral	–	16	1-2	Mohr, 1965
Cavia	–	sitting, lying	open, exposed area	ventral	occasionally	3-4	2.1	Rood, 1972
Microcavia	6	sitting, lying	open, exposed area	ventral	yes	3-4	2.8	Rood, 1970, 1972
Galea	–	sitting, lying	open, exposed area	ventral	yes	5	2.7	Rood, 1972
Dolichotis	–	sitting, lying	open, exposed area	lateral	no	8-12	1-2	Mohr, 1949
Pediolagus	5	sitting	open, exposed area	lateral	no	>4	2	D. G. Kleiman & S. Wilson, unpublished
Hydrochoerus	–	sitting	open area	?	no	16	4	Ojasti, 1971; Zara, 1973
Dasyprocta	2	lying, sitting	open, exposed area	ventral	no	12	1	Smythe, 1970a; Roth-Kolar, 1957

Myoprocta	4	lying, sitting	open, exposed area	ventral	no	8–12	2	Kleiman, 1972
Lagidium	–	–	open, exposed area	lateral	–	>4<16	2	Pearson, 1948
Myocastor	–	sitting	open area or nest	lateral	–	8	5–6	Newson, 1966; Klapperstück, 1954
Capromys	–	sitting	open area	lateral	–	22	2	Taylor, 1970
Octodon	25	huddle	nest	ventral	yes	5–6	5	S. Wilson & D. G. Kleiman, unpublished
Octodontomys	25	huddle	nest	ventral	–	5–6	2	S. Wilson & D. G. Kleiman, unpublished
Proechimys	25	huddle	nest	ventral	–	6	2.8	Maliniak & Eisenberg, 1971, unpublished; Enders, 1935
Abrocoma	–	sitting, lying	nest	ventral	–	4	2–3	J. F. Eisenberg, unpublished

Source: From D. G. Kleiman, 1974, reproduced with permission of Academic Press.

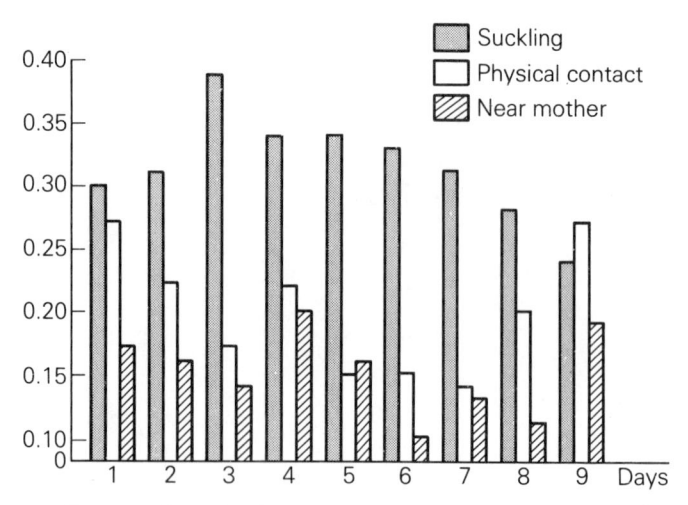

Figure 1 Frequencies of observed suckling, physical contact and near responses during the first 9 days of observation. All responses refer to both mothers present and not just an infant's biological mother. The maximum possible total for the three measures per day is 1.00. (Reproduced with permission from C. Fullerton, J. C. Berryman and R. H. Porter, *Behaviour* 48 (1974), 145-56)

Within the rodent 'Families of Uncertain Affinities' the mole rat (*Heterocephalis glaber*) is included (see Eisenberg, 1981). These rodents display a type of reproductive behaviour quite unlike that found in other rodents. Jarvis (1981) reported that these animals live in large colonies (of about 40 members) in which only one female breeds, whilst the remaining animals constitute two or three castes: frequent workers, infrequent workers, and non-workers. Only the breeding female suckles the young (up to 12 per litter) but other castes sleep with them, keep them warm, carry them about and bring food to the weanlings. Thus there is communal care of young by both sexes. Little published work is available on these animals, but it is clear that their behaviour is quite unlike that of other rodents.

COMMUNICATION

In the preceding sections of this chapter there has already been some discussion of the way in which pup stimulation can affect maternal behaviour in rodents. This section will extend this account by covering in more detail communication between mothers and infants – with the greater emphasis on the mother's behaviour in this context. A very considerable number of studies have looked at communication between mothers and young in the myomorphs and related families (see Noirot, 1972b; Sales and Pye, 1974), but once again very much less is available on other types of rodent.

This section will deal with communication in its broadest sense as defined by Marler (1967), Rowell (1972) and Scott (1968), since the more strict definitions such as that of Deag (1980), or Otte's (1974) notion of 'functional signals', do not permit the inclusion of chemical signals determined by dietary factors in the context of communication.

Rodent species, as has been noted, are found throughout the world and occupy a very wide range of habitats, however the majority of rodent species are small, many are nocturnal, and some fossorial. These factors have tended to reduce the reliance on visual communication in many species, and for burrowing animals it is the tactile, chemical, and auditory stimuli which are most effective in influencing the behaviour of conspecifics. In rodent species which have altrical young, sensitivity to tactile and chemical signals are often the only fully functional senses at birth, and thus this determines the nature of maternal communication.

Tactile and thermal signals

Close physical contact between parents (usually mothers) and young, and between siblings when parents are absent, is characteristic amongst rodents. Contact other than suckling is common even when young rodents are precocial and homeothermic. However since many rodent species are poikilothermic at birth and are thus highly dependent on parents and nest for warmth, it is not surprising that thermal and tactile sensitivity is present at birth. The rat shows tactile sensitivity several days before birth, and the precocial guinea pig is well advanced, responding to thermal and tactile stimuli very early in gestation (Gottlieb, 1971; Carmichael and Lehner, 1937).

Leon et al. (1978) have shown that the nesting behaviour of the mother rat is influenced by both pup temperature and ambient temperature. By huddling with her pups, the pups' temperature rises, but this increases the mother's body temperature eventually causing her to leave the nest and thus terminate the nesting bout. This occurs when her ventral temperature reaches a critical value. At higher ambient temperatures nest bouts are shorter, but at cooler temperatures the termination of the nest bout is not accounted for solely in terms of an increase in maternal temperature. High levels of ultrasonic calling by pups, which may be stressful to the mother (Elwood and Broom, 1978), may also be a factor in influencing the length of the nest bout.

Hofer (1978) reported several studies which indicate that warmth from the mother is responsible for physiological changes in the rat pup during the sucking period. Pups which have had several days' exposure to room temperature in the absence of the mother, show profound motor deficits, as well as slowing biological maturation of the brain, internal organs and skeleton (Stone et al., 1976). Plaut and Davis (1972) found that pups which were separated from their mothers had a higher mortality rate than those housed with a non-lactating adult female at 12–13 days of age. It was suggested that the warmth provided by the non-maternal animal promoted survival of pups.

Tactile stimulation of the young has a vital role in their survival. Porter (1983) has suggested that cleaning the neonate may be important in facilitating the onset of breathing, and the licking of the anogenital region of the young is essential in the stimulation of defaecation and urination in both precocial and altrical rodents. Retrieval of the young by use of the mouth is characteristic of many rodent species when the young become separated from the nest. In mice, rats and hamsters a dorsal grip is common (Ewer, 1968), whereas in hystricomorphs retrieval is less often found, although the green acouchi (Kleiman, 1972) holds her infant, by the mid-section. The young themselves may take an active role in staying attached to their mother by climbing on her back, as in the capybara when swimming (MacDonald, 1981) or by clinging to her neck, as in *Sciurus hoffmanni* (Lanf, 1925), whilst the mother squirrel holds them by the skin of the ventral surface.

Chemical signals

A wide range of scent producing glands have been identified in rodents (Brown, 1979) and so it is not surprising to discover that chemical signals can have a variety of effects in influencing parent–young relations. Chemical signals have the quality of being effective in the absence of the signaller and they play a vital role in identifying individuals, groups, nests or territories.

The ability to perceive chemical stimuli is functional in neonatal altrical rodents and consequently a number of researchers have explored the extent to which infants can discriminate between olfactory stimuli. Several studies have indicated that rodents will show a preference for home-cage nest odours. This has been demonstrated in infant rats from 12 days (Gregory and Pfaff, 1971) and in rat mothers from just before birth until the second week postpartum (Bauer, 1983). Although the preference for nest odour does not indicate the nature of the chemical stimuli salient in determining these preferences, evidence is accumulating which suggests that lactating females produce a chemical signal which is attractive to the young and possibly also to themselves.

Leon and Moltz (1971) have termed this chemical signal from the lactating female the maternal pheromone and have talked of the 'pheromonal

Parental Behaviour in Rodents 63

bond' (1972); this appears to exist in a number of rodent species including: the rat (Leon and Moltz, 1971); spiny mice (Porter et al., 1978); mice (Breen and Lesher, 1977); and probably gerbils (Gerling and Yahr, 1982). The infant rodents do not necessarily show a preference for their *own* mother over another lactating female however, although this does not necessarily mean that they cannot make this distinction (Spencer-Booth, 1970). In view of the considerable quantity of research on this topic only that on the rat will be detailed here.

Leon (1974, 1975) has shown that the maternal pheromone is contained in cecotrophe, an unformed anal excrement produced in large quantities by lactating rats. Leon and Moltz (1972) found that the period over which pups are responsive to maternal pheromone is closely synchronized with the stage postpartum during which the female releases these attractive odours: this is between about 14 and 27 days postpartum (Gregory and Pfaff, 1971; Leon and Moltz, 1971, 1972), and the onset of pheromonal emission is linked with the stage when rat pups are capable of moving around and hence may become separated from the mother. Emission of the pheromone is under endocrine control, in particular prolactin (Leon and Moltz, 1973), however stimulus characteristics of the young are also a factor in the onset time of pheromonal emission. If a female's own young are replaced continuously by neonatal pups during the first two weeks postpartum, pheromonal emission is prevented, but if older pups are substituted for the mother's own younger pups emission cannot be hastened (Moltz and Leon, 1973). It has been shown that the properties of anal excrement are a direct function of dietary factors (Leon, 1975; Galef, 1981; Clegg and Williams, 1983) and rats can be raised on a diet which inhibitis maternal pheromone production (Leon et al., 1977); this has led to the use of the term pheromone being questioned by some researchers (Clegg and Williams, 1983). A discussion of the original definition (Karlson and Lücsher, 1959) and of the problems surrounding its current usage, are found in Beauchamp et al., (1976) and Brown (1979). It is clear however that chemical cues from the mother have attractive qualities for pups at a stage when they are mobile but still dependent. Since diet influences these chemical cues this has the added benefit of attracting pups only to those females which have identical diets. In the wild state it is likely that this limits the pups to a very few and probably only one female: their mother.

Apart from attracting the separated young rats to their mother, or another lactating female, and thereby enhancing their chances of receiving parental care, there is evidence that attraction to maternal faeces and the consumption of them, has adaptive benefits for the pups. Moltz and Lee (1981) have evidence to suggest that by consuming lactating females' faeces, the young ingest substances which play a part in the immunocompetance of the gut and in brain development. The authors suggested that the pheromone is 'exquisitely designed to promote the survival of the young' (305) by first inducing the pups to approach and then consume what the mother provides anally. For further details on the existence of the maternal pheromone in other myomorph rodent species Porter (1983) has reviewed these in some detail.

Studies of olfactory communication in sciuromorphs and hystricomorphs are rather more limited, although some research is available in the latter. Kleiman (1974) recorded that a number of genera show enurination over their young by the male (e.g. *Myoprocta, Cuniculus, Octodontomys* for example), and several studies on the guinea pig reveal that mothers and infants probably use olfactory cues in recognition. An artificial odour (for example ethyl benzoate) which has been applied to their young will be preferred by mother guinea pigs in a choice test between this odour and an unfamiliar one; research has also implicated natural olfactory cues in maternal recognition by infants (Porter et al., 1973; Fullerton et al., 1974). Work by Carter and Marr (1970) demonstrated that infants will show a preference for a familiar (artificial) odour, when given a choice between this and a natural or unfamiliar one, thus indicating that chemical cues may be important in mother-recognition.

Finally, olfaction plays a role in nipple location by the young. Olfactory bulbectomy in mice was found to result in impaired nipple location (Cooper and Cowley, 1976) and it led, in some cases, to inadequate nursing and death in rats (Tobach, 1967). Research has shown that the mother's behaviour is not disrupted by this operation, indicating that inability to find the nipple is due to changes in the pup postoperatively. When pups were placed in direct contact with the nipple they were able to attach, and so it appears that the disruption was to nipple location from a distance (Teicher et al., 1978). Nursing was also found to be disrupted if 2 week old rat pups were treated with zinc sulphate intranasally (Hofer, 1976) and a similar result was found in pups with females which have had chemical lavage of the nipples (Teicher and Blass, 1976). Apparently nipple attachment in the rat is initially elicited by the amniotic fluid contained in the mother's saliva which is deposited during grooming at parturition on the females nipples; thereafter nipple attachment is determined by the saliva deposited on the nipples by the pups themselves (Blass and Teicher, 1980).

Acoustic signals

Maternal behaviour in rodents is greatly influenced by the acoustic signals emitted by the offspring and the females frequently direct a variety of vocalization specifically to their young at this time, the chief function of which appears to be to maintain proximity with the latter. For parents of altrical young, true acoustic communication between them and their offspring is not possible since the infants typically do not respond to sound until some days after birth; rats, mice and deermice begin to respond at about 10 days of age (Gottlieb, 1971). Nevertheless neonates of these and other myomorphs emit various ultrasonic vocalizations which are important in attracting the mother or both parents, but which the young themselves cannot hear (Noirot, 1972b; Sales and Smith, 1978).

Early ultrasonic calling has been linked with the infant's inability to thermoregulate. Homeothermy is not achieved in altrical rodents until some days or weeks after birth (see Okon, 1970, 1972; Dieterlin, 1961). In various species of rodent, for example the house mouse, ultrasonic calls begin to be emitted as the body temperature of the young drops below 33 °C (Okon, 1970). Both rate and sound pressure level of the calls vary with age and temperature. In many rodents studied (reviewed by Smith and Sales, 1980) calling in response to mild cold stress (22°C) ceases at about the time that homeothermy is achieved; in the wild house mouse for example calling ceases at about 13 days in mild cold stress; but at 19–20 days in more severe cold stress (23°C) the more precocial spiny mouse pup ceases to call much earlier, at about 7 days (Sales and Smith, 1978). Thus this calling is associated with the stage when infant rodents are poikilothermic.

Infant mice of various species also call (independently of the cold) in response to tactile stimulation such as handling, or during retrieval (Sewell 1968). However in rats, tactile stimulation appears to suppress ultrasonic calling in the first few days after birth, although later more calls are elicited by handling than by cold alone (Sales and Smith, 1978). Rat pups will increase their rate of ultrasonic calling when placed on clean bedding, relative to their rate on soiled bedding (Oswalt and Meier, 1978).

Calls produced by these infant rodents appear to attract the mother and initiate retrieval by her; Smith and Sales (1980) suggested that rodent ultrasounds are not easily localizable and the calls serve to attract the mother, rather than assist her to locate the young. The authors also suggested that calls may have the effect of maintaining a high level of maternal behaviour through an arousal effect.

Amongst sciuromorphs and related groups a variety of audible calls have been recorded, for example in the fox squirrel (Zelly, 1971) and in the Columbian ground squirrel (Steiner, 1973), but no detailed reports of acoustic behaviour between mother and infants appear to have been published.

In the hystricomorphs some research is available. Kleiman (1974) recorded that contact calls are found between mothers and infants, and Eisenberg (1974) reported this in *Myoprocta pratti, Dasyprocta punctata, Cuniculus paca, Octodon degus* and the domestic guinea pig *Cavia porcellus*. Vocalizations of the domestic guinea pig have been extensively studied (for example, Coulon, 1973; Arvola, 1974; Berryman, 1970,

1974, 1976, 1981). Research on the guinea pig shows that lactating females, unlike virgin females, respond vocally to calls made by isolated infants and exhibit searching behaviour in response to such calls (Berryman, 1974, 1981), but lactating females do not appear to be able to differentiate between their own and alien young by sound alone (Berryman, 1981). Females' responsiveness to infants' calls wanes over the suckling period (Berryman, 1974). The incidence of contact and mild distress calls (chuts, chutters, low whistles and whistles) is initially high among lactating females briefly separated from their young, but it declines gradually over the four postpartum weeks. This decline is correlated with other changes in the female's behaviour, for example latency to approach her young. Pettijohn (1977), however, found that mothers only respond strongly to infant's distress calls during the first postpartum week. Distress calling was low in infants in the mother's presence, relative to the level recorded in isolated infants.

Rood (1972) has suggested that domestication may have resulted in guinea pigs showing a lower threshold for vocalization than their wild relatives of the genera *Cavia*, Microcavia and *Galea*, since these wild rodents, despite having a range of audible vocalizations, are less vocal than the domestic guinea pig.

Contact calls of the hystricomorphs are typically short and repetitive and thus localization of these sounds is enhanced; this is in contrast to the ultrasounds of the myomorphs. It would appear that in the precocial rodents, where separation between mother and young is more likely because of the degree of infant mobility, easy localization of mothers' and infants' calls may be adaptive.

Visual signals

Finally, visual communication will be considered only briefly because it does not seem to be a major channel of communication in rodents' parental behaviour. In rodents where young are altrical, vision is usually the last of the major sensory systems to become functional (Gottlieb, 1971) and its use is limited, or non-existent, in the early relationship between mothers and infants.

In the precocial species, for example the hystricomorphs, vision may have a slightly more important role. Guinea pig mothers do appear to recognize their young by distal cues (Porter et al., 1974) and visual cues are thought to be important in recognition. Infant guinea pigs will develop preferences for visual stimuli, as a result of brief exposure, in an imprinting-like situation (Shipley, 1963; Sluckin, 1968; Sluckin and Fullerton 1969; Gaston, et al., 1969), thus visual signals may also play a role in maintaining mother–infant proximity during the period of infant dependency.

PATERNAL BEHAVIOUR

Paternal behaviour in rodents is not always precisely defined. Many studies report the incidence of maternal-like behaviour in males, that is, behaviour typical of mothers with their young, but the notion that fathers may make their own unique contribution to infant care is less often considered. Indeed adult males are sometimes *more* responsive to infants than are non-maternal females as has been observed in the domestic guinea pig (Berryman and Fullerton, 1976).

It is possible that the male's parental role may be a protective one even where no clear care-taking activities are shown. Elwood (1983b), in his review of paternal care of rodents, observed that the majority of reports of parental care from laboratory studies (see table 2), with few observations of this nature in the field. Laboratory housing conditions may

Table 2 Species in which paternal care has been observed in captive animals

Species	Reference
Mus musculus	Leblond, 1936
Rattus rattus	Strozik and Korda, 1977
Rattus fuscipes	Horner and Taylor, 1969
Microtus californicus	Hatfield, 1935
Microtus pennsylvanicus	
Microtus ochrogaster	Hartung and Dewsbury, 1979
Microtus montanus	
Baiomys taylori	Blair, 1941
Onychomys torridus	Horner and Taylor, 1969
Onychomys leucogaster	Ruffer, 1965
Peromyscus gracilis	
Peromyscus maniculatus	Horner, 1947
Peromyscus leucopus	
Peromyscus californicus	Dudley, 1974
Peromyscus eremicus	Hatton and Meyer, 1973
Meriones persicus	Eibl-Eibesfeldt, 1951
Meriones tameriscinus	
Meriones vinogradovi	Fiedler, 1973
Meriones crassus	
Meriones unguiculatus	Elwood, 1975
Dicrostonyx groenlandicus	Mallory and Brooks, 1978
Mesocricetus auratus	Marques and Valenstein, 1976
Myoprocta pratti	Kleiman, 1972

Source: Reproduced with permission of John Wiley and Sons; from R. W. Elwood (ed.), in *Parental Behaviour of Rodents*, Chichester, John Wiley and Sons, 1983.

impose a highly artificial degree of proximity between males and females when the latter are producing young and this may increase the male's interaction with the young as well as the incidence of care-taking responses, which might in natural conditions never arise. Similarly laboratory conditions may separate the sexes and artificially reduce male–young interactions. These problems have been discussed by a number of authors (Hatfield 1935; Eibl-Eibesfeldt, 1951).

The type of social organization found amongst rodents gives some initial clues to the likely occurrence of paternal care. Relatively few rodents are truly solitary as was indicated in the introduction, but of these animals paternal care is non-existent by definition, since the mother must bear the young. However amongst the majority of rodents some degree of social contact beyond that between mothers and pre-weanlings, and between the sexes during mating, does occur (Elwood, 1983b), and within these the nature of contact between males and young is clearly worthy of investigation.

Paternal care has been observed in captive animals in a wide range of species, but chiefly, to date, amongst the myomorphs and related families (Elwood, 1983b) and to some degree in the hystricomorphs (Kleiman, 1974).

In *Mus musculus* there is strong evidence of maternal care in the males of some domestic strains (Leblond, 1936; Noirot, 1964c); males presented with a live one-day-old pup showed a high incidence of retrieval, licking and nest-building, and their behaviour was not significantly different from that of naïve female mice. However, male 'maternal' behaviour towards a drowned pup has also been studied (Noirot, 1964c) in an experiment in which both males and females were tested and it was noted that 'a very small proportion' of animals (sex unspecified) reacted to the drowned pup by eating it and these subjects were eliminated from the experiment (442); thus the true incidence of this behaviour was not recorded.

Other studies reveal that male mice may show an enormous range of responses towards newborn young: from retrieval, to indifference, to pup killing (Gandelman, 1973; vom Saal, 1983). The presence or absence of testosterone appears to affect this behaviour, since castration increases the proportion of animals which show retrieval, and decreases the proportion of animals showing pup killing (Gandelman and vom Saal, 1975). Similarly, exposure to testosterone will induce pup killing in virgin female mice, which otherwise typically show retrieval (Davis and Gandelman, 1972; Gandelman and vom Saal, 1975).

In laboratory studies of other muroid rodents (Superfamily *Muroidea*) Hartung and Dewsbury (1979) found that males of four *Microtis* species (*californicus, ochrogaster, montanus* and *pennsylvanicus*), and two *Peromyscus* species (*maniculatus,* and *luecopus*) exhibited 'appreciable amounts' of sitting on the nest, licking pups and manipulation of nest materials. The *Microtis* species also showed pup retrieval and manipulation of the pups within the nest. The authors found that, in general, sex differences in parental behaviour were quite small. Studies of the influence of parity on male parental behaviour all showed little effect.

Among *Meriones unguiculatus* Elwood (1975) found extensive paternal care in laboratory cages, but he reported other unpublished data (Elwood, 1983, 238) which indicated that the provision of a nest box, or the location of the nest in a burrow system, lessened the contact between the young and father. Elwood (1983), like Hartung and Dewsbury (1979), emphasized that laboratory reports of paternal care cannot be taken as evidence for this behaviour under natural conditions, similarly an amicable pair living in the laboratory cannot be seen as a sign that animals live in pairs under natural conditions.

Despite evidence that males may show paternal care towards their offspring, there is evidence that males of a variety of altrical rodent species will respond to young of their own species, but not normally their own offspring, by infanticide. This has already been discussed in mice, but is also found in hamsters (Rowell, 1961), rats (Rosenberg and Sherman, 1975) and gerbils (Elwood, 1977).

Paternal care has been studied less extensively in the sciurmorphs, hystrimcomorphs and rodent 'Families of Uncertain Affinities'. In the former group some work on the species of *Marmota* is available however. Barash (1974c) found that the hoary marmot *Marmota caligata* may be highly colonial. Colonies consist of a male, one or several females and their offspring. Males occupy a single burrow or share one with a parous female and her yearlings. The males do not appear to play any obvious role in the care of the young and females with newborn young defend their burrows against all colony members including other males. Hoary marmots which live in more isolated colonies (Barash, 1975) were however found to show a rather different social organization; a male and a female share a single burrow and more paternal care is evident. Barash (1975) suggested that there is an inverse relationship between paternal behaviour and the frequency of social interaction between adult marmots. He argued that amongst marmots which have more opportunities for social interactions, opportunities for the male to copulate with a neighbouring female will have an evolutionary advantage, particularly since a strange male may also inseminate his female(s), and thus paternal care will have a low priority. Where animals live in relative isolation males are 'free' to devote more time and attention to paternal behaviour 'because the need for watchfulness and the recompense for roguery are vastly reduced' (615). Barash (1975) also suggested that the male's involvement may also be a factor in reducing predation of the young.

Amongst hystricomorphs the male is known to be tolerant towards the young in a number of species (Kleiman, 1974) and infanticide is not typical in precocial rodent species. A variety of forms of contact have been observed including approaching and following young, marking the young and sleeping with them. 'Courtship' of infants by adults of both sexes is commonly observed in the domestic guinea pig; this behaviour involves following, sniffing and licking the genitalia, and performing the rhumba whilst emitting the characteristic purr vocalization (Berryman, 1976; and own observation). Kleiman (1974) notes that paternal behaviour in these groups seems to be correlated with the degree of physical contact occurring between males and females outside the context of reproduction. Species showing little or no contact outside this time (for example, *Pediolagus*) show little paternal care, whereas in species where pairs show much contact-promoting behaviour (*Octodon* for instance) the opposite is true. Nevertheless males may be tolerant, if not active, in involvement with the young.

The extent to which males and females of the *Caviinae* associate, other than for mating, has been much debated. Rood (1972), in his study of three genera of Argentine cavies, indicated that permanent social bonds are not formed. However, more recent work on the domestic guinea pig by Jacobs (1976) indicated that the same male tends to associate with the same female over several successive pregnancies, and Sacher's (1983) research on guinea pigs showed evidence for long-lasting attachments in guinea pigs housed in groups (this was more evident among stronger or more dominant males). In line with Kleiman's (1974) hypothesis, this closer association might also be expected to reveal more paternal involvement with the young but less evidence for this is available. Rood (1972) does not report it, nor do other studies of groups of guinea pigs (Kunkel and Kunkel, 1964; King 1956); Nevertheless tolerance of infants by males is well documented and research by Berryman and Fullerton (1976) revealed that unfamiliar males (that is, males which are not the fathers of the young) but which are housed with the mother are more interactive with infants than are virgin females. In this study males and females housed in pairs huddled together with the young and males neither avoided nor acted aggressively towards young as did virgin females. Pettijohn (1979) showed that young guinea pigs isolated from the home cage show a reduction of distress vocalizations in the presence of their father, which was less than the reduction produced by the mother's presence, but greater than that produced by the presence of a sibling or an unfamiliar female. Evidently a familiar male adult is effective in reducing distress.

In the Patagonian hare, or mara (*Dolichotis patagonum*), Genest and Dubost (1974) reported that whilst a male does not take part in rearing the young, his continual presence close to the female may facilitate successful rearing. Males appear to be protectors and guards and may 'assist a female in difficulty' (161). Thus the male's role here may be a vital one for the female and young in maintaining isolation from conspecifics, and ensuring safety from predators.

Finally, in the mole rat *Heterocephalus glaber* (Jarvis, 1981), males are observed to take part in caring for the young of the single breeding female. Apart from suckling, which is carried out by the biological mother, other aspects of care are shared between adults of both sexes.

THE INFLUENCE OF THE MALE ON THE DEVELOPMENT AND SURVIVAL OF YOUNG

Apart from studying the participation of the male in care of the young, a number of researchers have investigated the effects of the presence of the male on pup development and survival. Several authors have suggested that pups will benefit from the extra warmth that males may provide, particularly in species where young are altrical. Dudley (1974) studied the development of *Peromyscus* pups reared with or without the male, and with the female present or removed from the nests for periods during the day. He found that pups caged with the father in the absence of the female, developed faster than those reared without the father, and that more early-weaned pups survived if caged with their father. These authors suggested that the beneficial effect of the father lies in the additional warmth which he can provide. Other studies lend support to the notion that warmth provided by any other adults (whether male or female) may be beneficial (for example, Plaut and Davis, 1972). However, Elwood and Broom (1978) found no difference in the rate of pup survival when the young were reared by one or both parents. Clearly, ambient temperature and the degree to which females must leave the nest for food, will influence the extent to which the males presence has a beneficial effect.

Accelerated development in the young has been shown to occur in some rodent species when the male is present with, or near, the female and her young. Fullerton and Cowley (1971) found that the presence of male's urine and faeces (but no direct contact with the male) had the effect of enhancing development and weight-gain in mice, but Priestnall and Young (1978) found no differences between mice reared with or mice reared without males. In the gerbil *Meriones unguiculatus* (Elwood and Broom, 1978) earlier eye-opening was observed in pups reared by both parents. In this case it was suggested that the accelerated development might be attributed to the additional warmth provided by the male, but warmth cannot account for the first example. The accelerated onset of sexual maturity in female mouse pups reared in the presence of the male (Fullerton and Cowley, 1971) has been attributed to a pheromone in the urine of adult males.

The presence of the male also has effects on the female's behaviour. In their work on gerbils, Elwood and Broom (1978) found that females showed *less* nest-building activity when rearing their young with a male present, and less gnawing of the bars of the cage (the latter is an activity which is thought to be stress related). These authors suggest that the presence of the male enables pups' temperature to be kept up without the amount of nest-building required by females housed singly, thus his presence reduces her work load and the resultant stress which this causes.

In summary it appears that the presence of the male may both directly, if he is involved in paternal care, or indirectly, by providing warmth by his passive presence, have the effect of reducing the level of maternal care necessary for pup survival.

CONCLUDING REMARKS

The parental behaviour of rodents has been shown to be chiefly the concern of the female and so this conclusion will consider her behaviour first. Whilst generalizations are dangerous in relation to a group as diverse as the rodents, on the whole (in those species studied) adult females are non-maternal towards their young prior to their first pregnancy. Females of the myomorphs may be 'sensitized' by prolonged exposure to infants and thus induced to show maternal care, but these studies, whilst revealing important information about the factors eliciting maternal care, are misleading if they are interpreted as showing that, in their natural habitats, females are likely to be maternal prior to their own first pregnancy. Domestic mice have been shown to be highly maternal prior to pregnancy, but this animal's behaviour does not appear to be characteristic of *Mus musculus* as a whole.

The considerable volume of research on the hormonal correlates of maternal behaviour in the rat has revealed that the onset of maternal behaviour during pregnancy and near term is regulated by the hormones of pregnancy. A rise in oestrogen is the trigger for this behaviour at the end of pregnancy combined with a decline in progesterone at this time. There is no uniformity in the hormonal basis of maternal behaviour in the other myomorph rodent species studied. The onset phase of maternal behaviour in the hamster is similar to that of the rat, but further research on the hamster, gerbil and mouse is needed.

The *maintenance* of maternal behaviour in the rat appears to be dependent on pup stimulation, and the role of hormones is chiefly in relation to lactation. Pup stimulation is also a factor in the maintenance phase of hamster maternal behaviour, but the role of hormones at this time is not clear from the existing research. In certain mouse strains postpartum maternal behaviour is *not* dependent on continuous pup stimulation since virgin animals show maternal responsiveness.

In the rat there is a third phase, the *transition* phase, between the *onset* and *maintenance* of maternal behaviour, which occurs during the shift between prepartum regulation and postpartum regulation of maternal behaviour. Pup contact with the female around parturition seems to be essential for this shift to occur; if pups are removed from the mother immediately postpartum, maternal responsiveness over the next few days declines rapidly. Some evidence in support of a similar process in the hamster is available.

Hormonal correlates of maternal behaviour in other rodent types have been assessed from indirect evidence and further research is needed on the sciuromorphs and hystricomorphs.

The behaviour shown by the mother postpartum is influenced by the maturity of her offspring at birth, in particular their ability to thermoregulate. In altrical species, where homeothermy is achieved only after several weeks, maternal care involves long periods in the nest, the maintenance of the nest structure and a responsiveness to signs from the infants that they are cold. In precocial species the role of the mother is rather different; whilst infants may still require warmth early in life, their dependence on the mother for their sanitary needs and supply of milk, parallels those of altrical species, but protection from predators for an extended period seems to be a very important aspect of parental care of these young. This is observed more clearly in naturalistic studies, and laboratory workers have tended to overlook the parents' essential role in this aspect of infant survival in species which appear to be independent at an early age.

Parental communication has evolved to an extraordinarily high degree amongst rodents. The 'pheromonal bond', described in detail in the rat, and the great repertoire of sounds used by almost all rodents studied, exemplify perfectly the adaptations which have produced close synchrony in behaviour of parents and infants. Although signals are shaped by natural selection to benefit the signaller, the parent–infant context has been described by Porter (1983) as unique because any genetic advantage gained by the parent usually accrues to the infant because the former's benefit is mediated through the latter's increased fitness.

The available knowledge concerning rodent paternal behaviour is very limited relative to that documented for females. It is generally thought that the typical mammalian pattern of reproductive behaviour involves mate desertion by the male, leaving the female to rear her young alone or in association with other females. The suggestion that male parental care has arisen as a secondary adaptation has been considered in this chapter. It is clear that a variety of factors may be conductive to his involvement with the young, such as difficult or harsh environments jeopardizing infant survival, or a low population density; even within a genus involvement with the young may be markedly different as a function of these factors. Male behaviour towards the young is not necessarily a mirror image of that shown by the female, although research has tended to explore behaviour in terms of male 'maternal behaviour'. It is unfortunate that many patterns of parental care have the label 'maternal' because with this label goes the underlying assumption that this behaviour is the 'natural' province of the female. The use of the term parental behaviour applied to both the male and the female would avoid this connotation.

Parental behaviour shown by non-parental animals has been touched on briefly in this chapter, indeed in some rodent species 'communal' or 'promiscuous' nursing has been reported, whilst other species have a highly exclusive relationship with their own offspring. Until recently the apparently altruistic nature of this form of behaviour has been a puzzle to those studying behaviour; the latter fits the 'selfish gene' notion (Dawkins, 1976; Dawkins and Krebs, 1978) whereas the former appears to be a wasted investment in unrelated young. Trivers (1971) has indicated, however, that parental care towards unrelated offspring may be to the genetic advantage of the 'giver', providing that such behaviour is returned in kind. This is an extreme position since communal parental care is more usually found amongst related animals and thus each parent is caring for offspring which do indeed share their genes; thus caring for offspring of relatives may indeed enchance the individual's 'inclusive fitness'. The factors which influence the occurrence of a non-exclusive relationship between a mother and her offspring need to be more fully researched.

In conclusion, rodent parental behaviour in the female has been researched in depth in certain myomorph rodents, the rat in particular. Parental behaviour in other rodents is very unevenly explored and more further study of sciuromorphs and hystricomorphs would be of great value, as would more research on all rodents in their natural habitats. The role of the male in the life of the young is relatively unexplored, but the research currently available suggests that this may be greater than was once thought.

REFERENCES

Ader, R. and Grota, R. J. 1970: Rhythmicity in the maternal behaviour of *Rattus norvegicus*. *Animal Behaviour*, 18, 144–50.

Arvola, A. 1974: Vocalization in the guinea-pig, *Cavia porcellus* L.. *Annales Zoologici Fennici*, 11, 1–96.

Avery, G. T. 1925: Notes on reproduction in guinea pigs. *Journal of Comparative Psychology*, 5, 373–96.

Barash, D. P. 1974a: The evolution of marmot societies: a general theory. *Science*, 185, 415–20.

—— 1974b: Mother–infant relations in captive woodchucks (*Marmota monax*). *Animal Behaviour*, 22, 446–8.

—— 1974c: The social behaviour of the hoary marmot (*Marmota caligata*). *Animal Behaviour*, 22, 256–61.

—— 1975: Ecology of paternal behaviour in the hoary marmot (*Marmota caligata*): an evolutionary interpretation. *Journal of mammalogy*, 56, 613-18.
Bauer, J. H. 1983: Effect of maternal state on the responsiveness to nest odours of hooded rats. *Physiology and Behavior*, 30, 229-32.
Beauchamp, G. K., Doty, R. L., Moulton, D. G. and Mugford, R. A. 1976: The pheromone concept in mammalian chemical communication: a critique. In R. L. Doty (ed.), *Mammalian Olfaction, Reproductive Processes, and Behaviour*, London: Academic Press.
Beniest-Noirot, E. 1958: Analyse du comportement dit 'maternel' chez la souris. *Monographie Français de psychologie No 1, Centre National des Republicans Sociaux, Paris*.
Berryman, J. C. 1970: Guinea pig vocalizations. *Guinea-pig Newsletter*, 2, 9-18.
—— 1974: A study of guinea pig vocalizations: with particular reference to mother-infant interactions. Unpublished PhD thesis, University of Leicester.
—— 1976: Guinea pig vocalizations: their structure, causation and function. *Zeitschrift für Tierpsychologie*, 41, 80-106.
—— 1981: Guinea pig responses to conspecific vocalizations: playback experiments. *Behavioural and Neural Biology*, 31, 476-82.
Berryman, J. C. and Fullerton, C. 1976: A developmental study of interactions between young and adult guinea pigs (*Cavia porcellus*). *Behaviour*, 59, 22-39.
Biggers, J. D., Finn, C. A. and McLaren, A. 1962: Long-term reproductive performance of female mice. II. Variation of litter size with parity. *Journal of Reproduction and Fertility*, 3, 313-30.
Birch, H. G. 1956: Sources of order in the maternal behaviour of animals. *American Journal of Orthopsychiatry*, 26, 279-84.
Blass, E. M. and Teicher, M. H. 1980: Suckling. *Science*, 210, 15-22.
Blair, W. F. 1941: Observations on the life history of *Baiomys taylori subator*. *Journal of Mammalogy*, 22, 378-83.
Breen, M. F. and Leshner, A. I. 1977: Maternal pheromone: a demonstration of its existence in the mouse (*Mus musculus*). *Physiology and Behavior*, 18, 527-9.
Bridges, R. S. 1975: Long-term effects of pregnancy and parturition upon maternal responsiveness in the rat. *Physiology and Behavior*, 14, 245-9.
—— 1977: Parturition: its role in the long term retention of maternal behavior in the rat. *Physiology and Behavor*, 18, 487-90.
Bridges, R. S., Feder, H. H. and Rosenblatt, J. S. 1977: Induction of maternal behaviors in primigravid rats by ovariectomy, hysterectomy, or ovariectomy plus hysterectomy: effect of length of gestation. *Hormones and Behavior*, 9, 156-69.
Bridges, R. S., Rosenblatt, J. S. and Feder, H. H. 1978: Stimulation of maternal responsiveness after pregnancy termination in rats: effect of time of onset on behavioral testing. *Hormones and Behavior*, 10, 235-45.
Brown, K. 1979: Chemical communication between animals. In K. Brown and S. J. Cooper (eds), *Chemical Influences on Behavior*, London: Academic Press.
Carmichael, L. and Lehner, G. F. J. 1937: The development of temperature sensitivity during the fetal period. *Journal of Genetic Psychology*, 50, 217-27.
Carter, C. S. and Marr, J. N. (1970): Olfactory imprinting and age variables in the guinea pig, *Cavia porcellus*. *Animal Behaviour*, 18, 238-44.
Clegg, F. and Williams, D. I. 1983: Maternal pheromone in *Rattus norvegicus*. *Behavioral and Neural Biology*, 37, 223-36.

Cooper, A. J. and Cowley, J.J. 1976: Mother-infant interaction in mice bulbectomized early in life. *Physiology and Behavior*, 16, 453-9.

Cosnier, J. 1963: Quelques problemes poses le comportement maternel provoque chez la ratte. *Psychophysiologie Compte Rendu des Séances de la Société de Biologie*, 157, 1611-13.

Cosnier, J. and Couturier, C. 1966: Comportement maternel provoqué chez les rattes adultes castrées. *Compte Rendu des Séances de la Société de Biologie*, 160, 789-91.

Coulon, J. 1973: Le repertoire sonore du cobaye domestique et sa signification comportementale. *Revue Comportement Animal*, 7, 121-32.

Daly, M. and Daly, S. 1975: Behaviour of Psammoys obesus (*Rodentia: Gerbillinae*) in the Algerian Sahara. *Zietschrift für Tierpsychologie*, 37, 298-321.

Daly, M. and Wilson, M. 1978: *Sex, Evolution and Behavior*. North Scituate, Massachusetts: Duxbury Press.

Davis, D. E. and Hall, O. 1951: The seasonal reproductive conditions of female Norway (Brown) rats in Baltimore, Maryland. *Physiological Zoology*, 24, 9-20.

Davis, P. G. and Gandelman, R. 1972: Pup killing produced by the administration of testosterone propionate to adult female mice. *Hormones and Behavior*, 3, 169-73.

Dawkins, R. 1976: *The Selfish Gene*. Oxford: Oxford University Press.

Dawkins, R. and Carlisle, T. R. 1976: Parental investment, mate desertion and a fallacy. *Nature*, 262, 131-3.

Dawkins, R. and Krebs, J. R. 1978: Animal signals: information or manipulation? In J. R. Krebs and N. B. Davies (eds), *Behavioural Ecology*, Oxford: Blackwell Scientific Publications.

Deag, J. M. 1980: Social behaviour of animals. *Studies in Biology*, 118, London: Edward Arnold.

Dieterlen, F. 1961: Beiträge zur biologie der Stachelmaus, *Acomys cahirinus dimidiatus*, Cretzschmar. *Zietschrift für Säuge tierkunde*, 26, 1-3.

Dieterlen, F. 1962: Geburt und Geburtshilfe bei der Stachelmaus, *Acomys cahirunus*. *Zeitschrift für Tierpsychologie*, 19, 191-222.

Dollinger, M. J., Holloway, W. R. and Denenberg, V. H. 1980: Parturition in the rat (*Rattus norvegius*): normative aspects and the temporal pattern of behaviour. *Behaviour Processes*, 5, 21-37.

Dudley, D. 1974: Paternal behavior in the California mouse *Peromyscus californicus*. *Behavioral Biology*, 11, 147-252.

Eibl-Eibesfeldt, I. 1951: Gefangenschaftsbeobachtungen au der persischen Wustenmaus (*Meriones persicus persicus* Blandford): Ein Bietrag zur vergleichenden Ethologie der Nager. *Zietschrift für Tierpsychologie*, 8, 400-23.

Eisenberg, J. F. 1974: The function and motivational basis of hystricomorph vocalizations. In I. W. Rowlands and B. J. Wier (eds), *The Biology of the Hystricomorph Rodents*, Symposia of the Zoology Society of London, 34, London: Academic Press.

—— 1981: *The Mammalian Radiations: An Analysis of Trends in Evolution, Adaptation and Behaviour*. London: The Athlone Press.

Elwood, R. W. 1975: Paternal and maternal behaviour of the Mongolian gerbil. *Animal Behaviour*, 23, 766-72.

—— 1977: Changes in the response of male and female gerbils (*Meriones unguiculatus*) towards test pups during pregnancy of the female. *Animal Behaviour*, 25, 46–51.

—— (ed.) 1983a: *Parental Behaviour of Rodents*. Chichester: John Wiley.

—— 1983b: Paternal care in rodents. In R. W. Elwood (ed.), *Parental Behaviour of Rodents*, Chichester: John Wiley.

Elwood, R. W. and Broom, D. M. 1978: The influence of litter size and parental behaviour on the development of the Mongolian gerbil pups. *Animal Behaviour*, 26, 438–54.

Enders, R. K. 1935: Mammalian life histories from Barro Colorado Island, Panama. *Bulletin of the Museum of Comparative Zoology at Harvard College*, 78, 383–502.

Ewer, R. F. 1968: *Ethology of Mammals*. London: Logos Press.

Festing, M. 1968: Some aspects of reproductive performance in inbred mice. *Laboratory Animals*, 2, 89–100.

Fiedler, U. 1973: Observations on the biology of some Gerbillinae (jirds) especially of *Gerbillus* (*Dipodillus*) dasyurys in captivity. I. Behaviour. *Zietschrift für Säugetierkunde*, 38, 321–40.

Fleming, A. and Rosenblatt, J. S. 1974: Maternal behavior in the virgin and lactating rat. *Journal of Comparative and Physiological Psychology*, 86, 957–72.

Fullerton, C. and Cowley, J. J. 1971: The differential effect of the presence of adult male and female mice on the growth and development of the young. *Journal of Genetic Psychology*, 119, 89–98.

Fullerton, C., Berryman, J. C. and Porter, R. H. 1974: On the nature of mother–infant interactions in the guinea pig (*Cavia porcellus*). *Behaviour*, 48, 145–56.

Galef, B. G. Jr. 1981: Preference for maternal odours in rat pups: implications of a failure to replicate. *Physiology and Behavior*, 26, 783–6.

Gandelman, R. 1973: The development of cannibalism in male Rockland-Swiss mice and the influence of olfactory bulb removal. *Developmental Psychobiology*, 6, 159–64.

Gandelman, R. and vom Saal, F. S. 1975: Pup killing in mice: the effects of gonadectomy and testosterone administration. *Physiology and Behavior*, 15, 647–51.

Gaston, M. G., Stout, R. and Tom, R. 1969: Imprinting in guinea pigs. Psychonomic Science, 16, 53–4.

Genest, H. and Dubost, G. 1974: Pair-living in the mara (*Dolichotis patagonum*). *Mammalia*, 38, 155–62.

Gerling, S. and Yahr, P. 1982: Maternal and paternal pheromones in gerbils. *Physiology and Behavior*, 28, 667–73.

Gilbert, A. N., Pelchat, R. J. and Adler, N. T. 1980: Postpartum copulatory and maternal behaviour in Norway rats under semi-natural conditions. *Animal Behaviour*, 28, 989–95.

Gottlieb, G. 1971: Ontogenesis of sensory function in birds and mammals. In E. Tobach, L. R. Aronson and E. Shaw (eds), *The Biopsychology of Development*, New York: Academic Press.

Goy, R. W., Hoar, R. M. and Young, W. C. 1957: Length of gestation in the guinea pig with data on the frequency and time of still birth. *Anatomical Record*, 128, 747–57.

Gregory, E. H. and Pfaff, D. W. 1971: Development of olfactory-guided behavior in infant rats. *Physiology and Behavior*, 6, 573–6.
Grota, L. J. and Ader, R. 1969: Continuous recording of maternal behaviour in *Rattus norvegicus*. Animal Behaviour, 17, 722–9.
Hartung, T. G. and Dewsbury, D. A. 1979: Paternal behaviour in six species of muroid rodent. *Behavioural and Neural Biology*, 26, 466–78.
Hatfield, D. M. 1935: A natural history of *Microtus californicus*. *Journal of Mammalogy*, 116, 261.
Hatton, D. C. and Meyer, M. E. 1973: Paternal behavior in cactus mice (*Peromyscus eremicus*). *Bulletin of the Psychonomic Society*, 2, 330 (abstract).
Herrenkohl, L. R. and Campbell, C. 1976: Mechanical stimulation of mammary gland development in virgin and pregnant rats. *Hormones and Behavior*, 7, 183–98.
Hofer, M. A. 1976: Olfactory denervation: its biological and behavioural effects in infants rats. *Journal of Comparative and Physiological Psychology*, 90, 829–38.
—— 1978: Hidden regulatory processes in early social relationships. In P. P. G. Bateson and H. Klopfer (eds), *Perspectives in Ethology. 3: Social Behaviour*, New York: Plenum Press.
Horner, B. E. 1947: Parental care of young mice in the genus *Peromyscus*. *Journal of Mammalogy*, 28, 31–6.
Horner, B. E. and Taylor, J. M. 1969: Paternal behaviour in *Rattus fuscipes*. *Journal of Mammalogy*, 50, 803–5.
Ishii, O. 1920: Observations on the sexual cycle of the guinea pig. *Biological Bulletin*, 38, 237–50.
Jacobs, W. W. 1976: Male–female associations in the domestic guinea pig. *Animal Learning and Behavior*, 4, 77–83.
Jakubowski, M. and Terkel, J. 1982: Infanticide and caretaking in non-lactating *Mus musculus*: influence of genotype, family group and sex. *Animal Behaviour*, 30, 1029–35.
Jarvis, J. U. M. 1981: Eusociality in a mammal: Cooperative breeding in naked mole-rat colonies. *Science*, 212, 571–3.
Karlson, P. and Lüscher, M. 1959: 'Pheromones': a new term for a class of biologically active substances. *Nature*, 183, 55–6.
King, J. A. 1956: Social relations of the domestic guinea pig living under seminatural conditions. *Ecology*, 37, 221–8.
Klapperstück, J. 1954: Der Sumpf biber (*Nutria*) *Neue Brehm Bücherei*, No. 115.
Kleiman, D. G. 1972: Maternal behaviour of the green acouchi (*Myoprocta pratti pocock*) a South American caviomorph rodent. *Behaviour*, 43, 48–84.
Koller, G. 1952: Der Nestbau der weissen Maus und seine hormonale Auslösung. *Verhandlungen der Deutschen zoologischen Gesellschaft Frieburg*, 160–8.
—— 1956: Hormonale und psychische steuerung bein Nestbau weiser Mäuse, *Zooligischer Anzeiger* 19, Supplement, (*Verhandlungen der Deutschen zoologischen Gesellschaft,* 1955) 123–32.
Kristal, M. P., Peters, L. C., Franz, J. R., Whitney, J. F., Nishita, J. K. and Steuer, M. A. 1981: The effect of pregnancy and stress on the onset of placentophagia in Long-Evans rats. *Physiology and Behavior*, 27, 591–5.
Kunkel, P. and Kunkel, I. 1964: Beiträge zur Ethologie Hausemeer-

schweinchens, *Cavia aperea F. porcellus* L. Zietschrift für Tierpsychologie, 21, 603-41.
Labriola, J. 1953: Effect of caesarian delivery upon maternal behavior in rats. *Proceedings of the Society of Experimental Biology and Medicine*, 83, 556-7.
Lang, H. 1925: How squirrels and other rodents carry their young. *Journal of Mammalogy*, 6, 18-24.
Leblond, C. P. 1936: Extra-hormonal factors in maternal behaviour. *Proceedings of the Society for Experimental Biology and Medicine*, 38, 66-70.
Leon, M. 1974: Maternal pheromone. *Physiology and Behavior*, 13, 441-53.
—— 1975: Dietary control of maternal pheromone in the lactating rats. *Physiology and Behavior*, 14, 311-19.
Leon, M. and Moltz, H. 1971: Maternal pheromone discrimination by pre-weanling albino rats. *Physiology and Behavior*, 7, 265-7.
—— 1972: The development of the pheromonal bond in the albino rat. *Physiology and Behavior*, 8, 683-6.
—— 1973: Endocrine control of the maternal pheromone in the postpartum female rat. *Physiology and Behaviour*, 10, 65-7.
Leon, M., Croskerry, P. G. and Smith, G. K. 1978: Thermal control of mother-young contact in rats. *Physiology and Behavior*, 21, 793-811.
Leon, M., Galef, B. G. Jr. and Behse, J. H. 1977: Establishment of pheromonal bonds and diet choice in young rats by odor-pre-exposure. *Physiology and Behavior*, 18, 387-91.
Lisk, R. D., Pretlow, R. A. and Freidman, S. A. 1969: Hormonal stimulation necessary for elicitation of maternal nest building in the mouse. *Animal Behaviour*, 17, 730-7.
Lott, D. F. and Rosenblatt, J. S. 1969: Development of maternal responsiveness during pregnancy in the rat. In B. M. Foss (ed.), *Determinants of Infant Behaviour*, vol 6, London: Methuen.
Louttit, C. M. 1927: Reproductive behaviour of the guinea pig: the normal mating behaviour. *Journal of Comparative Psychology*, 1, 247-63.
MacDonald, D. W. 1981: Dwindling resources and the social behaviour of capybaras (*Hydrochoerus hydrochaeris*) (*Mammalia*). *Journal of Zoology*, 194, 371-91.
MacNamara, M. C. 1981: Kind of cottontail but no kind of rabbit. *Animal Kingdom*, 84, 22-5.
Mallory, F. F. and Brooks, R. J. 1978: Infanticide and other reproductive strategies in the collared lemming, *Dicrostonyx groenlandicus*. *Nature*, 273, 144-6.
Marler, P. J. 1967: Animal communciation signals. *Science*, 157, 769-74.
Marques, D. M. and Valenstein, E. S. 1976: Another hamster paradox: more males carry pups and fewer kill and cannibalize young than do females. *Journal of Comparative and Physiological Psychology*, 90, 653-7.
Mohr, E. 1949: Einiges vom Grossen und vom Kleinen Mara (*Dolichotis patagonum* Zimm. und *salinicola* Burm). Zoologische Garten. Lepizig, 16, 111-33.
Mohr, E. 1965: Altweltliche Stachelschweine. *Neue Brehm Bücherei. No. 350.*
Moltz, H. and Lee, T. M. 1981: The maternal pheromone of the rat: identity and functional significance. *Physiological Behavior*, 26, 301-6.

Moltz, H. and Leon, M. 1973: Stimulus control of the maternal pheromone in the lactating rat. *Physiology and Behavior*, 10, 67-71.

Moltz, H. and Wiener, E. 1966: Effects of ovariectomy on maternal behavior of the primiparous and multiparous rat. *Journal of Comparative and Physiological Psychology*, 62, 382-7.

Moltz, H., Levin, R. and Leon, M. 1969: Differential effects of progesterone on the maternal behavior of primiparous and multiparous rat. *Journal of Comparative and Physiological Psychology*, 67, 36-40.

Moltz, H. Robbins, D. and Parks, M. 1966: Caesarian delivery and maternal behavior of primiparous and multiparous rats. *Journal of Comparative and Physiological Psychology*, 61, 455-60.

Naaktgeboren, C. 1962: Die Geburt des Meerschweinschens. *Acta Morphologica Neerlando-Scandinavica*, 5, 195-6.

Naaktgeboren, C. 1963: Die Geburt des Meerschweinschens. *Archives Néerlandaises de Zoologie*, 15, 379.

Nagasawa, H., Tôzaki, T., Shôda, Y. and Naito, M. 1960: Lactation curve of the guinea pig. *Japanese Journal of Zootechnical Science*, 31, 195-9.

Newson, R. M. 1966: Reproduction in the feral coypu (*Myocastor coypus*). *Symposium of the Zoological Society of London*, 15, 323-34.

Noirot, E. 1964a: Changes in responsiveness to young in the adult mouse. I. The problematical effect of hormones. *Animal Behaviour*, 12, 52-8.

—— 1964b: Changes in responsiveness to young in the adult mouse. II. The effect of external stimuli. *Journal of Comparative and Physiological Psychology*, 57, 97-9.

—— 1964c: Changes in responsiveness to young in the adult mouse. IV. The effect of an initial contact with a strong stimulus. *Animal Behaviour*, 12, 442-5.

—— 1965: Changes in repsonsiveness to young in the adult mouse. III. The effects of immediately preceding performances. *Behaviour*, 24, 318-25.

—— 1966: Ultrasounds in young rodents. I. Changes with age in albino mice. *Animal Behaviour*, 14, 459-620.

—— 1972a: The onset of maternal behaviour in rats, hamsters and mice. In D. S. Lehrman, R. A. Hinde and E. Shaw (eds), *Advances in the Study of Behaviour*, vol. 4, New York: Academic Press.

—— 1972b: Ultrasounds and maternal behavior in small rodents. *Developmental Psychobiology*, 5, 371-87.

Numan, M., Rosenblatt, J. S. and Komisaruk, B. R. 1977: The medical peroptic area and the onset of maternal behavior in the rat. *Journal of Comparative and Physiological Psychology*, 91, 146-64.

Ojasti, J. 1968: Notes on the mating behaviour of the capybara. *Journal of Mammalogy*, 49, 534-5.

Okon, E. E. 1970: The effect of environmental temperature on the production of ultrasounds by isolated non-handled albino mouse pups. *Journal of Zoology*, 162, 71-83.

—— 1972: Factors affecting ultrasound production in infant rodents. *Journal of Zoology*, 168, 139-48.

Ostermeyer, M. C. 1983: Maternal aggression. In R. W. Elwood (ed.), *Parental Behavior of Rodents*, Chichester: John Wiley.

Oswalt, G. L. and Meier, G. W. 1975: Olfactory, thermal and tactual influences on infantile ultrasonic vocalization in rats. *Developmental Psychobiology*, 8, 129-35.

Otte, D. 1974: Effects and functions in the evolution of signalling systems. *Annual Review of Ecology and Systematics*, 5, 385-417.

Pearson, O. P. 1948: Life history of mountain viscachas in Peru. *Journal of Mammalogy*, 29, 345-74.

Pettijohn, T. F. 1977: Reaction of parents to recorded infant guinea pig distress vocalization. *Behavioural Biology*, 21, 438-42.

—— 1979: Attachment and separation distress in the infant guinea pig. *Developmental Psychobiology*, 21, 73-81.

Plaut, S. H. and Davis, J. H. 1972: Effects of mother-litter separation on survival growth and brain amino acid levels. *Physiology and Behavior*, 8, 43-51.

Porter, R. H. 1983: Communication in rodents: adults to infants. In R. W. Elwood (ed.), *Parental Behavior of Rodents*, Chichester: John Wiley.

Porter, R. H., Doane, H. M. and Cavallaro, S. A. 1978: Temporal parameters of responsiveness to maternal pheromone in *Acomys cahirinus*. *Physiology and Behavior*, 21, 563-6.

Porter, R. H., Fullerton, C. and Berryman, J. C. 1973: Guinea pig maternal-young attachment behavior. *Zietschrift für Tierpsychologie*, 32, 489-95.

Priestnall, R. 1972: Effect of litter-size on the behaviour of lactating female mice (*Mus musculus*): an observational study. *Animal Behaviour*, 20, 386-94.

—— 1983: Postpartum changes in maternal behaviour. In R. W. Elwood (ed.), *Parental Behaviour of Rodents*, Chichester: John Wiley.

Prilusky, J. 1981: Induction of maternal behavior in the virgin rat by lactating rat brain extracts. *Physiology and Behavior*, 26, 149-52.

Read, J. M. 1912: Observation on the suckling period in the guinea pig. *University of California Publications in Zoology*, 9, 341-51.

Reisbick, S., Rosenblatt, J. S. and Mayer, A. D. 1975: Decline of maternal behavior in the virgin and lactating rat. *Journal of Comparative and Physiological Psychology*, 89, 722-32.

Rheingold, H. L. 1963: Introduction. In H. L. Theingold (ed.), *Maternal Behaviour in Mammals*, New York: John Wiley.

Richards, M. P. M. 1966: Maternal behaviour in the golden hamster: responsiveness to young in virgin, pregnant and lactating females. *Animal Behaviour*, 14, 310-13.

—— 1967: Maternal behaviour in rodents and lagomorphs: a review. In A. McLaren (ed.), *Advances in Reproductive Physiology*, vol. 3, London: Logos Press.

—— 1969: Effects of oestrogen and progesterone on nest building in the golden hamster. *Animal Behaviour*, 17, 356-61.

Rodriguez-Sierra, J. and Rosenblatt, J. S. 1977: Does prolactin play a role in estrogen-induced maternal behavior in rats: apomorphine reduction of prolactin release. *Hormones and Behavior*, 9, 1-7.

Rood, J. P. 1970: Ecology and social behaviour of the desert cavy (Microcavia australis). *American Midland Naturalist*, 83, 415-54.

—— 1972: Ethological and behavioural comparison of three genera of Argentine cavies. *Animal Behaviour Monographs*, vol. 5, part 1.

Rosenberg, K. M. and Sherman, G. F. 1975: The role of testosterone on the

organization, maintenance and activation of pup killing in the male rat. *Hormones and Behavior*, 6, 173-9.
Rosenblatt, J. S. 1967: Nonhormonal basis of maternal behavior in the rat. *Science*, 156, 1512-14.
Rosenblatt, J. S. and Lehrman, D. S. 1963: Maternal behavior of the laboratory rat. In H. L. Rheingold (ed.), *Maternal Behavior in Mammals*, New York: John Wiley.
Rosenblatt, J. S. and Siegel, H. I. 1975: Hysterectomy-induced maternal behaviour during pregnancy in the rat. *Journal of Comparative and Physiological Psychology*, 89, 685-700.
—— 1981: Factors governing the onset and maintenance of maternal behavior among nonprimate mammals. The role of hormonal and nonhormonal factors. In D. J. Gubernick and P. H. Kopfer (eds), *Parental Care in Mammals*, New York: Plenum Press.
—— 1983: Physiological and behavioural changes during pregnancy and parturition underlying the onset of maternal behaviour in rodents. In R. W. Elwood (ed.), *Parental Behaviour of Rodents*, Chichester: John Wiley.
Rosenblatt, J. S., Siegel, H. I. and Mayer, A. D. 1979: Progress in the study of maternal behavior in the rat: Hormonal, nonhormonal, sensory, and developmental aspects. In J. S. Rosenblatt, R. A. Hinde, C. Beer and M. C. Busnel (eds), *Advances in the Study of Behaviour*, vol. 10, New York: Academic Press.
Roth, L. L. and Rosenblatt, J. S. 1967: Changes in self-licking during pregnancy in the rat. *Journal of Comparative and Physiological Psychology*, 63, 397-400.
—— 1968: Self-licking and mammary development during pregnancy in the rat. *Journal of Endocrinology*, 42, 363-78.
Roth-Kolar, H. 1957: Beitrage zu einem Aktionssystem des Aguti (*Dasyprocta aguti aguti* L.). *Zeitschrift für Tierpsychologie*, 14, 363-75.
Rowell, T. E. 1960: On the retrieving of young and other behaviour in lactating golden hamsters. *Proceedings of the Zoological Society of London*, 135, 265-82.
—— 1961: Maternal behavior in non-maternal golden hamsters. *Animal Behaviour*, 9, 11-15.
—— 1972: *Social Behaviour of Monkeys*. Harmondsworth: Penguin Books.
Ruffer, D. G. 1965: Sexual behaviour of the northern grasshopper mouse (*Onychomys leucogaster*). *Animal Behaviour*, 13, 447-52.
Rugh, R. and Wohlfromm, M. 1967: The reproductive performance of the laboratory mouse: maternal age, litter size and sex ratios. *Proceedings of the Society for Experimental Biology and Medicine*, 126, 685-7.
Sachser, V. N. 1983: Soziale Beziehungen, räumliche Organisation und Verterlung agonistischer Interaktionen in einen Gruppe von Hausmeer-schweinschen (*Cavia aperea f. porcellus*). *Zietschrift für Säugetierkunde*, 48, 100-9.
Sales, G. and Pye, D. 1974: *Ultrasonic Communication by Animals*. London: Chapman & Hall.
Sales, G. D. and Smith, J. C. 1978: Comparative studies of the ultrasonic calls of infant murid rodents. *Developmental Psychobiology*, 11, 595-619.
Scott, J. P. 1968: Observation. In T. A. Sebeok (ed.), *Animal Communication: Techniques of Study and Results of Research*, Bloomington & London: Indiana University Press.

Seitz, P. F. D. 1954: The effects of infantile experiences upon adult behaviour in animal subjects. I. Effects of litter-size during infancy upon adult behaviour in the rat. *American Journal of Psychiatry*, 110, 916–27.
—— 1958: The maternal instinct in animal subjects: I. *Psychosomatic Medicine*, 20, 214–26.
Seward, J. P. and Seward, G. H. 1940: Studies on the reproductive activities of the guinea pig. I. Factors in the maternal behaviour. *Journal of Comparative Psychology*, 29, 1–24.
Sewell, G. D. 1968: Ultrasound in rodents. *Nature*, 217, 682–3.
Shipley, W. U. 1963: The demonstration in the domestic guinea pig of a process resembling classical imprinting. *Animal Behaviour*, 11, 470–4.
Siegel, H. I. and Greenwald, G. S. 1975: Prepartum onset of maternal behavior in hamsters and the effects of estrogen and progesterone. *Hormones and Behavior*, 6, 237–45.
Siegel, H. I. and Rosenblatt, J. S. 1975: Hormonal bases of hysterectomy-induced maternal behavior during pregnancy in the rat. *Hormones and Behavior*, 6, 211–22.
—— 1978: Duration of estrogen stimulation and progesterone inhibition of maternal behaviour in pregnancy-terminated rats. *Hormones and Behavior*, 11, 12–19.
Simpson, G. G. 1945: The principles of classification and a classification of mammals. *Bulletin of the American Museum of Natural History*, 85: 1–16, 1–350.
Simpson, G. G. 1974: Chairman's introduction: taxonomy. In I. W. Rowlands and R. J. Weir (eds), *The Biology of Hystricomorph Rodents*, Symposia of the Zoological Society of London, 34, London: Academic Press.
Sluckin, W. 1968: Imprinting in guinea-pigs. *Nature*, 220, 1148.
Sluckin, and Fullerton, C. 1969: Attachments of infant guinea pigs. *Psychonomic Science*, 17, 179–80.
Smart, J. L. 1983: Undernutrition, maternal behaviour, and pup development. In R. W. Elwood (ed.), *Parental Behaviour of Rodents*, Chichester: John Wiley.
Smith, J. C. and Sales, G. D. 1980: Ultrasonic behaviour and mother–infant interactions in rodents. In R. W. Bell and W. P. Smotherman (eds), *Maternal Influences and Early Behaviour*, New York: Spectrum.
Smythe, N. 1970: Ecology and behaviour of the agouti (*Dasyprocta punctata*) and related species on Barro Colorado Island, Panama. PhD Thesis, University of Maryland.
Spencer-Booth, Y. 1970: The relationship between mammalian young and conspecifics other than mothers and peers: a review. *Advances in the Study of Behaviour*, 3, 119–94.
Steiner, A. L. 1973: Self- and allogrooming behaviour in some ground squirrels (*Sciuridae*), a descriptive study. *Canadian Journal of Zoology*, 51, 151–61.
Stern, J. J. and Bronner, G. 1970: Effects of litter size on nursing time and weight of the young in guinea pigs. *Psychonomic Science*, 21, 171–2.
Stone, E., Bonnet, K. and Hofer, M. A. 1976: Survival and development of maternally deprived rats: role of body temperature. *Psychosomatic Medicine*, 38, 242–9.
Strozik, E. and Korda, P. 1977: Behaviour of rats in pairs (parents) towards their offspring. *Zwierzeta Laboratoryjne*, 14, 83–96.
Taylor, R. H. 1970: Reproduction, development and behaviour of the Cuban

Hutia conga, *Capromys p. pilorides*, in captivity. M. S. thesis, University of Puget Sound.

Teicher, M. and Blass, E. M. 1976: Suckling in newborn rats: eliminated by nipple lavage, reinstated by pup saliva. *Science*, 193, 422–5.

Teicher, M. H., Flaum, L. E., Williams, M., Eckert, S. T. and Lumia, A. R. 1978: Survival, growth and suckling behavior of neonatally bulbectomized rats. *Physiology and Behavior*, 21, 553–61.

Terkel, J. and Rosenblatt, J. S. 1968: Maternal Behavior induced by maternal blood plasma injected into virgin rats. *Journal of Comparative and Physiological Psychology*, 65, 479–82.

Tobach, E., Rouger, Y. and Schnierla, T. C. 1967: Development of olfactory function in the rat pup. *American Zoologist*, 7, 792–3.

Trivers, R. L. 1971: The evolution of reciprocal altruism. *Quarterly Review of Biology*, 46, 35–57.

—— 1972: Parental investment and sexual selection. In B. Campbell (ed.), *Sexual Selection and the Descent of Man*, Princetown, New Jersey: Princetown University Press.

Voci, V. E. and Carlson, N. R. 1973: Enhancement of maternal behavior and nest building following systemic and diencephalic administration of prolactin and progesterone in the mouse. *Journal of Comparative and Physiological Psychology*, 83, 388–93.

vom Saal, F. S. 1983: Variation in infanticide and parental behavior in male mice due to priot intrauterine proximity to female fetuses: elimination by prenatal stress. *Physiology and Behavior*, 30, 675–81.

Waring, A. and Perper, T. 1979: Parental behaviour in the Mongolian gerbil (*Meriones unguiculatus*). I. Retrieval. *Animal Behaviour*, 29, 1091–7.

Weisner, B. P. and Sheard, N. M. 1933: *Maternal Behaviour in the Rat*. London: Oliver and Boyd.

Wise, D. A. and Pryor, T. L. 1977: Effects of ergocornine and prolactin on aggression in the postpartum golden hamster. *Hormones and Behavior*, 8, 30–9.

Wood, A. E. 1965: Grades and clades among rodents. *Evolution*, 19, 115–30.

Young, W. C. 1969: Psychobiology of sexual behaviour in the guinea pig. In D. S. Lehrman, R. A. Hinde and E. Shaw (eds), *Advances in the Study of Behavior*, vol 2, London: Academic Press.

Zara, J. L. 1973: Breeding and husbandry of the capybara. *Hydrochoerus hydrochaeris*, at Evansville Zoo. *International Zoo Yearbook*, 13, 137–9.

Zelly, A. R. 1971: The sounds of the fox squirrel *Sciurus niger rufiventer*. Journal of Mammalogy, 52, 597–604.

4 Parental Behaviour in Ungulates

D. G. M. Wood-Gush, K. Carson,
A. B. Lawrence and H. A. Moser

GENERAL INTRODUCTION

In this chapter attention is focused on the maternal behaviour of the common domesticated ungulates. The reasons for this are threefold. First, the underlying and concurrent physiological changes and accompanying physiological mechanisms are relatively well-known. Second, there are sometimes detailed records of individual variability and the factors underlying these differences can often be identified. (This information might be useful in understanding any individual differences found in species in the wild.) Finally, the effects of artificial selection on domesticated ungulates together with changes recognizably due to environmental factors can be useful indicators of the flexibility of homologous behaviour in closely related species. In this chapter we deal mainly with maternal behaviour. An important aspect of the dam's role in the care of her offspring is lactation and the nursing behaviour of the young in some ungulate species is well reported. To avoid confusion, nursing behaviour is used here to refer to all teat seeking and sucking behaviour of the young, while sucking is used only to refer to the actual event of milk intake. For those interested in paternal behaviour in ungulates examples are cited by Lent (1974).

BOVIDAE

Capridae: sheep (*Ovis aries*) and goats (*Capra hircus*)

The maternal behaviour of the *Capridae* has attracted much interest because of the long lasting mother–young bond that they form rapidly at birth. Attention has also been directed towards understanding the nature of maternal behaviour in the postpartum period up to weaning. However, little is understood of the process of weaning and the integration of the offspring into the wider social group.

Geist (1971) reports that mountain sheep withdraw from the group 2-3 weeks before parturition and select isolated rugged cliffs to lamb, whilst Grubb (1974) found that Soay sheep also separated from the group but stayed within the home range. In domestic sheep Arnold and Morgan (1975) found that 66 per cent of ewes lambed in isolation from the flock and Alexander et al. (1979) indicated that pregnant ewes tended to lamb away from the main concentration of non-pregnant ewes. The variation in seeking isolation may be related to the degree of predation, given that mountain sheep are exposed to a greater risk of predation than Soay or domestic sheep, and they appear to seek isolation more commonly than the others (Arnold and Dudzinski, 1978). However isolation may also facilitate strengthening of the ewe–lamb bond (Fraser, 1968: Geist, 1971).

Compounded in the preparturient period with the phenomenon of seeking isolation is the act of seeking shelter. Hunter (1954) reported that Blackface sheep were more sensitive to climatic conditions before parturition than a month earlier and Winfield et al. (1969) found that Welsh Mountain ewes sought shelter at lambing in wet and windy conditions. Experiments carried out on shorn and unshorn Merino ewes indicate that shelter-seeking by ewes is a direct response to chilling and is not motivated by maternal behaviour (Alexander et al., 1980; Stevens et al. 1981). Wild and feral sheep do not appear to actively seek out shelter to lamb (Geist, 1971; Grubb, 1974). Kilgour and Ross (1980) reported that in feral goats complete withdrawal by females from the group was rare; they usually gave birth in a bush area near the grazing site.

There are several descriptions of the behaviour of the ewe at lambing (Collias, 1956; Smith, 1965; Poindron and Le Neindre, 1975; Bareham, 1976). Ewes generally lie down to give birth but may stand at the point of lambing (Bareham, 1976) after which they immediately start to lick the young, emitting low pitched grunts, concentrating on the head region whilst the lamb remains lying down (Bareham, 1976; Geist, 1971; Grubb, 1974). The licking of the young and orientation of the ewe to the lamb facilitate teat-seeking for the young, but these actions are not essential for the new-born lamb to find the teats (Alexander and Williams, 1964). The new-born lamb appears to be motivated to find the teats irrespective of how hungry it is (Alexander and Williams, 1966), and it would appear to use thermal cues to aid it in finding the teats (Vince, 1984). Stephens and Linzell (1974) found that kids rather easily located the teat on a mammary gland which had been transplanted from the inguinal area to the neck, possibly because the temperature at the neck is similar to that found in the inguinal area.

Poindron and Le Neindre (1980) have shown that hormones play an important role in the onset of maternal behaviour in the ewe (see figure 1), and indeed this behaviour can be induced in non-pregnant ewes by injections of progesterone and oestrogen. However, Poindron and Le Neindre

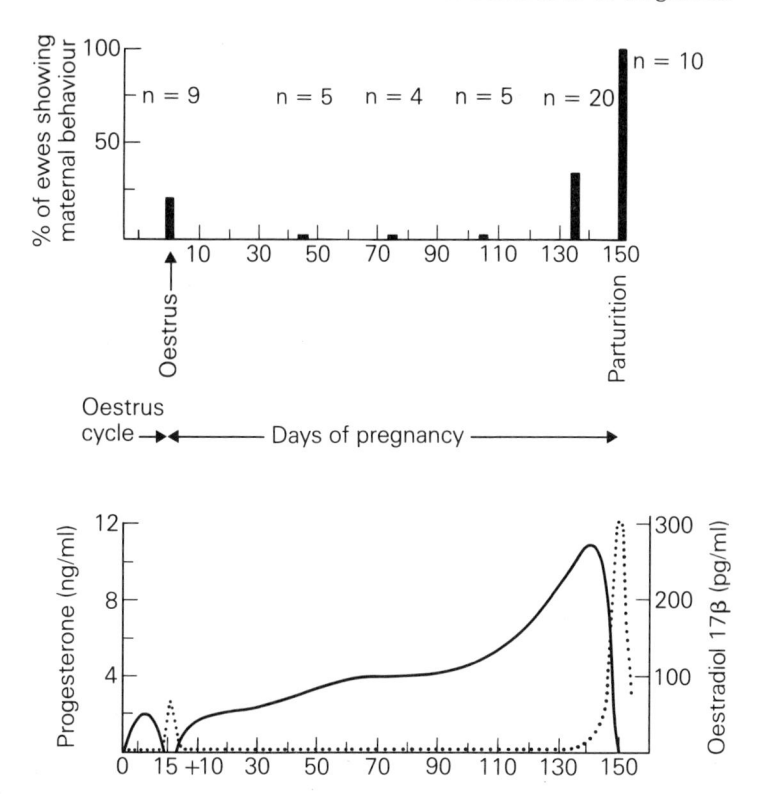

Figure 1 Maternal responsiveness of ewes towards new-born lambs at various stages of reproductive cycle with parallel evolution of progesterone and oestradiol 17β in maternal blood (from P. Poindron and P. Le Neindre, 1979a, Hormonal and behavioural basis for establishing maternal behaviour in sheep. In L. Zichella and P. Pancheri (eds), *Psychoneuro-endocrinology in Reproduction*, Amsterdam: Elsevier, 121-8)

never obtained a 100 per cent induction of maternal behaviour in non-pregnant ewes and so concluded that other factors must also be involved. Keverne et al. (1983) showed that vaginal-cervical stimulation immediately invoked the full complement of maternal behaviour in non-pregnant ewes primed with oestrogen and progesterone, illustrating that the physical act of giving birth also plays an important role in the onset of maternal behaviour in the ewe.

It has been recognized for many years that there exists a critical period during which the ewe is imprinted on her lamb(s), and this period has been defined by Poindron and Le Neindre (1979a) as the time during which a ewe is able to remain maternal when separated from her lamb at parturition. Smith et al. (1966) found that ewes separated from their lambs at birth would accept them up to 8 hours afterwards. Poindron and Le Neindre (1979a) have shown that the critical period can be extended in the ewe if she is induced to lamb with an injection of oestradiol benzoate. This lengthening of the critical period seems related to the high levels of oestrogen in the blood 24 hours after lambing. It would appear therefore that oestrogen is involved in the control of maternal responsiveness – whether it is the direct initiator that acts on the brain or merely an indirect component interacting with other hormones is not known (Poindron and Le Neindre, 1980).

During the critical period the newborn lamb provides the dam with an experience which allows her to remain maternal beyond the initial limits of the sensitive period. Poindron and Le Neindre (1979a) have shown that suckling and licking of the newborn are not necessary for the maintenance of maternal responsiveness but that perception of olfactory cues are, and these results are supported by olfactory bulb ablation experiments (Baldwin and Shillito, 1974). However it is not known if sight and/or hearing are also needed for the establishment of maternal behaviour.

Goats have also been found to have a critical period. Klopfer et al. (1964) reported that only five minutes of contact with at least one of her kids immediately after birth was sufficient for a doe to establish a maternal bond. However recent results suggest that maternal imprinting in goats may take longer than previously thought and that in addition the mother may 'label' her kids by licking them or by passing cues through her milk (Gubernick et al, 1979; Gubernick, 1980; Gubernick, 1981). A 'labelled' kid may be recognized and accepted by its own mother, but rejected by other mothers. (Interestingly, 'labelling' does not appear to occur in sheep.) Alexander and Stevens (1982) have shown that ewes exhibit no interest in their own milk, saliva or inguinal gland secretion but show significant interest in swabs rubbed around the rump, tail and anal region of their lambs. Further work may elucidate why two, otherwise quite similar species, have developed rather different methods with which to recognize their offspring after birth and establish a lasting mother–young bond.

Experience has been found to have a profound effect on maternal behaviour at parturition. Alexander (1960) found that behaviour detrimental to the lamb was more frequent in primiparous ewes, and Poindron and Le Neindre (1980) found that aggressive behaviour towards lambs was more common in primiparous than multiparous ewes. In addition, they found that injections of progesterone and oestrogen failed to induce maternal behaviour in primiparous ewes. Similarly Lickliter (1982) found that primiparous goat mothers were less likely to accept their kid back after a 2-hour postpartum separation than multiparous mothers. It seems then that maternal experience plays an important role in the manifestation of adequate maternal behaviour at parturition.

Beyond the immediate postpartum period sensory information provided by the lamb plays a major role in the regulation of maternal behaviour; hormonal manipulations appear to have little effect upon short-term maternal behaviour at this stage (Poindron and Le Neindre, 1980). Although, as described above, recognition of lambs at birth appears to be due to odour and newborn characteristics such as birth fluids, the mechanisms involved in recognition could change as lambs become mature (Gonyou, 1984). Alexander and Shillito (1977a) indicated that visual and, to a lesser extent, auditory cues were important in maternal discrimination over the short distances tested, whilst olfaction was useful only at distances of less than 0.25 metres. Visual recognition appears to depend heavily on cues from the lamb's head (Alexander and Shillito, 1977b).

Experiments on auditory recognition (Shillito-Walser et al., 1982) have shown that Dalesbred ewes have a greater range of 'voice types' than Jacobs ewes, thus enabling Dalesbred lambs more easily to locate their mothers. Dalesbred are a hill breed of sheep, and it is suggested that vocal recognition may be of great importance in allowing ewe and lamb to locate each other in bad weather conditions. Lenhardt (1977) has shown by acoustical analysis that mother goats fail to recognize individually their kids' cries until the fourth day of life because all kids' cries are essentially identical up to this time, but beyond the fourth day individuality of 'voice type' occurs.

As mentioned previously, olfaction is important in maternal recognition at close quarters and seems to be important in organising the behaviour of the lamb at nursing when the lamb invariably passes near the front of the dam before reaching the udder (Poindron and Le Neindre, 1980) and usually the ewe smells her young at this time and again during nursing. Alien lambs are denied access to the udder (Ewbank, 1967). Nursing frequencies decrease from one nursing or more per hour in the first week, to one or two nursing periods per 6 hours by the twelfth week postpartum (Munro, 1956; Ewbank, 1964; Fletcher, 1971). (Nursing bouts are always terminated after the first week by the ewe stepping over the lamb (Munro, 1956; Ewbank, 1964). The relationship between nursing frequency and milk intake is unclear; Ewbank (1967) found that singles that were nursed more put on more weight, and yet twins that had light birth weights nursed more yet put on less weight than heavier twins. Fletcher (1971), however, found no relationship between nursing frequency and weight gain.

Morgan and Arnold (1974) reported in the field that the Merino ewe and lamb remained in close contact for the first four weeks of life, the largest distances between them occurring when the ewe was grazing and the lamb playing. Hewson and Verkaik (1981) found that in Scottish Blackface the largest distances occurred when the lamb was lying and the ewe grazing. Very little is written about the mechanisms, other than the powers of recognition, that enable the ewe and lamb to remain in contact under dynamic field conditions. Preliminary results with Blackface (Lawrence and Wood-Gush, in preparation) suggest that the lamb may be operantly conditioned, through reward from sucking, to approach the ewe with the intention of sucking whenever the ewe stands in the alert anti-predator posture. With the lamb at close quarters the ewe may allow it to suck or she may move away with the lamb following. Ewbank (1967) similarly noticed that ewes would initiate nursing by vocalization and would walk away immediately the lamb approached; he postulated that this may be a method by which a ewe attempts to keep close contact with her lambs in the latter part of lactation when the lambs tended to get separated from the ewe.

Again, very little is known about the process of weaning under field conditions. Arnold et al. (1979) found that weaning occurred when milk production declined to less than 1,000 ml per day. However, Geist (1971) found that weaning was a gradual process in mountain sheep, with the ewe and lamb finally disassociating when the lamb was about one year old. Contrary to this, the work of Hunter and Milner (1963), Hunter (1964) and Grubb and Jewell (1966) suggested that the long lasting nature of the maternal bond in domestic and Soay sheep was responsible for the formation of the matriarchal social groups found in these sheep. Recent work by Lawrence and Wood-Gush (in preparation) has found that the situation in Blackface at least is equivalent to that found in mountain sheep, as the ewe and lamb cease to associate in any significant manner from about the ninth month after lambing. This suggests that factors other than the maternal bond influence the continuing existence of social groups in sheep, a hypothesis supported by studies on home range behaviour (Lawrence and Wood-Gush, 1981).

Finally, there is increasing evidence of the effects that maternal influence can have on the post-weaning behaviour of lambs. Key and MacIver (1980) have found that lambs cross-fostered onto ewes of a different breed, showed the grazing preferences of their adopted mothers after weaning. Lynch et al. (1983) have illustrated the importance of maternal influences in determining the subsequent acceptance of a wheat supplement by lambs. These results raise questions as to the nature and extent of maternal influences not only on diet selection but also on other behavioural processes of the growing lamb.

CATTLE (BOS TAURUS)

The maternal behaviour of the cow comes under diverse pressures as a result of artificial selection. On the one hand, in the dairy cow, there is selection for generalization of maternal behaviour involving easy separation from the calf, which is generally removed from its dam shortly after birth, and for ease of milk let-down in response to the milking machines. In beef cows, on the other hand, selection is essentially for retention of the wild type. Clearly these divergent selection pressures, as well as others arising from differences in the husbandry of the two types, might then be expected to result in differences in maternal behaviour within the species.

There do indeed appear to be differences in the behaviour of cows before calving. Edwards and Broom (1982) report that in the Friesians watched by them, some cows removed themselves from the herd, while Kiley-Worthington and de la Plain (1983) found that in suckler beef cows such withdrawal is a rare event. Generally the cow gets left behind. From the latter study it appears that some three to four hours before calving cows give the low 'mm' call which is usually given to the calf, and that once the amniotic fluid is shed the cow smells it and eats the grass and earth on which it has fallen. Some of the cows lie down for the severe contractions and stand for the later stages (Selman et al., 1970; Kiley-Worthington and de la Plain, 1983), whilst the Friesians watched by Edward and Broom (1982) remained recumbent.

After a normal birth the cow will lick the calf and for the first hour this is her predominant activity (Edwards and Broom, 1982), and indeed, most of the licking is confined to the first three hours postpartum (Kiley-Worthington and de la Plain, 1983). Again, breed differences appear to exist in the amount of licking performed, Friesians licking their calves less than cows belonging to the Saler breed, an 'unimproved' dual-purpose breed (Poindron and Le Neindre, 1979).

Generally, the cow calls frequently in the first three hours after calving and Kiley-Worthington and de la Plain (1983) suggest that this may serve to auditorally imprint the calf. However bellowing may sometimes be performed (Selman et al., 1970a). Once standing the calf proceeds to seek the teat and Selman et al. (1979b) suggest it is seeking a recess – in their study the teat was more quickly reached when the dam was a beef cow with a primitive udder conformation than in the case of Ayrshire calves whose dams had pendulous udders. Edwards (1982) likewise reported that the possession of a pendulous udder hindered the calf in finding the teat, and in these cases the calf may end up searching in the area of the cow's brisket. Also, in both these studies, when calving was indoors some calves directed their teat-seeking towards the pen walls or manger. Over the first three hours in the field, suckler cows will stay within one metre of the calf which, by the age of two to three hours, will be making its first forays in exploration under field conditions (Kiley-Worthington and de la Plain, 1983). In that study the average time spent in teat-seeking was 12.9 minutes (range 5–24.01 minutes).

Apart from the existence or otherwise of a pendulous udder, other factors can affect the teat seeking behaviour of the calf. Edwards (1982) has made a detailed study of 133 Friesians which shows the problems that can occur. In cases in which there has been a difficult calving or in which the dam is a primiparous heifer or, particularly, if she has given birth to a large cross-bred calf, the heifer may remain lying for some time. In these cases the calf may not be able to find the teat and may direct its seeking towards environmental objects. Furthermore, when several cows are calving in the same area it may be 'adopted' by a preparturient cow dominant to its dam. Older cows in their third or fourth parities may approach their calves less frequently than younger, inexperienced cows. Some heifers, on the other hand, appear to be frightened of their calves (Edwards, 1982) and may butt and kick them. In addition, Edwards found that there was also a paternal genetic effect in the time taken to nurse as well as a seasonal effect, for calves born in January took longer to nurse than those born in August. Failure to suck within the first six hours after birth will deprive the calf of colostrum and it will thus fail to obtain its normal levels of serum immunoglobulins. In addition, its ability to absorb colostrum declines with age (Edwards, 1982). In all about 60 of the calves in the study failed to suck within six hours. Poor nutrition will also affect the mothering ability of the cow at this early stage of maternity and in agriculture it is generally accepted that under-nourished cows will remain maternally inadequate, but whether this inadequacy involves any aspects of maternal behaviour apart from milk yield has not been investigated.

As stated earlier, some preparturient cows give the 'mm' call, but all cows give this call after calving and the number of calls increases dramatically in the hours after birth, the average being 83 per hour in the study of Kiley-Worthington and de la Plain (1983). Three hours after calving the frequency begins to decline and by the end of the first week it is down to 15 per hour. The calf rarely calls during the first three hours and when it does its calls are of the 'men' type, given with mouth open rather than the purring 'mm' type. It has been suggested that these calls of the calf are given when there is a change in the environment leading to a novel situation or to discomfort of the calf.

How the cow recognizes her calf has not yet been rigorously tested. It is thought that initially the smell of the calf is important and that later visual and auditory cues are employed. Le Neindre and Garel (1976) attempted to establish whether there is a sensitive period in which she can learn to identify her calf. In a test using primi- and multiparous dairy cows, they separated them from their calves at birth. The cows were divided into four groups and each cow in each group was presented successively with a strange calf of her breed and then with her own calf. The four groups underwent these tests at 1, 6, 12 and 24 hours after calving and during each presentation, which was ten minutes long, the cow's behaviour was recorded. Those presented with the calves at 1 hour postpartum rarely showed any aggression towards any calf, whereas 77 per cent of those separated at 12 hours were aggressive towards both calves. The authors conclude that the sensitive period is about 3 hours which would coincide with period of intense reactivitiy of the cow towards the calf reported by Kiley-Worthington and de la Plain (1983). Postpartum endocrine profiles have been established for L. H., prolactin, progesterone, cortisol, oestradiol 17 and oestrone (Arije et al., 1974; Le Voie and Moody, 1976; Wetteman et al., 1978; Kaltenbach and Dunn 1980; Gauthier et al., 1983), and some of these are shown in figure 2. However, no study has yet tried to relate these profiles to the various stages of the maternal behaviour of the cow, as has been done in the case of the ewe; the main interest in these studies has centred on the factors governing the length of anoestrus following calving. It can be seen in Figure 2 that postpartum oestrogens decrease after calving and remain low until near the postpartum oestrus, while progestins, too, remain low over the postpartum period. Prolactin increases 4–2 days before parturition and remains high until 20 days after calving. The presence of the calf can affect basal levels of prolactin (Akers and Lefcourt, 1982), but limited access to the calf may depress prolactin levels (Goodman et al., 1979). Milking and nursing, which are discussed below, have temporary effects on prolactin and oxytocin levels.

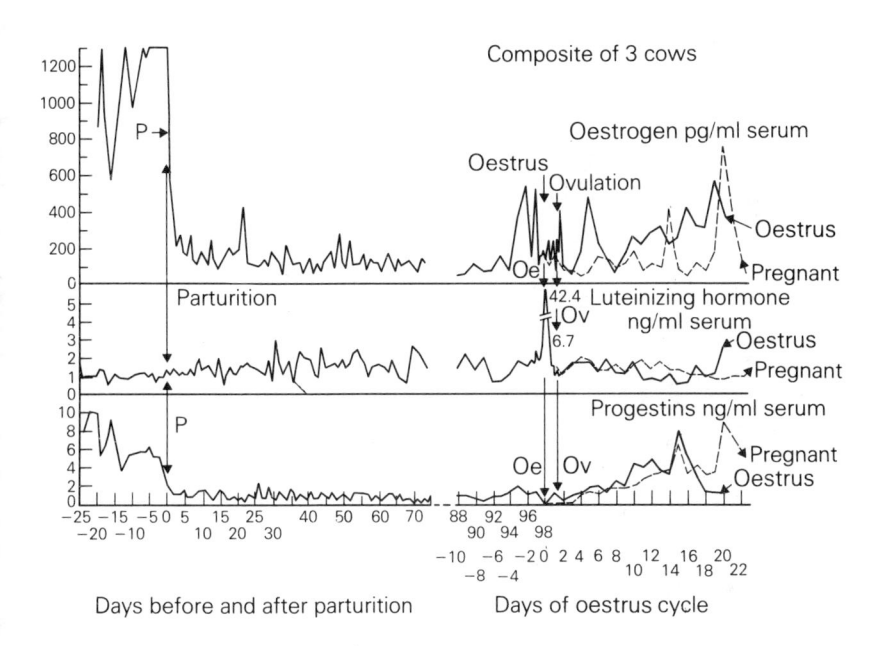

Figure 2   Serum levels of oestrogen, luteinizing hormone and progestins before parturition until the second postpartum oestrus or pregnancy in three beef cows (from Arije et al., 1974)

Table 1 The nursing and sucking frequency and duration in cattle as reported by several groups of workers

Authors	Length of observation period (hours)	Breed of dam	Breed of sire	Conditions	Mean frequency/24 hours and period of observation
Walker	24	AA, Her, AA x Jersey	AA	field	5-3.7 (from calving to 6 months)
Hafez and Lineweaver	24	Her, Hol (no cross-bred calves)	Her, Hol	housed	7.5-6.25 (1-6 wks)
Ewbank	12, 24	Her	Her	field	3.9 (at 1 wk), 2.3 (at 12 wks)
Nicol and Sherafeldin	dawn to dusk	AA, AA x Her, AA x Shorthorn	AA or F	field	5.6 (at 7 days), 3.5 (at 24 days), 3.5-3.0 (at 120 days)
Somerville and Lowman	24	Her, F Blue grey Blue grey	Charolais Charolais Charolais	field field housed	5.5 (at 4.5 wks) 5.5 (at 7-8 wks) 9.1 (at 2-3 wks)
Lewandski and Hurnik	24	Beef breed	Beef breed		4.8 (over 100 days)

AA = Aberdeen Angus
F = Friesian
Her = Hereford
Hol = Holstein

Parental Behaviour in Ungulates 95

Nursing or suckling has been studied by a number of authors (Walker, 1962; Hafez and Lineweaver, 1968; Ewbank, 1969; Nicol and Sherafeldin, 1975; Somerville and Lowman, 1979; Lewandski and Hurnik, 1981). The average frequency of suckling as reported by them is variable, as is shown in table 1. Furthermore, within their average frequencies is hidden further variability between and within data from individual females. Also with increasing age there is a decline in the number of sucklings by the calf. Hafez and Lineweaver (1968), Ewbank (1969) and Somerville and Lowman (1979) report that sucklings were fairly evenly distributed between day and night but Lewandski and Hurnik (1981) found only 35 per cent of sucklings to be at night. It can be deduced from these sources that the duration of a suckling event is approximately between six and ten minutes, but none of these workers accurately defines what they meant by a suckling event.

Milk let-down is a neuro-endocrine reflex and is controlled by oxytocin. The lactating female becomes conditioned to visual and tactile stimuli associated with suckling or milking and this conditioning induces the release of oxytocin. It then acts on the myoepithelial cells surrounding the aveoli in the mammary gland, and the contraction of these cells puts pressure on the aveoli which displaces milk into the duct system of the mammary gland (Reeves, 1980).

Stimulation of the udder leads to a three- to ten-fold increase in plasma prolactin lasting for up to one hour, but the magnitude of the response appears to be related to the stage of lactation of the cow (Fell et al., 1971; Goodman et al., 1979) and to the presence or absence of the calf. Holstein cows tested for prolactin concentrations at milking or nursing, on either the second or third day after calving showed small increases in prolactin concentrations at milking or sucking if their calf was present, but if they had been housed without their calves there was a significant rise in prolactin concentrations. However the increase was from different basal levels (Akers and Lefcourt, 1982). Oxytocin levels, too, may be affected by the presence of the calf (Akers and Lefcourt, 1982). These responses are less in the case of milking than for calves sucking, but they do not appear to be entirely due to the type of tactile stimulation, for the presence of the calf alone can lead to increased milk flow from cannulated teats in certain cows and heifers (Peeters et al., 1973).

Normally the calf adopts the inverse parallel position for nursing, and in 37 per cent of nursings witnessed by Le Neindre and Garel (1977) the calf passed in front of the cow before adopting this position, and in 21 per cent of these cases it was licked. However other positions can be taken by the calf and these are common in fostered calves, although the results of fostering are variable (Le Neindre et al., 1978). The calf can be effectively adopted, in which case it nurses in the inverse parallel position with the other calf (as in the case of twins) and is allowed to start nursing before the arrival of the legitimate calf. Some cows, however, allow the fostered calf to take up the inverse parallel position but will not allow it to nurse first. Alternatively, the calf will only be tolerated, in which case it can only nurse when the legitimate calf is already nursing and it cannot use the inverse parallel position. In this case the foster calf nurses from the rear or from the 'double position' in which it stands parallel to the legitimate calf in the inverse parallel position, but distally from the cow so that it is hidden by the legitimate calf. Occasionally a cow may adopt a second calf and refuse her own calf. Adoption is more likely to occur if the fostered calf is presented at calving (Le Neindre et al., 1978, Kiley-Worthington and de la Plain, 1983), if this is the case the latter authors state that it should be rubbed in the amniotic fluid of the foster cow. Blindfolding the cow over parturition may also help fostering. Both groups of workers agree that the behaviour of the calf at presentation is important and Kiley-Worthington and de la Plain (1983) state that a cow with experience of double nursing in one lactation will more readily adopt a calf in subsequent lactations. Furthermore Poindron and Le Neindre (1979) found breed differences in the propensity of cows to adopt rather than tolerate the foster calf.

After the intense preoccupation with her calf over the first three hours postpartum in suckler beef cows, the distance between the cow and calf gradually increases; by the end of the first week it is sometimes as great as 30 metres and by the end of the second week can be as great as 50 metres. However, the amount of time spent in contact with the dam does not change over this period (Kiley-Worthington and de la Plain, 1983). The calf begins to lie with other calves as the dam grazes at considerable distances away and its interactions with peers increase from 0.22/30 minutes in the first week to 1.73/30 minutes in the ninth week. Concomitantly its interactions with the dam decline over the same period from 1.45 to 0.15 (Wood-Gush et al., 1984). As in the case of the sheep, little attention has been paid by researchers to the process of natural weaning.

CERVIDAE

Reindeer (*Rangifer tarandus*)

The reindeer has not been subject to such intense artificial selection as the cow but it is, nevertheless, the most domesticated of all cervids. Its maternal behaviour has been reported on by Espmark (1971a, and 1980).

One to two days before calving the reindeer cow shows restlessness and her behaviour becomes asynchronous with the rest of the herd. In addition, like some cattle, she may at this time give the calls normally given to her calf. She eventually isolates herself and, again similarly with cattle there is variation in this aspect; Forest reindeer isolate themselves and mountain reindeer give birth in the immediate vicinity of the herd. As labour increases so the female may alternate between lying and standing but she lies while giving birth (however Espmark feels there is individual variation in this). As in cattle the placenta is eaten. For the first 25–30 minutes after the birth the cow licks the calf nearly continuously, but only sporadically thereafter, and the calf takes between 10–40 minutes to stand. In the first stages of teat-seeking the calves generally direct their behaviour towards incorrect parts of the cow's body, particularly the chest and throat region, as in the case of many cattle calves. The first succesful nursings witnessed by Espmark were from that position as the calf nosed its way backwards along the belly of the cows. The cow usually initiates departure from the place of birth to rejoin the herd between 3–30 hours postpartum.

In the early stages, nursing appears to be initiated mostly by the dam but after about two weeks of life the calf is usually the initiator. The most common posture for nursing, as in cattle, is the inverse position with the body of the calf forming an acute angle with, or parallel to, the body of the dam. Other positions included nursing from behind or nursing while the cow was lying, but the incidence of these last two positions was much lower. Nursing with cows other than the dam was seen in most calves in the study, and was generally done when the legitimate calf was nursing. During the first week the calves nursed successfully approximately three times per hour while at three months of age the rate had declined to one successful nursing every five hours. Nursing attempts occurred mostly in clusters, the number in a cluster declining from four in the first week to one at three months of age. The duration of each nursing declined during the first two months and then stabilized, the longest bouts being in the second week when they averaged about 70 seconds. The body care received by the calf from the dam was nearly always connected with nursing. Licking was of very short duration and confined to the genital area, but this declined with age and no licking was seen in connection with nursing after two months of age. Espmark (1980) investigated the effect of undernourishment during pregnancy on maternal behaviour and found that calves from undernourished females tended to take longer to stand after birth and were less tolerated by their dams.

Reindeer calves are followers. Initially the dam and calf remain very close together with the calf making occasional exploratory forays two to three metres from the dam. Later, when the dam joins the herd, she will stop if the calf lags behind and will greet it with nose to nose contact when it comes and allow it to suck before she moves on, thus reinforcing its following. When she is grazing, the calf usually lies beside her and as she moves on it will join her. (However, some cows are less attentive to their calves in this phase than others.) By the time the calf is three weeks old its distance from the dam is no different from the general pattern of spacing in the herd, except that, when resting, calves lie very close to the mother until they are two months of age. Both the cow and the calf can recognize one another's calls (Espmark 1971b), and once the calf rejoins its mother she sniffs it. Thus, it would appear that, as in cattle, olfactory cues are important in recognition of the calf. (The latter has yet to be tested experimentally.) It is not known whether there is a critical period for the reindeer cow to learn this recognition. Weaning takes place about six months after birth, but the calf remains with its dam.

Red deer (Cervus elaphus)

In the case of red deer, there are descriptions of hinds calving in the wild (Clutton-Brock and Guiness, 1975; Clutton-Brock et al., 1982) and indoors in intensive-breeding conditions (Arman, 1974). In the wild the hind moves away from the matriarchal group, generally 2–12 hours before calving which takes place in a peripheral area of the home range. For successive calvings a hind may use the same general area but not the same place. In intensive conditions some hinds calved standing and one calved lying, while Clutton-Brock and his colleagues report that in the wild the hind may lie during severe labour and stand when the calf is exposed. After the birth, the hind licks the calf and then rests, licks it again and then nurses it. In three of the hinds watched in intensive conditions the calf found the teat while the hind was still lying and it nursed while she lay. As the calf nurses the hind licks the perineal area of the calf; it then urinates and defaecates and she drinks the urine and eats the faeces. In the wild the calf chooses the lying site and is visited by the hind. Over the first day or two it is nursed every two to three hours, but by the end of the first week the frequency drops to eight times per 24 hours and to four to five times per 24 hours when the calf is three months old. (There is a tendency for male calves to nurse more frequently than female calves.) If the hind conceives again, weaning is at five to seven months, but if not, she might nurse the calf until the following summer or autumn, the frequency of which falls to a low level in the winter and early spring and rises again in the summer. Over the first day or two after birth calves nurse weakly and inefficiently and the bouts are protracted, but by the time the calves are three weeks old the average bout is roughly 100 seconds in duration, this declining to about 60 seconds and 30 seconds at 10 and 25 weeks postpartum. The data of Clutton-Brock et al. (1982) suggest that the duration of nursing bouts is longer in the calves of young (three and four year old) and old (12 and 13 year old) hinds than in the calves of middle-aged hinds. By the age of 10–15 days the hiding phase changes for the calf spends progressively more time with the hind and by three weeks of age it is seldom more than 100 metres from the hind.

EQUIDAE

There are six species of wild *Equidae* alive today and two domesticated species, all of which show similar maternal behaviour. The wild species are the Plains or Burchell's Zebra (*E. quagga*), the Mountain Zebra (*E. Zebra*), Grevy's Zebra (*E. grevyi*), the wild horse (*E. przewalski*), the African wild ass (*E. africanus*) and the Asiatic wild ass (*E. hemionus*). Equine domestication has led to our present day domestic horse (*E. przewalski f. caballus*) and the donkey (*E. africanus f. asinus*), which are the domesticated forms of the wild horse and the African wild ass respectively (Carson and Wood-Gush, 1983a).

The offspring of all these equid species show the 'follower' type of relationship with their dam, as defined by Walther (1961). These 'follower' species maintain close and frequent contact between foal and dam from a few hours postpartum and the foal travels with the herd from the first day of life. The equine foal's and the dam's maternal behaviour are specialized to protect the health and safety of the new-born foal in the open grassland habitat in which the wild species live, and the 'following' strategy is part of this specialization. The foal must therefore be 'fit' at birth for running with the herd.

The equine foetus is physically active from three months of age, with the frequency of movements increasing towards parturition. These movements have been likened to isometric exercises which are necessary for good muscular devlopment and functioning of the joints to enable the new-born foal to make its early adaptations to life (Fraser et al., 1975). As pregnancy advances the mare becomes physiologically and behaviourally ready to respond to her new-born foal (Shillito-Walser, 1977). Most mares, domestic and feral, give birth during the hours of darkness and, if necessary, are able to delay the onset of labour until environmental conditions are favourable, thus protecting the vulnerable neonate from predators (Campitelli et al., 1982/83; Klingel, 1969). Herd-living mares sometimes seclude themselves from the herd to give birth (Tyler, 1972). Some mares may seek the foal even before the foetus is expelled and, like many other ungulates, the mare shows considerable attraction to her own amniotic fluid (Fox, 1968) sometimes even neglecting the foal in this way (Beaver, 1981). Mares normally lie down for the final stages of labour and for the birth (Klingel and Klingel, 1966; Rossdale, 1967) and delivery of the foal while standing is considered abnormal (Rossdale, 1968). The afterbirth is normally passed 60-90 minutes after parturition and is not normally eaten by the mare (Campitelli et al., 1982; Klingel, 1966), again a feature of the 'follower' type of ungulate which moves away from the birth place soon after parturition; thus the presence of the afterbirth in attracting predators is not a serious risk.

The early development of behaviour in the foal is similar in both wild and domesticated species but it is interesting that the wild species are significantly faster in achieving the early goals (such as standing and sucking) and have presumably retained this rapid adaptation to extra-uterine life as an anti-predator mechanism. A new-born Plains zebra takes only 11 mintues to stand and 67 minutes to start sucking postpartum (Klingel and Klingel, 1966), whereas a thoroughbred foal takes, on average, 57 minutes to stand and 111 minutes to suck (Rossdale, 1967). The suck-reflex is apparent in foals even before standing and the primary aim of the neonate after standing is to look for the udder and find the milk supply. The dam gets to her feet soon after the foetus has been expelled, zebra mares stand as soon as 13 minutes after parturition (Klingel, 1969) whereas thoroughbred mares take on average 40 minutes to stand (Rossdale, 1967). The mare licks her new-born foal thoroughly and this, as in other ungulates, is thought to stimulate respiration and improve muscle tone. Licking also familiarizes the dam with her foal and teaches her to distinguish her foal from others.

Mares are able to distinguish their foals almost immediately after parturition (Waring, 1970) and are very attentive to the foal, approaching it and sniffing it and keeping it in view all the time (Francis-Smith, 1979). They are quick to protect their foal from any moving object and herd-living mares keep themselves between the foal and conspecifics during the first few days postpartum. The new-born foal takes two to three days to identify its dam and so the dam's behaviour, in keeping herself as the primary object in the foal's field of view, is to ensure correct imprinting (Klingel, 1969; Tyler, 1972; Powell, 1978). Young wild asses have difficulty in identifying their dam at this age (Rashek, 1976). In domestic mares, isolated from conspecifics, this 'drive' is redirected towards other moving objects such as humans, dogs and even sparrows (Francis-Smith, 1979). After this critical period the frequency of mare–to–foal interactions decreases and foals take most of the initiative in maintaining contact with their dam, especially when pastured (Francis-Smith, 1979).

In recognizing one another mares and foals largely use the visual sense, and despite olfactory cues also being important, when visual cues are eliminated, mares and foals are hindered in their efforts to find one another (Wolski et al., 1980). Mares are able to identify the calls of their foals whereas foals respond to the reaction of a mare to their approach to identify their dam from other mares. Mutual sniffing appears to act as final identification between a mare–foal pair.

Foals look for the udder by reaching up and along the underside of the dam's belly with their muzzle, moving it from side to side and grasping parts of the mare with the lips and sucking, often confusing the pectoral region with the pelvic. Early teat-seeking is probably directed at the mare's belly but the activity seems to involve a certain element of luck in that foals grasp and suck the dam's tail and legs as well as the udder and teats during the first day postpartum (Carson and Wood-Gush, 1983b). Having grasped a teat, foals suck for several seconds and then change to the other teat, swopping teats several times during one bout. Foals initiate all nursing activity, often approaching the dam and walking along one side of her, pushing under her neck and walking down the other side to stand in a reverse parallel position which makes the mare stand still if she is walking or grazing (Tyler, 1972). Dams help their foals nurse by standing still and flexing the hind leg on the side away from the foal to angle the teats towards the foal's mouth. During the first week postpartum foals nurse frequently – 7 times per hour – with each nursing bout lasting a mean of 105 seconds, but not all nursing attempts achieve sucking (Carson and Wood-Gush, 1983b). The frequency of nursing decreases as the foal grows older until 12 weeks postpartum after which they nurse approximately once an hour.

In contrast to the decreasing nursing frequency of foals the mare's milk production increases to a peak at roughly the 10th week of lactation (Bouwman and Schee, 1978) and foals take advantage of this extra milk by being able to swallow more quickly while nursing as they grow older (Rogalski, 1973).

Nursing behaviour consists of two behaviour patterns, teat-seeking or 'nosing', as it has been called, and sucking. Once the foal has grasped a teat it thrusts its muzzle into the udder several times before sucking; this pushing or bunting is a common feature of nursing behaviour and encourages milk flow by stimulating the release of oxytocin from the neurophypophysis (Fraser, 1968). Oxytocin apparently acts in the same way as in the cow in the expulsion of the milk from the alveoli and small ducts of the mammary gland. Bunting is a vigorous activity and at times seems painful to the mares. In domestic mares and foals the frequency of bunting reaches a peak in the second week postpartum and simultaneously the mares become unwilling to let their foals nurse. Emotional factors can inhibit milk let-down (Tindal, 1966) and the mares, being wary of the painful bunting, may withhold milk ejection, thus setting up a vicious circle with the foals having to bunt more to stimulate milk ejection. The mares may be quite violent in their efforts to stop their foals nursing by running away and kicking or biting (Francis-Smith, 1979; Tyler, 1972). It is not known why nursing activity should be more painful in the second and third week postpartum and such behaviour has only been recorded in the domestic horse.

Feral mares become less tolerant of their foals nursing behaviour as the foals grow older (Rashek, 1976; Martin-Rossett et al., 1978) but do not usually wean their foals until shortly before the next parturition.

Weaning then occurs as a sudden and complete rejection of the foals nursing attempts by the mare (Tyler, 1972). If she is not pregnant, the mare may allow a foal to continue nursing as a yearling and there are reports of mares allowing a foal and a yearling to nurse at the same time (Feist and McCullough, 1975).

Even after weaning foals maintain a close contact with their dams and remain in their family group and the social bonds between dam and offspring are broken only when the young reach puberty. At this time the family stallion chases away the adolescent males from the family group whilst young females in their first oestrous are usually abducted by other stallions to join a different family group within the same herd. Puberty usually occurs when the young are between one and four years (Klingel, 1975; Tyler, 1972).

There are two distinct types of social organization amongst the *Equidae* (Klingel, 1972a) and these affect the role of the stallion as a father. In the territorial species (Grevy's zebra, the wild ass, Asiatic wild ass and the feral donkey) the stallions have no relationship with their offspring since the mares migrate through the stallions territory and move on after mating. The non-territorial species live in family groups consisting of a family stallion, several mares and their offspring. Amongst these species the stallion defends his mares and foals from predators and conspecifics (Feist and McCullough, 1975) and may stay close to the mare during birth in order to protect the mare and foal (Klingel and Klingel, 1966).

SUIDAE

The domestic pig

The members of the pig family are the only true multiparous ungulates, and seasonal breeding is the common mode of reproduction in natural environments. Synchronous birth in herbivorous and omnivorous species has an obvious advantage since sufficient food, in many geographic areas, may only be available for limited periods. Another advantage of synchronous breeding is that a large group of young born at the same time acts as an anti-predator strategy. Synchrony of reproduction between sows has been reported in the wild boar (Meynhardt, 1981) and in the domestic pig (Stolba, 1982).

Sows are able to conceive during lactation (Rowlinson et al., 1975; Stolba, 1982). However, much of the physiological and behavioural causation of lactational oestrus is still unknown. Although hormonal changes of considerable magnitude occur during pregnancy, they do so gradually until just before parturition (Catchpole, 1969). No evidence is available, yet, on hormonal mechanisms of the onset of maternal behaviour in pigs, but the general outline of the mechanisms of the pig is unlikely to differ from that of the ewe or cow.

Sows observed in a semi-natural environment graze more intensively than the non-pregnant females (Hafez, 1962). About three weeks before farrowing, the udder of a pregnant sow starts to swell and then two weeks later her teats and the vulva do likewise. Grundlach (1968) reported that the female feral pig becomes more active and investigates her environment during the week before parturition. This activity increases until, after a few days, she decides on the new nest-site, preferring a site on a forest ridge or in a burrow on a hill. The same features are to be seen in the domestic sows studied in a semi-natural enclosure at Edinburgh. Most nests are established on sheltered sites; 40 per cent are at least partially covered by a roof of dense branches of trees or bushes and 89 per cent are sheltered on at least one side. Pigs apparently like an open view from their nests, for they select the nest-site on ridges from which they can see up and downhill, and behind which they can easily hide from intruders. (Stolba, personal communication). The same behaviour can be seen in spacious pens with many ecological features where the sow due to farrow chooses a nest-area, builds a nest with all material available and defends it against other conspecifics (Stolba, 1982). So it can be said that the modern sow shows much ancestral nest-building behaviour around the time of parturition (Grundlach, 1968; Signoret et al., 1975). If the nest material carried by the domestic sow is removed, she will show great tenacity in retaining the site and carrying back all the nesting material (Grauvogel, 1958). However, in the United Kingdom 85 per cent of all sows farrow in small, bare, confined crates (M. R. Baxter, personal communication) and as the sow's nesting behaviour is most likely determined by her endocrine state, welfare considerations arise from her restraint. (Agricultural Committee, 1981; The Brambell Report, 1965).

As a wild boar's birth takes place in a nest or a burrow (Sambraus, 1978), there are still very few observations of the birth process. Normally the young are born while the female is lying on her side. The neonates are covered by a quickly drying transparent foetal membrane. The umbilical cord is detached by the repeated struggle of the young to reach the teats, and the sow, unlike most other ungulate females, does not lick her young or give any help.

In the semi-natural enclosure at Edinburgh and in spacious ecologically enriched pens the sow lies down for contractions (Stolba, 1980, 1981), in most cases leaning against a solid object – either a wall, a rail, or in nature, towards the highest ridge of the nest (Moser, personal observation). The litter is delivered within two to three hours and most sows eat the after-birth. They may even eat any dead foetus (Craig, 1981). The early hours of darkness appear to be the most likely time for farrowing in pigs (Alexander, 1974). Fraser (1968) postulated that adrenalin, produced in aroused animals, may be responsible for inhibiting birth during periods of greater activity or excitement.

Domestic piglets are extremely precocial at birth in comparison with other domestic animals, and, like the piglets of the wild boar, fend for themselves. After struggling to their feet within a minute or two of birth, they explore the body of the sow and usually locate a teat within the first 15 minutes. Piglets compete for teats and those arriving late in large litters might be at a disadvantage and may fail to gain access to a productive teat or, in some cases, to any teat at all (Craig, 1981). The piglets form a teat order in which each piglet suckles and defends a particular nipple (Donald, 1937).

Young piglets do not initially follow the mother and are 'hiders' for the first five days (Hutton, 1983). During this time the young are forced to stay in close contact with their mother and/or litter-mates in order to conserve body heat, and the less the environmental cover, the more essential is this 'huddling together'. At this stage the female very seldom leaves the young.

Whatson and Bertram (1983) reported that naso-nasal contacts between a sow and her young are frequent in the first five minutes after a successful suckling and suggested that this behaviour acts to re-establish the identity of the piglets nursed by the sow at a time when the sow is placid and receptive. Meese and Baldwin (1975) found that sows in farrowing crates either attacked strange piglets or examined them vigorously with their snout, while bulbectomized sows showed no aggression towards them, indicating that an olfactory component is implicated in the mother's recognition of her young. However piglets may be moved from sow to sow at least on the first two days after farrowing. Meynhardt (1980), reporting on vocal recognition in the feral swine, reported that the vocalizations of different litters differ in their sonograms, but the vocalization of piglets and their mother are very similar. Playback experiments show that the mother recognizes the calls of her piglets. Olfactory recognition seems to work over a very short distance, while recognition at a distance is through vocalisation; visual recognition is very poor. Kiley (1972) reports that there is an overall facilitation of calling in the piglets which functionally would be expected to increase the cohesion of the piglet group.

Pigs (like camels) differ considerably from other domesticated mammals in the mother's early response to the young for they are not cleaned by their dam at birth. The new-born pig is very much on its own and, as stated, most piglets stand within two minutes of delivery and suck on the sow, moving about exploring her udder until a teat is found and colostrum is ingested. This, as in other young ungulates, gives the piglet passive immune protection. Suckling bouts during early and mid-lactation are at about hourly intervals, but these intervals tend to lengthen in later stages of lactation to one and a quarter to two hours (Barber et al., 1955). The interactions between the sow and her piglets during suckling have been studied by Whittemore and Fraser (1974) and

Fraser (1975a,b,c,d). Although piglets might hold a nipple in their mouth for long period, especially in early development, milk let-down is relatively brief (20–25 seconds) and occurs only after vigorous 'nosing' of the udder. The amount of nosing required to stimulate a sow to let her milk down gradually increases; from about a minute during the first weeks to about two minutes by the eighth week of lactation. Typically the nursing sow gives a series of rapid grunts for about 25 seconds before milk ejection and stimulation of the anterior teats is more important for milk ejection than stimulation of the posterior teats (Fraser, 1973). When three or fewer piglets are present, sows may stop lactating altogether, possibly because the amount of udder stimulation is below a critical threshold (Pomeroy, 1960). Removal of those on posterior teats has a less marked effect, suggesting that the period of rapid grunting coincides with physiological events which lead to milk ejection. Hutton (1983) has described the behaviour of the neonate piglets in semi-natural conditions. Over the first few days it is the sow which initiates suckling bouts by grunting, approaching and sniffing her piglets, rousing them from sleep and then lying on her side with her teats well exposed while grunting rhythmically. Usually there is no clear-cut termination of a sucking bout at this age, as the sow tends to remain on her side after milk let-down and the piglets continue to nose and suck the teats until they fall asleep in a huddle against the warm udder (Hutton, 1983).

Although, over the first few months the sow often leads her piglets into a nest to nurse them, she increasingly nurses away from the nest during periods of activity, especially when the weather is warm and dry. Furthermore, under these conditions, slightly more than half of all the nursings were synchronized, which is more than could be expected by chance. It seems that sows respond to the sucking vocalization of other sows and their piglets as much as to the stimulation of their udders by their young (Hutton, 1983). Communal sucking is reported to occur in wild and feral pigs (Kurz and Marchinton, 1972; Lent, 1974). However, Hutton's observations suggest that the piglets suck from other sows in a opportunistic way, taking advantage of a teat in the absence of its mother. True communal sucking did not occur among piglets under semi-natural conditions.

Meynhardt (1981) reports that there is no fixed teat order in the feral swine in the first fortnight, but that after three weeks every piglet has established its own teat. The nursing rhythm in the first few days is determined by the sow; later on the hunger calls of the young are important, socially facilitating other young and the mother to nurse. Amazingly Meynhardt, in over a hundred farrowings, never saw any piglet crushed in the nest.

The teat order of new-born piglets develops progressively over the first two weeks of life. However, when the litter size is ten or more there is a marked decrease in the stability of the teat order (Hemsworth, 1976). In the semi-natural conditions used by Hutton (1983) the piglets' growth rates were not consistently influenced by the position of its teat along the udder, and Hutton concluded that the concept of dominance at the udder is not justified. Crowding at the udder is probably an important factor prompting piglets to seek milk and solid food elsewhere. Piglets are mainly responsible for locating and defending their own teat from others and the sow plays only a minor role in determining which piglet suckles from her (Hutton, 1983).

As the piglets become older, the attachment to the nest gradually declines (Fraedrich, 1971; Hutton, 1983). The mean age at natural weaning is 82 days in domestic stock (Stolba, 1982). However, before weaning the young animals must learn to ingest solid food if the transition is to be gradual and not associated with a nutritional setback when separated from their dams. Piglets start to nibble at solid food five days after birth but the first appreciable intake of solid food occurs only at an age of ten days (Moser, personal observation). Age at weaning is not determined by the degree of aggression shown by the mother, as this varies between sows and can start either a long or short time before the completion of weaning (Hutton, 1983). The maternal aggressiveness of the domestic sow compared to that of the wild boar sow has probably been reduced through artificial selection aimed at enabling nursing sows to remain calm in the presence of humans. In fact, the time taken to achieve weaning may depend on the amount of satisfactory food available to the piglet, for with less suckling the sows may go dry sooner. Weaning does not disrupt the social bond between the mother and her offspring, and females of the Wild Boar (Meynhardt, 1980) and the domestic sows (Stoba, 1982) retain their matrilinear bonds into adult life.

GENERAL DISCUSSION

The study of individual variation is useful in understanding the causal mechanisms underlying maternal behaviour. For example, it can illustrate the effect of early experience, nutritional state and age on maternal behaviour. In general it would seem that writers on the behaviour of wild species have tended to ignore individual variation, whilst some have described variation in major behavioural traits. For example, Altmann (1952) writing on the maternal behaviour of the elk (*Cervus canadensis Nelsoni*) states that some females took their still-wobbly calves to pasture, sometimes with the umbilical cord still turgid, while others enforced the 'hiding' system by pounding on the calves with nose and front feet when they started to get up. Gosling (1969) describes differences in the distance between the females of Cokes Hartebeest and their calves. Further work may illustrate more examples of individual variation in maternal behaviour in wild ungulates.

Parental Behaviour in Ungulates 107

There is evidence in cattle of variation between breeds, probably reflecting genetic differences and it seems likely that some genetic variation in maternal behaviour exists within wild species. The genetic variation in domestic species is likely to be greater because of the relaxation of certain selection pressures and the possible pleiotropic effects of some genes concerned with production traits.

Furthermore, some variation may be due to early experience, and again this might have a greater effect on domesticated species which are probably subjected to a wider range of rearing conditions than their counterparts in the wild. In the case of both sheep and cattle, descriptions were given of primiparous females avoiding or attacking their young. In the wild this could result in desertion of the young which has been reported in red deer (Clutton-Brock et al., 1982) and Soay sheep (Grubb, 1974). Aberrant behaviour of this sort in domesticated ungulates might be due to lack of experience with the young of their species, for in modern agriculture animals tend to be reared in single age/single sex groups. An alternative hypothesis is that the physiological mechanisms for maternal behaviour may not be fully functional when the primiparous female is mated, with the result that a critical threshold for the relevant hormones may not be reached. For example, the lack of co-ordination of different components of reproductive behaviour in domesticated female ungulates is well known in the form of the 'silent' oestrus. Interestingly, this has been reported in Wildebeest (Watson, 1969). Prolonged labour in ewes and cows is often related to poor nutrition and, as we have seen, it is often later connected with difficulties in maternal behaviour, and similar effects might therefore be expected in wild species. Finally the impact of early learning on subsequent maternal behaviour has to be considered for, as has been shown in other species, experience as an infant can affect the maternal behaviour of the adult female.

Variation in the time taken by the calf to find the teat has undoubtedly been influenced by selection for milk production in cattle, for pendulous udders result in many calves failing to suckle within six hours. Many of these calves are described as searching in the chest area of the dam and this behaviour has also been reported in Cokes Hartebeest (Gosling, 1969). Problems with nursing as reported in mares have also been reported in ewes (J. Doney, personal communication) and Clutton-Brock et al. report that three calves in a population of red deer in the wild died as a result of 'suckling difficulties'. It can be concluded, therefore, that the variation and difficulties in teat-seeking recorded in the domesticated species are not unique to these species, although they may be more common.

For the elucidation of theories of natural selection, such as those dealing with maternal investment, domesticated species are unlikely to yield much information. But for the study of the effects of maternal behaviour on the later food and pasture preferences of the young and on their home range behaviour, these species provide promising material. However, their major contributions to the study of maternal behaviour will be from the physiological mechanisms underlying the behaviour and from the nature of the mother–young bond – how it is formed and maintained.

REFERENCES

Agricultural Committee 1981: Animal welfare in poultry, pig and veal calf production. London HMSO.

Akers, R. M. and Lefcourt, A. M. 1982: Milking and suckling – induced secretion of oxytocin and prolactin in parturient dairy cows. *Hormones and Behaviour*, 16, 87-93.

Alexander, G. 1960: Maternal behaviour in the Merino ewe. *Proceedings of the Australian Society of Animal Production*, 3, 105-14.

Alexander, G., Lynch, J. J. and Mottershead, B. E. 1979: Use of shelter and selection of lambing sites by shorn and unshorn ewes in paddocks with closely or widely spaced shelters. *Applied Animal Ethology*, 5, 51-69.

Alexander, G., Lynch, J. J., Mottershead, B. E. and Donnelly, J. B. 1980: Reduction in lamb mortality by means of grass wind breaks: Results of a five year study. *Proceedings of the Australian Society of Animal Production*, 13, 329-32.

Alexander, G. and Shillito, E. E. 1977a: The importance of odour, appearance and voice in maternal recognition of the young in Merino ewes ('Ovis aries'). *Applied Animal Ethology*, 3, 127-35.

—— 1977b: Importance of visual clues from various body regions in maternal recognition of the young in Merino sheep ('ovis aries'). *Applied Animal Ethology*, 3, 137-43.

Alexander, G., Signoret, J. P. and Hafez, E. S. E. 1974: Sexual and maternal behaviour. In E. S. E. Hafez (ed.), *Reproduction in Farm Animals*, 3rd edn, Philadelphia: Lea & Febiger.

Alexander, G. and Stevens, D. 1982: Odour cues to maternal recognition of lambs: an investigation of some possible sources. *Applied Animal Ethology*, 9, 165-75.

Alexander, G. and Williams, D. 1964: Maternal facilitation of suckling drive in new born lambs. *Science*, 146, 665-6.

—— 1966: Teat-seeking activity in lambs during the first hours of life. *Animal Behaviour*, 14, 166-78.

Arije, G. R., Wiltbank, J. N. and Hopwood, M. L. 1974: Hormone levels in pre- and post-parturient beef cows. *Journal of Animal Science*, 39, 338-47.

Arman, P. 1974: A note on parturition and maternal behaviour in captive red deer (*Cervus elaphus* L.). *Journal of Reproduction and Fertility*, 37, 87-90.

Arnold, G. W. and Dudzinski, M. L. 1978: *Ethology of free-ranging domestic animals*. Amsterdam: Elsevier Scientific Publishing Company.

Arnold, G. W. and Morgan, P. D. 1975: Behaviour of the ewe and lamb at lambing and its relationship to lamb mortality. *Applied Animal Ethology*, 2, 25-46.

Arnold, G. W., Wallace, S. R. and Maller, R. A. 1979: Some factors involved in natural weaning processes in sheep. *Applied Animal Ethology*, 5, 43–50.

Baldwin, B. A. and Shillito, E. E. 1974: The effects of ablation of the olfactory bulbs on parturition and maternal behaviour in Soay sheep. *Animal Behaviour*, 22, 221–4.

Barber, R. S., Brande, R. and Mitchell, K. G. 1955: Studies on milk production of large white pigs. *Journal of Agricultural Science*, Camb., 46, 97–118.

Bareham, J. R. 1976: The behaviour of lambs on the first day after birth. *British Veterinary Journal*, 132, 152–61.

Beaver, B. V. 1981: Maternal behaviour in mares. *Veterinary Medicine/Small Animal Clinician*, 76, 315–17.

Bowman, H. and Schee, W. 1978: Composition and production of milk from Dutch warm blooded saddle horse mares. *Zeitschrift für Tierphysiologie, Tierenährung und Futtermittelkunde*, 40, 39–53.

Brambell, F. W. R. 1965: *Report on the technical committee to inquire into the welfare of animals kept under intensive livestock husbandry systems.* London: HMSO, Cmnd 2836.

Campitelli, S., Carenzi, C. and Werga, M. 1982/83: factors which influence parturition in the mare and development of the foal. *Applied Animal Ethology*, 9, 7–14.

Carson, K. and Wood-Gush, D. G. M. 1983a: Equine behaviour: I. A review of the literature on social and dam–foal behaviour. *Applied Animal Ethology*, 10, 165–78.

—— 1983b: Behaviour of thoroughbred foals during nursing. *Equine Veterinary Journal*, 15, 257–62.

Catchpole, H. R. 1969: Hormonal mechanism during pregnancy and parturition. In H. H. Cole and P. T. Cupps (eds), *Reproduction in Domestic Animals*, 2nd edn., New York: Academic Press Inc.

Clutton-Brock, T. H. and Guiness, F. E. 1975: Behaviour of red deer (*Cervus elaphus* L.) at calving time. *Behaviour*, 55, 257–300.

Clutton-Brock, T. H., Guiness, F. E. and Albon, S. D. 1982: *Red deer. Behaviour and ecology of two sexes.* Edinburgh: Edinburgh University Press.

Collias, N. E. 1956: The analysis of socialization in sheep and goats. *Ecology*, 37, 228–39.

Craig, J. V. 1981: Domestic animal behaviour. New Jersey: Prentice-Hall Inc.

Donald, H. P. 1937: Suckling and suckling preference in pigs. *Empire Journal of Experimental Agriculture*, 5, 361–8.

Edwards, S. A. 1982: Factors affecting the time to first suckling in dairy calves. *Animal Production*, 34, 339–46.

Edwards, S. A. and Broom, D. M. 1982: Behavioural interactions of dairy cows with their newborn calves and the effects of parity. *Animal Behaviour*, 30, 525–35.

Espmark, Y. 1971a: Mother–young relationship and ontogeny of behaviour in reindeer (*Rangifer tarandus* L.). *Zeitschrift für Tierpsychologie*, 29, 42–81.

—— 1971b: Individual recognition by voice in reindeer mother–young relationship. Field observation and playback experiments. *Behaviour*, 40, 295–301.

—— 1980: Effects of maternal prepartum undernutrition on early mother–calf relationships in reindeer. In E. Reimers, E. Gaare, and S. Skjenneberg, (eds),

Proceedings 2nd International Reindeer/Caribou Symposium, Roros, Norway, 1979.

Ewbank, R. 1964: Observations on the suckling habits of twin lambs. *Animal Behaviour*, 12, 34–7.

—— 1967: Nursing and suckling behaviour amongst Clun Forest ewes and lambs. *Animal Behaviour*, 15, 251–8.

—— 1969: The frequency and duration of the nursing periods in single-suckled Hereford beef cows. *British Veterinary Journal*, 125, IX–X.

Feist, J. D. and McCullough, D. R. 1975: Reproduction in feral horses. *Journal Reproduction and Fertility Suppl.*, 23, 13–18.

Fell, L. R., Beck, C., Blockley, M. A., Brown, J. M., Catt, K. J., Cumming, I. A. and Godling, J. R. 1971: Prolactin in the dairy cow during suckling and machine milking. *Journal of Reproduction and Fertility*, 24, 144–5.

Fletcher, I. C. 1971: Relationships between frequency of suckling, lamb growth and post-partum oestrous behaviour in ewes. *Animal Behaviour*, 19, 108–11.

Fox, M. W. 1968: *Abnormal behaviour in animals*. Toronto: W. B. Sanders Co.

Fraedrich, H. 1971: A comparison of behaviour in the Suidae. In *The Behaviour of ungulates and its relations to management*. ISUSN Publication, new series, 124.

Francis-Smith, K. 1979: Studies on the feeding and social behaviour of domestic horses. PhD Thesis, University of Edinburgh.

Fraser, A. F. 1968: *Reproductive behaviour in ungulates*. New York: Academic Press.

Fraser, A. F., Hastie, H., Callicott, R. B. and Brownlee, S. 1975: An exploratory ultrasonic study on quantitative foetal kinesis in the horse. *Applied Animal Ethology*, 1, 395–404.

Fraser, D. 1973: The nursing and suckling behaviour of pigs. I. The importance of stimulation of the anterior teats. *British Veterinary Journal*, 129, 324–35.

—— 1975a: The nursing posture of domestic sows and related behaviour. *Behaviour*, 57, 51–63.

—— 1975b: The 'teat order' of suckling pigs. II. Fighting during suckling and the effects of clipping the eye teeth. *Journal of Agricultural Science*, Cambridge, 84, 393–9.

—— 1975c: The nursing and suckling behaviour of pigs. III. Behaviour when milk ejection is elicited by manual stimulation of the udder. *British Veterinary Journal*, 131, 416–26.

—— 1975d: The nursing and suckling behaviour of pigs. IV. The effect of interrupting the suckling stimulus. *British Veterinary Journal*, 131, 549–59.

Gauthier, D., Terqui, M. and Mauleon, P. 1983: Influence of nutrition on prepartum plasma levels of progesterone and total oestrogens and post-partum plasma levels of luteinizing hormone and follicle stimulating hormone in suckling cows. *Animal Production*, 37, 89–96.

Geist, V. 1971: *Mountain sheep: A study in behaviour and evolution*. Chicago: University of Chicago Press.

Gonyou, H. W. 1984: The role of behaviour in sheep production: a review of research. *Applied Animal Ethology*, 11, 341–59.

Goodman, G. T., Tucker, H. A. and Convey, E. M. 1979: Presence of the calf affects secretion of prolactin in cows. *Proceedings of the Society for experimental Biology and Medicine*, 161, 421–4.

Grubb, P. 1974: Social organization of Soay sheep and the behaviour of ewes and lambs. In P. A. Jewell, C. Milner and J. Morton Boyd (eds), *Island survivors: The Ecology of the Soay sheep of St Kilda*, Athlone Press, University of London.

Grubb, P. and Jewell, P. A. 1966: Social grouping and home range in feral Soay sheep. *Symposium of the Zoological Society of London*, 18, 179–210.

Granvogel, A. 1958: Ueber das Verhalten des Hausschweines unter besonderer Berücksichtigung des Fortpflanzungsverhaltens. Berlin: Dissertation.

Grundlach, H. 1968: Brutfürsorge, Brutpflege, Verhaltensontogenese und Tagesperiodik beim europäischeen Wildchwein. *Zeitschrift für Tierpsychologie*, 25, 955–95.

Gubernick, D. J. 1980: Maternal 'imprinting' or maternal 'labelling' in goats? *Animal Behaviour*, 28, 124–9.

—— 1981: Mechanisms of maternal 'labelling' in goats. *Animal Behaviour*, 29, 305–6.

Gubernick, D. J., Jones, K. C. and Klopfer, P. H. 1979: Maternal 'imprinting' in goats. *Animal Behaviour*, 27, 314–15.

Hafez, E. S. E. 1962: *The behaviour of domestic animals*. London: Bailliere Tindall.

Hafez, E. S. E. and Lineweaver, J. A. 1968: Suckling behaviour in natural and artificially fed neonate calves. *Zeitschrift für Tierpsychologie*, 25, 187–98.

Hemsworth, P. H., Winfield, C. G. and Mullaney, P. D. 1976: A study of the development of the teat order in piglets. *Applied Animal Ethology*, 225–33.

Hewson, R. and Verkaik, A. J. 1981: Body condition and ranging behaviour of Blackface hill sheep in relation to lamb survival. *Journal of Applied Ecology*, 18, 401–15.

Hunter, R. F. 1954: Some notes on the behaviour of hill sheep. *Animal Behaviour*, 2, 75–9.

—— 1964: Home range behaviour of hill sheep. In D. J. Crisp (ed.), *Grazing in Terrestrial and Marine Environments*, Oxford: Blackwell.

Hunter, R. F. and Milner, C. 1963: The behaviour of individual, related and groups of South Country Cheviot hill sheep. *Animal Behaviour*, 11, 507–13.

Hutton, R. 1983: *The development of social behaviour in pigs.* PhD Thesis, University of Edinburgh.

Kaltenbach, C. C. and Dunn, T. G. 1980: Endocrinology of reproduction. In E. S. E. Hafez (ed.), *Reproduction in farm animals*, 4th edn, Philadelphia: Lea & Febiger, 85–113.

Key, C. and MacIver, R. M. 1980: The effects of maternal influence on sheep: breed differences in grazing, resting and courtship behaviour. *Applied Animal Ethology*, 6, 33–48.

Keverne, E. B., Levy, F., Poindron, P. and Lindsay, D. R. 1983: Vaginal stimulation: an important determinant of maternal bonding in sheep. *Science*, 219 (4580), 81–3.

Kiley, M. 1972: The vocalisation of ungulates, their causation and function. *Zeitschrift für Tierpsychologie,* 31, 171–222.

Kiley-Worthington, M. and de la Plain, S. 1983: *The behaviour of beef suckler cattle (Bos taurus)*. Basel: Birkhauser.

Kilgour, R. and Ross, D. J. 1980: Feral goat behaviour – a management guide. *New Zealand Journal of Agriculture*, 141, 15–20.

Klingel, H. 1969c: Reproduction in the plains zebra, *Equus Burchelli Boehmi*:

Behaviour and Ecological Factors. *Journal of Reproduction and Fertility Supplement*, 6, 339–45.
—— 1972: Social behaviour of African *Equidae*. *Zoologica Africana*, 7, 175–85.
—— 1975: Social organisation and reproduction in Equids. *Journal of Reproduction and Fertility Supplement*, 23, 7–11.
Klingel, H. and Klingel, U. 1966: Die Geburt eines Zebras, equus quagga boehmi, Matschie. *Zeitschrift für Tierpsychologie*, 23, 72–6.
Klopfer, P. H., Adams, D. K. and Klopfer, M. S. 1964: Maternal imprinting in goats. *Proceedings of the National Academy of Science*, 52, 911–14.
Klopfer, P. H. and Gubernick D. J. (eds) 1981: *Parental Care in Mammals*. London: Plenum Press.
Kurtz, J. C. and Marchinton, R. L. 1972: Radiotelemetry studies of feral hogs in South Carolina. *Journal of Wildlife Management*, 36, 1240–8.
La Voie, Y. and Moody E. L. 1976: Suckling effect on steroids in postpartum cows. *Journal of Animal Science*, 43, 292–3.
Lawrence, A. B. and Wood-Gush, D. G. M. (in preparation): *Maternal behaviour and social organisation in Scottish Blackface sheep*.
—— (in press): Home range behaviour and social organisation in Scottish Blackface sheep. *Applied Animal Ethology*.
Le Neindre, P. and Garel, J-P. 1976: Existence d'une période sensible pour l'établissement du comportement maternel de la vache après la mise-bas. *Biology of Behaviour*, 1, 217–21.
—— 1977: Etude des relations mére-jeune chez les bovins domestiques: Comparaison des liaisons existant entre la mère et des veaux légitimes ou adoptés. *Biology of Behaviour*, 2, 39–49.
—— 1978: Allaitement de deux veaux par des vaches de race Salers II. Etude de l'adoption. *Annals Zootechnie*, 27, 553–69.
Lent, P. C. 1974: Mother–infant relationships in ungulates. In V. Geist and F. Walther (eds), *The behaviour of ungulates and its relationship to management*, vol 1, Switzerland: IUCN, Morges, 14–55.
Lewandski, N. M. and Hurnik, J. K. 1981: Nursing and cross nursing behaviour in beef cattle. *Journal of Animal Science*, 53, Supplement I, 129.
Lickliter, R. E. 1982: Effects of a postpartum separation on maternal responsiveness in primiparous and multiparous domestic goats. *Applied Animal Ethology*, 8, 537–42.
Lenhardt, M. L. 1977: Vocal contour cues in maternal recognition of goat kids. *Applied Animal Ethology*, 3, 211–21.
Lynch, J. J., Keogh, R. G., Elwin, R. L., Green, G. C. and Mottershead, B. E. 1983: Effects of early experience on the post-weaning acceptance of whole grain wheat by fine-wool Merino lambs. *Animal Production*, 36, 175–84.
Martin-Rossett, W., Doreau, M. and Cloix, J. 1978: Etude des activités d'un troupeau de poulinières de trait et de leurs poulains au patûrage. *Annals Zootechnie*, 27, 33–45.
McBride, G., James, J. W. and Wyeth, G. S. F. 1965: Social behaviour of domestic animals. VII. Variation in weaning weight in pigs. *Animal Production*, 7, 67–74.
Meese, G. B. and Baldwin, B. A. 1975: Effects of olfactory bulb ablation on maternal behaviour in sows. *Applied Animal Ethology*, 1, 379–86.
Meynhardt, H. 1980: Untersuchungen zur akustischen, olfaktorischen und

visuellen Kommunikation des Europäischen Wildschweines (Sus scrofa): ein Beitrag für die Forstwirtschaft. Berlin: Akadenic Verlag, Heft 2/80

Meynhardt, H. 1981: *Verhaltens-biologische Forchungen am Schwarzwild und ihre mögliche Bedeutung für die Forschung am Hausschwein. Schrift zum Tagungsbericht.* Rostock-Dummerstorf: D. T. Akademie Landwirtschaftliche Wissenschaft.

Morgan, P. D. and Arnold, G. W. 1974: Behavioural relationships between Merino ewes and lambs during the first four weeks after birth. *Animal Production*, 19, 169–76.

Munro, J. 1956: Observations on the suckling behaviour of young lambs. *Animal Behaviour*, 4, 34–6.

Nicol, A. M. and Sharafeldin, M. A. 1975: Observations on the behaviour of single suckled calves from birth to 120 days. *Proceedings of the New Zealand Society of Animal Production*, 35, 221–30.

Peeters, G., De Buysscher, E. and van de Velde, M. 1973: Milk ejection in primiparous heifers in the presence of their calves. *Zentralbald Veterinar Medizin A.*, 20, 531–6.

Phillips-Powell, R. 1978: Early post-natal behaviour in the Camargue foal. *Applied Animal Ethology*, 4, 294.

Poindron, P. and Le Neindre, P. 1975: Comparaison des relations mère–jeune observées lors de la tétée chez la brebis (*Ovis aries*) et chez la vache (*Bos taurus*). *Annals of Biology, Biochemistry and Biophysics*, 15, 495–501.

—— 1979a: Hormonal and behavioural basis for establishing maternal behaviour in sheep. In L. Zichella and P. Pancheri (eds) *Psychoneuro-endocrinology in Reproduction*, Amsterdam: Elsevier, 121–8.

—— 1979b: Les relations mère–jeunes chez les ruminants domestiques et leurs conséquences en production animale. *Bulletin technique du Departement de genetique animale*, 1979, 29–30.

—— 1980: Endocrine and sensory regulation of maternal behaviour in the ewe. In J. J. Rosenblatt, R. A. Hinde, C. Beer and M. C. Busnel (eds) *Advances in the Study of Behaviour II*, London: Academic Press.

Pomeroy, R. W. 1960: Infertility and neonatal mortality in the sow. III. Neonatal mortality and fetal development. *Journal of Agricultural Science*, 54, 31–56.

Rashek, V. A. 1976: Details of feeding and feeding behaviour in young wild asses on Barsa-Kelmes Island (Aralsea). *Zoologicheskii Zhurnol.*, 55, 784–7 (In Russian).

Reeves, J. J. 1980: Neuro endocrinology of reproduction. In E. S. E. Hafez, (ed.), *Reproduction in farm animals*. 4th edn, Philadelphia: Lea & Febiger, 114–29.

Rogalski, M. 1973: Behaviour of the foal at pasture. *Przeglad Hodowlany*, 41, 14–15. (In Polish).

Rossdale, P. D. 1967: Clinical studies on the newborn thoroughbred foal. I. Perinatal Behaviour. *British Veterinary Journal*, 123, 470–81.

—— 1968: Abnormal perinatal behaviour in the thoroughbred horse. *British Veterinary Journal*, 124, 540–53.

Rowlinson, P., Boughton, H. G. and Bryant, M. J. 1975: Mating of sows during lactation: observations from a commercial unit. *Animal Production*, 2, 233–41.

Sambraus, H. H. 1978: *Nutztierethologie*. Verlag: Paul Parly.

Selman, I. E., McEwan, A. D. and Fisher, E. W. 1970a: Studies on natural suckling in cattle during the first 8 hours post partum – I. Behavioural studies (dams). *Animal Behaviour*, 18, 276–83.
—— 1970b: Studies on natural suckling in cattle during the first 8 hours post partum – II. Behavioural studies (Calves). *Animal Behaviour*, 18, 284–9.
Shillito-Walser, E. 1977: Maternal behaviour in mammals. Symposium of the *Zoological Society of London*, 41, 313–31.
Shillito-Walser, E. E., Willadsen, S. and Hague, P. 1982: Maternal vocal recognition in lambs born to Jacob and Dalesbred ewes after embryo transplantation between breeds. *Applied Animal Ethology*, 8, 479–86.
Signores, J. P., Baldwin, B. A., Fraser, D. and Hafez, E. S. E. 1975: The behaviour of swine. In E. S. E. Hafez (ed.), *The behaviour of domestic animals*, 3rd edn, London: Bailliere Tindall.
Smith, F. V. 1965: Instinct and learning in the attachment of lamb and ewe. *Animal Behaviour*, 13, 84–6.
Smith, F. V., van Toller, C. and Boyes, T. 1966: The 'critical period' in the attachment of lambs and ewes. *Animal Behaviour*, 14, 120–5.
Somerville, S. H. and Lowman, B. G. 1979: Observations on the nursing behaviour of beef cows suckling Charolais cross calves. *Applied Animal Ethology*, 5, 369–73.
Stephens, D. B. and Linzell, J. L. 1974: The development of suckling behaviour in the newborn goat. *Animal Behaviour*, 22, 628–33.
Stevens, D., Alexander, G. and Lynch, J. J. 1981: Do Merino ewes seek isolation or shelter at lambing? *Applied Animal Ethology*, 7, 149–55.
Stolba, A. 1981: A family in enriched pens as a novel method of pig housing. In R. Ewbank (ed.) Alternatives to intensive husbandry systems. Proceedings of Symposium, UFAW at Wye.
—— 1982: designing pig housing conditions according to patterns of social structure. In *Proceedings: Perth Pig Conference*, 11–19.
Tindal, J. S. 1966: Studies on the neuroendocrine control of lactation. In G. E. Lamming and E. C. Amoroso (eds), *Reproduction in the female mammal*, London: Butterworths.
Tyler, S. J. 1972: The behaviour and social organisation of the New Forest ponies. *Animal Behaviour Monograph*, 5, part 2.
Vince, M. A. 1984: Teat-seeking or pre-sucking behaviour in newly born lambs: possible effects of maternal skin temperature. Animal Behaviour, 32, 249–54.
Walker, D. E. 1962: Suckling and grazing behaviour of beef heifers and calves. *New Zealand Journal of Agricultural Research*, 5, 331–8.
Walther, F. 1961: Einige Verhaltensbeobachtungen am Bergwild des Georg von Opel Freigeheges. *JHRB, G. Opel Freigehege, Tierforsch ungen*, 1960–1: 53–89.
Waring, G. 1970: Primary socialization of foals (*Equus caballus*). *American Zoology*, 10, 293.
Wettermann, R. P., Turman, R. D. Wyatt, R. D. and Totusek, R. 1978: Influence of suckling intensity on reproductive performance of range cows. *Journal of Animal Science*, 47, 342–6.
Whatson, T. S. and Bertram, J. M. 1983: Some observations on mother–infant interactions in pigs (Susscafa). *Applied Animal Ethology*, 9, 253–62.
Whittemore, C. and Fraser, D. 1974: The nursing and suckling behaviour of pigs.

II. Vocalisation of the Sow in Relation to Suckling Behaviour and Milk Ejection. *British Veterinary Journal*, 130 (4), 346-56.

Winfield, C. J., Brown, W. and Lucas, L. A. M. 1969: Sheltering behaviour at lambing by Welsh Mountain ewes. *Animal Production*, 11, 101-5.

Wolski, T. R., Houpt, K. A. and Aronson, R. 1980: The role of the senses in mare-foal recognition. *Applied Animal Ethology*, 6, 121-38.

Wood-Gush, D. G. M., Hunt, K., Carson, K. and Dennison, S. G. C. 1984: The early behaviour of suckler calves in the field. *Biology of Behaviour*, 9, 295-306.

5 Parental Care in Carnivores

O. A. E. Rasa

GENERAL CONSIDERATIONS

Prey-capture techniques and social systems

The carnivores are a relatively old group, phylogenetically speaking, having arisen in the early Tertiary period from Miacid stock which were thought to be small, genet-like animals, characterized by the presence of true carnassial or shearing teeth (Thenius, 1969). Since they specialized in protein consumption, which constitutes the richest food source, they had to overcome other problems, the major one being the capturing of their prey. Several independent techniques of prey-finding and prey-capture have been perfected during evolution and, within the carnivores, there are many variations in behaviour for the exploitation of food sources, from a tendency towards a vegetarian diet, on the one hand, to capture of prey animals as large, if not larger than, the predator itself, on the other. From the examination of fossil remains, however, it appears that the primitive carnivores, with few exceptions (for example, the ancestral bears), were feeders on small- to medium-sized warm-blooded prey which a single hunter could bring down unaided. From this basic form of predation, three main hunter types have developed and with the development of these types, changes have taken place in the social systems of the species involved; changes which have far-reaching effects on the type of parental care afforded to their young. These three main hunter types are:

1 Stalkers: these have retained the 'ancestral' prey-capture technique of creeping up on the prey and overcoming it. This technique is most effective when done alone, as there is then less chance of the prey sighting the predator and fleeing. Although the technique may be phylogenetically primitive, the morphological and behavioural adaptations evolved in some species are extremely advanced; for example, curved, retractable claws and powerful forelegs, long canines and powerful jaw musculature, an oriented killing bite, and so on. Typical examples of stalkers

are felids, many mustelids and viverrids. Almost all stalkers are solitary, territorial animals.

2 Gatherers: carnivores which have, for the most part, returned to a primitive 'collecting' technique as a secondary adaptation, and have specialized in the annexation of sources of small prey, primarily invertebrates, and carrion. Almost without exception, this group is capable of tackling larger prey animals, if encountered, but these do not form a primary part of the diet. The gathering technique reaches its most extreme form in omnivorousness, ranging in varying degrees from the occasional taking of fruit (some canids for example) to an entirely herbivorous diet (for instance, pandas). Gatherers show great variation in social structure, from solitary animals to some of the most complex multi-familial societies known amongst mammals.

3 Harriers who, as their name implies, literally run down their prey and specialize in the capture of large herbivores. This method of prey-capture is inefficient for a solitary animal and so most harriers are highly social, pack hunters. They have no specialized weapons for grasping and holding prey, as do the stalkers, and rely on trailing ability, speed and stamina for success. Examples are some of the canids and the spotted hyena.

Exceptions to these three main groups are the lions and cheetahs who combine both stalking and harrying techniques for the capture of large prey and the pinnepeds which, although typical harriers, do not appear to use co-operative hunting techniques.

Superimposed on this subdivision of the carnivores dependent on hunting strategy is the fact that many predators are also themselves preyed upon, and this factor also influences their social structure. There is a tendency for an increase in group size to be correlated with the adoption of diurnal hunting in open areas (viverrids, canids and felids). This tendency is associated not only with the protection of the adult hunters, but also of their young, and their captured prey. Many carnivores are opportunists and will kill and eat the young not only of other species, but of their own as well (Bygott, Bertram and Hanby, 1979) and food-stealing, amongst the larger carnivores at least, appears to be common (Estes, 1967).

With all these factors influencing tendencies towards a social mode of life in this order, it is no surprise to find that probably no other group of mammals has evolved such a wide variety of parental care and den-helper systems as are found in the *Carnivora*. There are, however, exceptions to the general rule which have evolved independently in response to environmental pressures brought to bear on the species concerned during evolution.

Population control systems

Carnivore populations can never be as large as that of their prey. Several factors combine in restricting population growth to prevent over-exploitation of the food sources involved, and these factors and their impact on parental care will be dealt with briefly here.

Spacing. Competition for a restricted and difficult to obtain food source is probably the reason for the majority of carnivores being solitary animals. Practically all of them are territorial and the most typical pattern of territoriality is that in which a single male defends a large territory which overlaps with one or more smaller female territories (Lockie, 1966; Erlinge, 1968). This pattern is found, with few exceptions, in all families comprising the *Carnivora*. Although, as will be dealt with in detail later, the majority of carnivore males take no active part in rearing the young, this system of territoriality can be considered as playing a passive but essential role in the successful transition of the young from altricial infants to active hunters, since the female is provided with a stable environment in which to raise the young (Ewer, 1973). This is of especial importance in species where infant-killing is common.

Restraints on reproduction. The majority of young carnivores are late in reaching sexual maturity, compared with herbivores. Apart from a few small mustelids, the majority of carnivores do not reach sexual maturity until they are at least a year old, many of them maturing far later (for example, bears, lions, some canids and pinnepeds). Related to maturation is litter size; the longer the young are dependent on the mother, the fewer the young per litter. The longest periods of maternal dependence amongst carnivores are associated with birth of one, or at the most, two young. In species with relatively large litters, however, maternal investment is relatively short-term and it is in these species that den-helpers are most commonly present, provisioning of the young after weaning being taken over, in part, either by the male or by other group members (Kleiman and Eisenberg, 1973). The majority of carnivores only have a single litter of young per year but in species where maternal care is prolonged, young may be borne only every 2-3 years. The period of juvenile dependence, when the cubs are weaned but still rely on the mother for food, appears to influence the onset of her next oestrus. From data available on large felids (Schaller 1967; Robinette et al., 1961; Adamson, 1969), there is a physiological mechanism underlying this which has not been investigated to date.

Juvenile mortality. For carnivores, there are two main periods of juvenile mortality, the first immediately after birth and the second following independence.

Little is known of the factors which result in a female failing to rear her young. Common causes are thought to be disturbance, departure from the 'normal' situation and maternal debilitation. Frequently, the young are eaten shortly after birth or simply abandoned. If the litter is lost, the female will usually breed again shortly afterwards. From the point of view of natural selection this behaviour is compensated for; although a replaceable litter may be lost, the mother survives after having contributed the minimum in maternal investment. The continuation of maternal care under adverse conditions could mean the loss of both mother and young. Efficient maternal care thus seems to be a balance between firmly fixed maternal responses and factors determining individual survival (Ewer, 1973).

The highest level of mortality amongst juvenile carnivores occurs following independence. Although few data are available on percentage death rate in terrestrial species, Kenyon et al. (1954) estimated that 72 per cent of northern fur seal pups (*Callorhinus ursinus*) died before reaching sexual maturity, which occurs at three years old. Deaths occurred through predation, accident, hunger or disease. Similar data are available for a small viverrid (*Helogale*) where up to 70 per cent of the young are predated before they are four months of age (Rasa, unpublished). Kruuk (1972) estimated that 32 per cent of spotted hyenas died as cubs and 30 per cent of the survivors died as yearlings – a total of 62 per cent. This high rate of juvenile mortality effectively prevents overpopulation and the over-exploitation of food sources.

BASIC PATTERNS OF CARNIVORE PARENTAL CARE

The basic pattern of carnivore parental care is founded on the almost total dependence of the young on the mother for survival. It is the least specialized form of rearing young and is found predominantly amongst the least specialized hunters of the order, the stalkers and gatherers. The mother is responsible for the selection and preparation (if any) of the birth site, which is usually a protected place within her living area. Only a few species construct burrows to be used specifically as birth sites, but most usually give birth in sheltered areas in the open, in caves or in burrows taken over from other animals. There is a tendency for species living in northern latitudes to provide a lined nest for the young, with fur pulled from the belly region of the mother for this purpose (for example, wolves). Suckling and, later, providing solid food for the cubs until they can hunt for themselves is, in most cases, entirely the domain of the mother. This basic pattern is found throughout the *Carnivora* and, to facilitate reference, parental care techniques in the various species will be reviewed on a Family basis, using a type species (where necessary) to illustrate the basic patterns involved. In addition, departures from the typical parental care system will be described.

Mustelids

To illustrate the parental care techniques typical of this group, the stoat or ermine (*Mustela erminea*) will be described in detail as exemplary for the entire family (Erlinge, 1977). The mustelids are almost all solitary, stalking hunters, the few exceptions being associated with the utilization of special food sources. The stoat is entirely solitary, except during mating, and the larger territories of the males overlap one or more female territories. Prey up to the size of a rabbit is taken, killed with a specialized bite to the neck. The young (3–9 to each litter) are born in the spring in a lined nest built by the mother, usually in a burrow she digs herself. During the first five weeks, when the young are still blind and naked, the mother rarely leaves them and will defend them vigorously, carrying them to another nest if disturbed. Suckling continues for approximately six weeks, but even before their eyes open the mother drags food to the youngsters. The adult coat is well developed before the young leave the nest. During the time they remain there, all excrement is ingested by the mother. As in most carnivores, defaecation and urination by the young is not involuntary but brought about by stimulation of the urinogenital/anal area through repeated licking by the mother. When their eyes open, the young attempt to leave the nest but are continually carried back by the mother, who lifts them in her mouth (usually by the neck) at which point the young fall into an almost comatose condition with the extremities curled up, the '*Tragestarre*', as it is known, which facilitates transportation. By the time they are eight weeks old, however, they start following her on her hunting excursions and participate in the killing. Social contact is restricted to the family group and, as in the majority of young carnivores, frequent social play takes place both between the siblings and their parents. This play includes many of the motor patterns used in hunting prey, as well as those typical of social contexts. The family remains together and hunts as a unit until autumn, at which time the young disperse.

One of the exceptions to this basic mustelid pattern is that of the European badger (*Meles meles*) which differs from most carnivores in that it is omnivorous with almost three-quarters of its diet being plant material, the rest consisting mainly of invertebrates, especially earthworms. Badgers mate for life and inhabit spacious dens ('setts') which can be used by the same family over decades. Kruuk (1978) is of the opinion that group life in this species is associated with efficient defence of a feeding territory. Unlike the majority of mustelids, badgers experience a period of lethargy during the winter and the young are born in February or March. There are two to five pups in each litter, their eyes opening four to five weeks of age. Although the mother lives in the den with her mate and other females, there is no evidence to date of these other adults helping to feed or tend the young. In contrast to *Mustela* the mother does not appear to bring food to the den for the young, who will obtain their first solid food when they start accompanying her on foraging expeditions. As in most mustelids, dispersion takes place in the autumn, but on occasion, a youngster will remain in the den over winter with its parents and is then driven off in spring.

A similar pattern is found in most representatives of the skunks (*Mephitinae*), which are also ominvores like the badgers. Here there is a stronger tendency for the young to spend the winter with the mother, dispersing the following spring but, unlike badgers, skunks do not pair for life – the male leaves after copulation, as in most Mustelids (Wight, 1931; Hamilton, 1937; Allen and Shapton, 1942; Yeager and Woloch, 1962).

Within the *Lutrinae*, the otters, which are aquatic and predominantly fish-eaters, many variations on the basic mustelid parental system exist. The European otter (*Lutra lutra*) appears to retain the typical solitary mustelid pattern of parental care (Erlinge, 1968) although the young may remain to hunt with the mother for a long period of time and there is some evidence of co-operation between them (Sheldon and Toll, 1964). The giant otter (*Pteronura brasiliensis*), however, has been observed to hunt in groups of up to 20 individuals (Hershkovitz, 1969).

Probably the best-known otter species is the sea otter (*Enhydra lutris*). The earliest records of parental care were made by Steller in his posthumously published paper (1751). He stated that they lived in family groups consisting of a male, a female, a single baby and older juveniles and reported that strong bonds existed between the parents and their youngster. Since Steller's time, however, the sea otter was hunted almost to extinction and it is only during the last 50 years that the population has recovered to a point where it is no longer endangered, as the result of a complete ban on hunting. Kenyon (1969) reports that the species now lives in large mix-sexed groups which tend to segregate, the males using different hunting and hauling-out areas to those of the females and juveniles. His data show no evidence of the family-based society described by Steller. Sea otters are unusual amongst the carnivores in that they feed almost exclusively on molluscs and echinoderms, breaking them open with their strong teeth or by a 'hammer and anvil' method, one of the few instances of tool-use in mammals. A stone held on the animal's belly as it floats on its back serves as an 'anvil' on which the food item is placed, the 'hammer' being another stone, which is held in the forepaws and hammered against the food item with stereotyped movements until its shell is cracked and the meat inside can be extracted.

The sea otter produces a single pup which, in contrast to all other mustelids, is born with the eyes open, a full set of milk teeth and is fully furred. Birth takes place on land in some protected site, usually in the open. The suckling young appears to be taught to swim by the mother who takes it into the water within a few days of birth, carrying it on her belly where she holds it with the forepaws. Even on land, the cub lies on the mother's chest where it is suckled, groomed and licked. During the first weeks of its life it will either be left on the land while the mother hunts or taken with her into the water. When she dives to capture prey, the baby remains floating on the surface, belly up. Should danger threaten, the mother dives, clasping the youngster to her. The suckling time for this species is unusually long for the mustelids, over a year, although the youngster starts taking solid food long before this. The pup is not permitted to take food from the mother at first, although it may try to. It is handed empty shells, however, from which the food has been extracted and mouths and manipulates these (Sandegren et al., 1973). The pup may remain with the mother when the next pup is born and does not reach sexual maturity until three years of age. Whether or not the older siblings help the mother rear the baby, however, is not known to date.

Procyonids and pandas

Most procyonids, such as the racoons (*Procyon lotor*), together with the pandas, have adopted an omnivorous way of life and almost all are solitary (Mech et al., 1966; Schneider et al., 1971). Their parental care pattern is similar to that of the mustelids with the exception that solid food does not appear to be brought to the young in the den. Only the coati (*Nasua narica*), the most social species, shows some divergence from this pattern (Kaufmann, 1962). Only adult males are solitary and territorial, with the adult females and juveniles living in diurnally active groups which wander through the males' territories. Breeding is synchronized over a period of about a month in early spring. The highly-pregnant females leave the group shortly before parturition, each building a rough nest in a tree and giving birth there to two to seven young. The young are relatively precocial but remain in the nest until they are about four weeks old, being retrieved by the mother if they attempt to stray. At five weeks of age, they join her in the female group, where they remain together as a family. Although the young only suckle for four months, they remain with the mother until the birth of her next litter. In the coati, instead of the young dispersing when independent, the females remain in the natal group and only the males leave on sexual maturity.

Little is known of parental care in the almost entirely herbivorous pandas and almost nothing is known of their life in the wild. The lesser panda (*Ailurus*) produces 1–4 young while the giant panda (*Ailuropoda*) produces a single youngster in an earth or tree den. Since both species are probably solitary it is likely that only the mother is involved in parental care but the details of this are completely lacking.

Bears

In essence, the bears (*Ursidae*) also have a typical solitary hunter parental care system. All species, except the polar bear (*Thalarctos*) are omnivorous, tending towards entirely herbivorous habits. Bears living in northern latitudes are unusual as carnivores in that the young are born during the period of hibernation and are disproportionately small at birth (400–500 g in comparison with 375 kg adult weight for the polar bear –Volf, 1963). Usually only 2–3 young are born and, during the winter months in which she remains with them in the den, the mother does not hibernate. By the time spring comes, the young are capable of following the mother, who defends them vigorously against predators. In the brown bear (*Ursus arctos*) the mother may even teach her young predation avoidance behaviour by chasing them up trees (Pedersen, 1957). In the polar bear, maternal defence consists of leading the young to ice floes within the first week of emergence, where the chances of predation are much reduced compared to those in mainland areas. She only returns to the den to hibernate with the cubs the following winter. In the first days after leaving the den, the mother kills seals for the young to feed on, so they are introduced to solid food almost immediately after emergence. Young bears mature slowly and remain with their mother, sharing her den, for at least a second winter and even a further one if she does not mate the subsequent spring. All young, however, disperse by the time they are three years old. During the time they spend with their mother, they share large kills made by her but forage for their plant food themselves. It is during their second summer that the cubs start actively copying their mother's hunting techniques and they are technically capable of fending for themselves at this time. Bear cubs are suckled for almost two years (21 months in the Polar Bear – Pedersen, 1957) and thus have the longest period of true maternal dependency known for carnivores.

One anomaly in this group is the more arboreal sloth bear (*Melursus ursinus*) which feeds mainly on plants and insects. In *Melursus* the young are carried on the mother's back until they are almost full-grown – probably the only carnivore in which this type of maternal transport of the young takes place (Norris, 1969).

Viverrids

The large majority of viverrids are solitary hunters or gatherers and many are nocturnal. Species with typical mustelid parental care patterns belong to the more cat-like *Viverrinae*, the civets, genets, linsangs and Palm civets. Here, the mother defends the young vigorously, even against their own father. Suckling continues for six weeks although, by this time, the young start following the mother on her hunts. The transition to solid food appears to be rapid and the weaning period short. The family remains together for several months before the young disperse.

The *Herpestinae*, or mongooses, show a great variety of parental care systems. These are all basically solitary hunters and gatherers, even though many species live in groups. The prey is usually small and there is no co-operative hunting, each individual hunting for itself. Roughly speaking, the mongooses can be divided into two main groups: the nocturnal, solitary species (for example, many *Herpestes* species, *Ichneumia, Bdeogale*) and the diurnal social species (for example *Helogale, Suricata, Mungos*), although some species are both solitary (or at the most pair-living) and diurnal (for example, *Herpestes ichneumon, H. sanguineus*). The solitary nocturnal species retain the typical carnivore parental care pattern but here, as in the *Viverrinae*, prey is not carried to the young in the nest, but when the young start accompanying the mother she feeds them on prey which she holds for them in her mouth.

It is amongst the social species of herpestids that the various 'denhelper' systems found in carnivores have reached a level of complexity paralleled only by the highly social canids. The simplest helper system is present in *Suricata* where the male helps the female defend the young and spends long periods of time playing with and grooming his offspring (Ewer, 1963; Wemmer, 1974). He does not, however, take any part in feeding them (Ewer, 1963).

The most recent data on the Dwarf mongoose (*Helogale*), the most intensively studied species to date, indicate that group life in this viverrid species, at least, evolved as a means of protection of the young against predation (Rasa, unpublished). This may also hold true for other social viverrids. Since the parental care system in these species represents such a departure from the solitary hunter type, it will be dealt with in detail here.

Den helpers in social viverrids. The term 'den', when used in association with *Helogale*, is a misnomer, since the mother constructs no nest or den in the sense of a permanent abode for the young. *Helogale* inhabits primarily the ventilation shafts of termite mounds (especially those of *Macro-* and *Odonto-termes*; Rasa, 1983), although other shelters can be used on occasion. Dwarf mongooses are mainly insectivorous and nomadic with a home range of approximately one square kilometre. This nomadism is retained, in restricted form, even when very small young are present. The animals live in groups averaging 12–14 individuals, nearly all of these being the offspring of the founding pair and representing young from several litters. Only the founding pair breed successfully, the young of subordinate females being killed at birth, probably by the alpha female (Rasa, 1979).

Litter size varies from two to seven, the average being four. There is evidence of chronism, some of the young being eaten shortly after birth, probably by the mother. This behaviour may be related to the extremely disparate birthweights found in this species (5.8–13.2g). The day after parturition, the task of warming, grooming and protecting the litter falls to the older siblings. The mother leads the remaining members of the group on the morning foraging excursion, the young being left behind in the termite mound with 2 or 3 'helpers' (usually a female and 2 males). They are suckled on the mother's return to the mound, during the heat of the day. During the afternoon activity period, the young are usually carried by the mother and older siblings to another termite mound in the vicinity where the morning pattern is repeated, the mother leaving with the group and the 'babysitters' remaining, usually different individuals to those which had performed this function in the morning. At dusk, the whole family unites again at the termite mound and spends the night there. This process is repeated during the next 18–24 days, by which time the young start accompanying the adults when they forage. At first they only forage in the immediate vicinity of the termite mound they are staying in.

The eyes open between the tenth and fourteenth days of life and from this time onwards the young keep attempting to crawl out of the mound and are constantly retrieved by the babysitters. They are carried by the neck and very readily fall into the 'Tragestarre' described for mustelids. The daily moving of the young from site to site probably has two functions: first it enables the group to continue its nomadic existence, thus ensuring sufficient food for all group members; second, it is a useful means of evading predation, especially of the young by snakes and larger mongoose species, and of the adults by birds of prey. Mongoose young, in contrast to most other carnivore species, are very vocal and active from birth onwards. When body contact is not maintained, either with adults or siblings, the young wander about and emit a continual penetrating vocalization which would readily attract predators. It is therefore necessary for at least one adult to remain with them in the nest during the period prior to them accompanying the group. Since the mother is lactating, she requires nourishment to produce milk for the offspring and since dwarf mongooses feed mainly on small invertebrates, such food cannot be collected easily or quickly. By other group members taking over the 'mothering' functions, both young and mother are given optimal survival chances.

Once the young start accompanying the group, they are fed solid food, mostly insects and the father is especially active in this respect. Usually a single youngster accompanies a single adult during foraging, running flank-to-flank with the adult when on the move but when danger threatens, they either dash under the adult's body or are picked up and carried to safety. By the time they are four to five months old, the youngsters are capable of feeding themselves, although they will still beg food from the adults and are frequently being rebuffed at this time. By six months of age food begging has stopped, but instead of dispersing the young remain with the group until well past sexual maturity (1½ years for males, 2 years for females) and help raise subsequent litters. For the dwarf mongoose, therefore, the energy expenditure necessary for reproduction is spread over the entire family group.

Although learning in this species plays little part in prey-capture techniques (Rasa, 1973), it plays an important part in other contexts. The role of 'teacher' can be taken over by any of the group members. Recent findings suggest that dwarf mongoose young are taught which are their future predators by observing and copying the warning behaviour of the adults (Rasa, 1981). In addition, 'guarding', one of the most important altruistic behaviour patterns, is traditional and is actively taught to youngsters between six and twelve months old by older male siblings (Rasa, 1977).

The only other social viverrid species studied to date, the banded mongoose (*Mungos mungo*), which lives in groups of up to 40 animals, differs from *Helogale* in that breeding is not suppressed in subordinate females and all females give birth at approximately the same time. Here the young are left in the care of the subordinate males of the group while the females forage (Rood, 1974).

Hyenas

The hyenas, although dog-like in body form, are more closely related to the viverrids than the canids. There are two main genera, the spotted hyena (*Crocuta*) and the striped and brown hyenas (*Hyaena vulgaris* and *H. brunnea*) which differ both in social structure and feeding habits. The striped and brown hyenas are fairly solitary and can be classed as gatherers, feeding mostly on carrion, garbage, small vertebrates and fruits. Three to four young are born in a den, usually an enlarged warthog hole, and are tended there by the mother. She may move them frequently (Lang, 1958). Their eyes open five days after birth and, in contrast to viverrids and spotted hyenas, food is regularly carried to the young (Kruuk, 1976), resulting in the surrounding area becoming scattered with bones. The male may also take part in providing the cubs with food (Kruuk, 1975).

Spotted hyenas, although they do feed on carrion, are pack hunting harriers, numbering up to 100 individuals, which run their large ungulate prey down. Two cubs, which have their eyes open and are fully haired at birth, are usually produced in a hollow or entrance hole near a communal den (Kruuk, 1972). They are vigorously defended against other females and especially males for the first two weeks of life. They then join the other cubs at a communal den which is not dug by the adults but taken over from smaller inhabitants and enlarged by the cubs themselves. Each den is used by several females at once so that up to a dozen cubs of varying ages can be present at one time. Each female suckles only her own cubs, though she may play with others. The young are not fed by regurgitation and only rarely are pieces of a kill brought to the den. The mother visits the cubs usually in the morning and evening to suckle them, but, when the cubs are still small, she may spend the entire day there. Suckling may be prolonged, up to four hours since, when the mother is following migratory prey, the visits to the den may be days apart. When the mother is absent, the cubs may remain inside the den or play with other cubs outside it, fleeing into the den should any large animal approach. The extreme protective behaviour shown by the mother in the first weeks of the cubs' life and their habit of burrowing deep inside the den which is too small and narrow for an adult to enter, are probably mechanisms to counteract the cannibalism prevalent in this species. When the cubs are a few months old, they start following the mother, at first for short distances but later these excursions gradually getting longer. Usually, however, they do not feed regularly on meat until they are seven to eight months old and their mother's milk remains the nutritional mainstay until they are almost full-grown, at about 16 months. They rarely hunt with the pack until they are a year old, usually bringing up the rear with the other youngsters. True hunting is therefore delayed until they are almost adult, the cubs going through a transitional phase in which they follow the hunt but do not participate.

In the spotted hyena, therefore, the solitary hunter parental care system is maintained with the difference that several females use the same den. There is no evidence of other females actively caring for cubs which are not their own but non-parturent as well as parturent females may remain at the den for long periods of time and may function as babysitters, protecting the young against predation (Lawick and Lawick-Goodall, 1971).

Felids

All felids are stalkers of fairly large prey in relation to their body size. With few exceptions (lion, *Panthera leo*; domestic cat *Felis catus* and bobcat, *Lynx rufus*) they are entirely solitary and territorial. The basis on which the social system found in *F. catus* and *P. leo* is founded is intrasexual tolerance. Both species have developed this to varying degrees, probably as an adaptation to the annexing and exploitation of a new food source. For the domestic cat, this adaptation was a result of an association with human settlements, which provided a concentrated and predictable foodsource (Liburg, 1980), while for the lion, it was communal hunting of large herbivores in an open habitat (Kleimann and Eisenberg, 1973). In both cases, the primary association is between females, the males remaining solitary in the cat, while in the lion they may join to form hunting groups.

Feral and domestic cats live in mixed societies, the females usually remaining in their natal areas and the males dispersing. Females form mutually tolerant groups and alloparenting is common (Liburg, 1981; McDonald and Apps, 1978; Laundré, 1977; Panaman, 1981). The 'helping' found between females consists of grooming, warming and retrieving kittens, later of bringing captured prey to them. Mothers which have lost their own litters attempt to steal the young from others. It is likely that males cannibalize small kittens and the mother will drive them away from her litter, sometimes being helped by other females. This cannibalism, even between felid species, appears to be widespread amongst the cat family (Bygott, Bertram and Hanby, 1979; Pienaar, 1969; Schaller and Lowther, 1969). Since the domestic cat is the best-known in respect of maternal care, it will be described here in detail as being representative for the Family as a whole (Schnierla, Rosenblatt and Tobach, 1963; Leyhausen, 1965; Baerends van Roon and Baerends, 1979).

Young cats are born in a den or protected place and come into the world furred but with the eyes still closed. The mother removes and eats the birth membranes and placenta and pushes the newborn youngster towards her nipples, where it attaches itself to one of them. In the pauses between consecutive births, the young are thoroughly licked and the mother remains with them continually for the first two days. On the third day, during a period in which the young are satiated and have fallen asleep, she leaves them briefly to feed, drink and defaecate. All urine and faeces produced by the young are ingested by the mother while they are still in the den. The period the mother spends away from the kittens increases with their age and, by the time they are four weeks old, she only joins them for suckling purposes, although these suckling periods may be prolonged (up to two hours). It is also during their fourth week of life that the mother starts bringing live food (usually mice) to the den, at first killing and eating them herself. The young gradually start feeding on meat and by the fifth week are tackling the live mouse for themselves and often succeed in killing it. By seven weeks of age, they leave the den on the approach of the calling mother to take the prey from her. The mother usually drops the prey in front of the kittens, and they sniff, lick, bite and tap it, chasing it if it is still alive and mobile. If the kittens do not kill the prey within a timespan varying from some minutes to half an hour, the mother will catch and kill it, dropping it afterwards in front of them. Once the prey, or a portion of it, is obtained by a kitten, it is defended vigorously with growls and the kitten slams with the forelegs, claws extended, both against the mother and siblings.

The first signs of hunting behaviour are shown during the fourth week of life when the kittens are attracted to moving objects. At this time their attention span is short but it increases during the fifth and sixth weeks.

Prey brought by the mother during this time may be bitten anywhere on the body but, by the eighth week of age, the stereotyped killing-bite to the neck is used. Baerends van Roon and Baerends (1979) are of the opinion that the kittens' period of high activity prior to the killing of prey objects, acts as a period of learning. Specific characteristics of the prey, the motor patterns involved in capture and killing, and their orientation are practised. Their experiments have shown that the presence of the mother is not essential to successful prey-capture in adult life if the young have been exposed to and have killed, mice in the early period of development (four to eight weeks) and if they have been raised in a litter. This finding throws some doubt on the mother felid as a 'teacher' of prey capture techniques, but it should be said that it is only through the mother's bringing of live prey to the young during this time that they are capable of being exposed to it during the critical period. In free-ranging cats, the 'teaching' role may be extended to include *where* prey can be found.

Kittens remain with the mother for varying periods of time, depending on whether a second litter is produced during the same year. Males usually all disperse by the time they are a year old and may be actively chased away by the mother, although daughters may remain far longer and, if no intervening litter has been born, come into oestrus at the same time as their mother the subsequent spring. They may then take part in caring for each others' litters, especially when one of the mothers is absent. Daughters with no litter of their own may help, even during parturition, and have been observed to eat the afterbirth (Baerends van Roon and Baerends, 1979). It is through this intra-sexual tolerance arising as an extension of the mother–daughter bond, that female groups in domestic cats probably evolved and, with them, the 'helper' system.

Lions differ from domestic cats in several ways. Firstly, the males, when leaving their natal pride, usually do so as a group. Secondly, the females, which tend to cycle almost simultaneously, after they have given birth to two to four young away from the pride, bring their cubs together (when they are about two weeks old) and they are raised in a communal crèche (Schaller, 1967). Although the mother primarily suckles her own young, she will permit other cubs to suckle as well and a cub may wander between several lactating females in order to obtain a full meal. When the pride is hunting, at least one female remains behind with the cubs to protect them against predation by leopards, hyenas and adult male lions. Young raised in a pride have a higher survival rate than those raised by single mothers. Lions are co-operative hunters, feeding mainly on large plains game, often far from the cubs' resting place. Carr (1962) stated that, if the kill is made not far away, the mother will carry food to the cubs in her mouth, otherwise, she swallows pieces of meat and will regurgitate them for the cubs on her return – a behaviour pattern not recorded for any other felid to date. This finding, however, is not substantiated by other workers (Schaller, 1967; Kruuk and Turner, 1967).

The cubs start accompanying the mother on her hunts at about three months old but do not attempt to hunt for themselves at this time and simply watch. Hunting in lions is done predominantly by the females in co-operation, but lion cubs which accompany the mother to the kill get no preferential treatment. The kill is first annexed by the males of the pride which, after they have eaten their fill, allow the females to feed; whatever remains is left for the cubs. In years when prey is scarce, this behaviour can lead to high cub mortality through starvation (Schaller, 1972).

Cubs are weaned at six to seven months and between the ninth and tenth months, their permanent teeth erupt – another time at which there is high mortality amongst the young as the process is associated with a great deal of pain and fever. The survivors may remain with the mother for up to two years, taking part in the hunts, or until she bears a second litter, at which point they are abandoned. By this time they are independent and form a juvenile group which remains together until sexual maturity (about three to four years of age). At this time the males disperse as groups, the females either remaining with their natal pride or dispersing as well. Although the pride males, which are very likely to be the fathers of the cubs, take no direct care of them, they are very tolerant towards the cubs, allowing the young to play with them and to try to steal pieces of food. Bygott, Bertram and Hanby (1979) found, however, that males taking over a female pride will kill any cubs present, this having the effect that females come into oestrus again almost immediately, enabling the males to copulate with them and produce cubs of their own and these, when born, are not killed. Exchanges of pride males take place approximately every four years.

The bobcat (*Lynx rufus*) has adopted a different parental care system more typical of canids than felids. Here, the male remains with the female after mating (Young, 1958). Before parturition, the female lines a den with moss and foliage and usually between two and four young are born there, their eyes opening when they are between three and nine days of age. During the period in which the young are suckled exclusively, the mother chases the male away from the immediate vicinity of the den. He is permitted to return when the young start taking solid food and helps to provide food for the developing youngsters. The young stay with the female and hunt with her until the following year, at which point they reach sexual maturity and disperse. A similar pattern of parental care has been suggested for the leopard (*Panthera pardalis*) and the cheetah (*Acinonyx jubata*) but definitive data are lacking.

With few exceptions, therefore, felids show little or no difference in their parental behaviour to that of other stalking solitary hunters. The only 'helper' systems which have evolved are based on intra-sexual tolerance, mainly between related females and associated with the annexation of special food sources, the bobcat being an exception to this rule. Typical of the felids is their small litter size in comparison to other carnivores (Portmann, 1965) and their relative precocity at birth (Ewer, 1973). These factors may be associated with the fact that the female usually has no aid in raising the young and that the period on which they are dependent on her for solid food is relatively long, especially amongst the larger cats (Schenkel, 1966; Schaller, 1967; Ewer, 1973). Small litter size and relative precocity would therefore augment survival for both cubs and mother.

Canids

The canids, in comparison to all other carnivores, appear to have departed from the typical social organization pattern in that here the basic group consists of a permanent or at least seasonal pair-bond (Kleiman and Eisenberg, 1973). These authors attribute this tendency to several factors including adaptation to a cursorial (running) life, a tendency towards omnivorous food habits and the presence of a large litter size, all of which encourage tolerance between sexes. Only the maned wolf (*Chrysocyon brachyurus*) appears to be entirely solitary (Kleiman, 1972). In the highly social species (for example wolf, dhole, hunting dog), co-operative hunting and group life appear to have evolved as an adaptation to coursing large prey which cannot be brought down by a single individual. In contrast to felids, which have a specialized killing bite, canids have none and large prey is bitten anywhere on the body and brought down as a result of blood loss and exhaustion – a process practically impossible for a solitary hunter. Packs appear to form through continued association of a family and not through the association of unrelated individuals.

Foxes, jackals and coyotes tend to live in permanent mated pairs (Storm, 1965; McPherson, 1969, van Lawick and van Lawick-Goodall, 1971; Eisenberg and Lockhart 1972; Moehlman, 1979) and the young are fed by both parents once they start taking meat. In the golden jackal (*Canis aureus*) older female siblings may return to the family during this time and also help to feed the litter (Moehlman, 1979). The young, in contrast to the majority of felids, are born in a den which is excavated, in part, by the mother. In canids, the litter size is large (1–13 with an average of five – Asdell, 1964) and the problem of providing food for the young is therefore increased. Wyman (1967) showed, however, that black-backed jackals (*C. mesomelas*) hunting as a pair had a far higher success rate than a jackal hunting alone, and so it seems the bonded-pair social system typical of this family has a definite advantage in terms of providing food for the cubs.

In canids living in pairs, the male is primarily responsible for bringing food to his mate and his cubs when they are still in the den. This has two advantages: is reduces the pressure on the female to provide food for herself and her offspring and, since she is thus enabled to remain with them, there is less chance of the young being predated (Kruuk and Turner, 1967; Muckenhirn and Eisenberg, 1973; Eloff, 1973).

Canid young are relatively altricial at birth, their eyes opening around the second week of age. They suckle for between six to eight weeks but are introduced to solid food before suckling ceases. In contrast to all other carnivores, the majority of canids bring solid food to the den in their stomachs and regurgitate it there when the pups beg. This begging consists of licking and butting at the corners of the adult's mouth. Only some of the small species (for example, the Arctic fox, *Alopex lagopus* – McPherson, 1969; the red fox. *Vulpes vulpes* – McDonald, 1977) carry small prey to the den in their mouths. This habit of regurgitation confers on the canids an advantage over other carnivores in that it allows the pups to remain in the protection of the den for a long period of time prior to following the adults on their hunts.

Maternal care in canids does not differ appreciably from that found in other carnivores and there is also little variation within the family itself. The differences found in canid parental care systems are based on the social structure and hunting habits of the species concerned. These variations will be illustrated by the two most social members of the canid family, the wolf (*C. lupus*) and the Hunting dog (*Lycaon pictus*) which show the greatest departures from the norm.

Wolves live in packs of about seven animals (Mech, 1970) in which each sex has a dominance hierarchy. In males this is linear while, in females, apart from the alpha and beta positions, no strict rank order appears to be present (Woolpy, 1968). Usually only the alpha female produces offspring and breeding between subordinates is actively inhibited (Zimen, 1971). A few days before the birth of the young, the female pulls fur from her belly and lines a hole she has previously dug or enlarged to use as a den. She also buries food in the immediate vicinity of the den as provision for the first few days following birth. (This habit of food caching is found throughout the canid family but does not appear to be typical of other carnivores, with few exceptions (McDonald, 1976).) After the birth of the young, the mother remains with them continually for several days and protects them vigorously against pack members. It is only later that pack females and subsequently, pack males, take part in the care and rearing of the youngsters (Muris, 1944; Schönberger, 1965) and in their protection against enemies. Although males do not lie with the young, they show all behaviour patterns typically associated with maternal care towards them, grooming, licking up and swallowing excrement, play and retrieval for instance (Zimen, 1971). When the pack is away hunting, the pups are often guarded by a subordinate female which remains behind in the den with them. By the time they are a year old, they are almost as large as their parents yet, although they accompany the pack on its hunts, do not participate actively at this time. During the first year of life, young wolves, in contrast to young lions, have preferential access to the kill and are allowed to feed first (Zimen, 1971). In their second year, however, they are either integrated into the rank order of the nuclear pack (Woolpy, 1968) or leave it shortly afterwards, either to take up a 'lone' existence or to attempt to join another pack (Mech, 1970).

The hunting dog (*Lycaon*) is probably the most social of the carnivores, living in tight packs of 2–40 animals (Kruuk, 1972). The packs have very large, overlapping home ranges and do most of their hunting during the day (Kühme 1965) feeding primarily on large ungulates which they run down as a pack. In contrast to wolves, there appears to be little or no dominance order within the pack, especially in the case of males, and pack membership remains relatively constant over long periods of time (Estes and Goddard 1967). Again, in contrast to wolves, there is little overt aggression between pack members, what little dominance there is being maintained mainly through friendly and submissive gestures (van Lawick 1974).

The relationship between females in the pack is difficult to determine in relation to reproduction, although a linear hierarchy is present (van Lawick and van Lawick-Goodall, 1971). Any female within the hierarchy can give birth and usually one or two litters averaging ten pups are produced yearly. The young are born in a den at a den-site used by the entire pack but, again in contrast to wolves, they are not defended against other pack members by the mother in the first days following birth. Non-parturent females take great interest in the pups, especially those of low-ranking mothers (van Lawick and van Lawick-Goodall 1971). If two females litter simultaneously, however, the higher-ranking mother will kill the pups of the other female (van Lawick 1974). Lactating females compete with one another in attempts to suckle pups although the mother usually retains first rights.

After they have killed their prey, all pack members regurgitate to the pups, their mother and any other adults remaining at the den. Prey is also shared even when there is insufficient food present to feed the entire pack adequately and, if the kill is made near the den, juveniles are given precedence over adults in feeding from it. Estes and Goddard (1967) describe a case where five-week-old pups were orphaned on the death of their mother and subsequently raised successfully by the eight remaining pack members, all males, as a result of this altruistic food-sharing.

Pups start accompanying the adults at the age of two to three months, as soon as they are capable of covering fairly long distances, but do not participate in hunts at this time. Since *Lycaon* is nomadic, this system of few but large, quickly mobile litters enables the group to resume its wanderings fairly rapidly and prevents drastic prey depletion near the den site. No data are available on how new packs form, but the pups usually remain with their natal group until well past maturity.

In both the wolf and the hunting dog, therefore, the concept of parental care must be expanded to one of familial care, as in the case of the dwarf mongoose, with which both species have many parallels. The golden jackal can be considered an intermediate stage since the association of older female siblings is only temporary. The system found in the wolf can be considered as more primitive than that in *Lycaon*, since the mother prevents access to the young by other pack members for some time following birth, thus delaying 'helping' behaviour. In both *Lycaon* and the dwarf mongoose, access is immediate, but in *Lycaon* the mother is enabled to remain in the den with the young because of the typical canid habit of food regurgitation, which is not present in viverrids.

Seals, sealions and walruses (pinnepeds)

The taxonomic status of this group is still not clear. They and the terrestrial carnivores almost certainly evolved from a common ancestor but the pinnepeds diverged into the marine habitat relatively early in carnivore evolution, probably from the original Canoid stock during the Miocene period (Ewer, 1973). Although some authors now consider the *Pinnepedia* to be a separate order, they were until recently accorded suborder status. Since so many of their behaviour patterns resemble those of typical carnivores and the morphological changes warranting their separation are all associated with adaptations to the marine habitat, they will be included here as examples of extreme specialization to an aquatic mode of life.

As in the majority of carnivores, seals and sealions are solitary hunters which pursue their prey (mainly fish and squid) in the same way as harriers. The majority of prey is small and requires no co-operation from other seals or sealions in its capture. Basically, the seals and sealions retain the typical carnivore parental care system in that only the mother provides for her young. The departures from this basic pattern are a result of the unique mating systems found within the order. Although aquatic and solitary for most of the year, seals and sealions gather on land during the breeding season and may form colonies of several thousand animals (for example, Steller's sealion – Imler and Sarber, 1947). Males have harems of females in their territories and parturition occurs within a few days of the females hauling out. Copulation takes place again within a few days (northern fur seal) or weeks (grey seal) following birth and delayed implantation of the embryo is typical (Scheffer, 1958).

In species forming close female groups or pods, the mother attempts to defend a small area around herself immediately prior to parturition. A single pup is born head first or in the breech position and the mother may tug at the placenta to hasten expulsion, but makes no attempt to eat it

(Slijper, 1956). The pups are precocial and able to move on land as soon as they are born and in some species (harbour seal *Phoca vitulina* and monk seals *Monachus* species, for example) pups may have to swim on the day of birth. In other species, such as the elephant seal (*Mirounga*) the mother remains on land continually for about 28 days, suckling the pup, after which she abandons it completely. The pups wander out of the female pod, form their groups and take to the water at about three months old (Rasa, 1971). Maternal care in this species is therefore reduced to protection of the pups against other pod females – which will attack and sometimes kill them – and the act of lactation. The fat content of the milk is so high that the pup quadruples its weight within a few weeks and lives on these fat reserves during the period it remains on land.

In the majority of otariids or sealions, however, as well as in a large number of seal species, the mother leaves the pup by the time it is two weeks old and returns to the sea to feed, rejoining her pup on land afterwards. In species such as the harbour seal, where the pups are relatively dispersed, recognition between mother and offspring is no great problem. In colony breeders, such as the fur seals (*Callorhinus*) and sealions (*Zalophus*) where up to 50 pups per harem can be present, the task is much more complicated. Mother–pup recognition appears to be first vocal and then olfactory; the mother calls when she reaches land and the pup answers. Pups are prepared to suckle from any female but the mother will suckle only her own, sniffing its muzzle beforehand in order to identify it (Maxwell, 1967; Sandegren 1968).

In the majority of seal species, females, after suckling their pups for up to six weeks, abandon them to their own devices as soon as they have moulted from the wooly pup coat to the sleek juvenile one. In fur seals and sealions, however, the mother–young bond is much longer; in *Callorhinus*, the northern fur seal, the female suckles the pup for three months, in *Arctocephalus*, the southern fur seal, up to six months or even a year. The best-studied species in this respect is the cape fur seal (*A. pusillus*) (Rand, 1955). In this case the mother remains with her pup continually for the first week after birth and the pup starts suckling within an hour of being born. The mother then returns to the sea for a few days before emerging to suckle the pup again. By the time the youngster is two months old, she remains absent for up to a week or more. At five months, the pup is supplementing its diet with fish and crustaceans which it catches in offshore pools, but a large proportion of what it ingests is inedible. The pup is weaned 12 months after birth and by this time can take up an independent existence. However, if the subsequent pup produced by the mother happens to die the yearling may continue suckling. During the first year of life the pup does not wander far from its birthplace, but it increases its range within the second year (this is especially true of bull pups).

In the majority of pinneped species, the male takes no part in the care of the young except in providing them with a relatively undisturbed area within his territory for their maturation. In the Galapagos sealion (*Zalophus californianus wollebaeki*), however, observations by Eibl-Eibesfeldt (1970) indicate that although the males paid little attention to the pups when they were on land they showed strong protective behaviour if the pups entered the water, heading them off if they swam out too far and driving them landwards. Frightened pups would rush towards the territorial male and nuzzle up to him and this nuzzling was reciprocated. The paternal role in sealions may thus extend to the pup as an individual and not simply be the provision of a breeding site for the mother.

The walruses (*Odobenidae*) are intra-Arctic migrants which, during the breeding season, spend a large proportion of their time hauled out on ice-floes. They feed primarily on molluscs, which they dig out of the sea-bottom with their tusks. Walruses appear to live in family groups (Pedersen, 1962a, b) consisting of an old bull, between one and three cows and a number of young not more than four or five years old, and these families will congregate into larger herds. The pup is covered with a woolly coat at birth, which gradually becomes sparser with age. Although the pups can swim as soon as they are born, they are usually transported by the mother who, swimming on her back, holds them to her chest with one or both fore-flippers and young females have been observed to do the same with orphan pups. Although, within the family, the mother is tolerant of other females approaching the pup, females from other families are viciously attacked. The pup continues suckling until it is 18 months old, by which time it is already taking solid food. Pups appear to remain with the family groups for at least a year following weaning. Males, at least, leave at three to four years of age and join bachelor herds. In the walrus, therefore, the harem system does not appear to be restricted to the breeding season alone, as in seals and sealions, but extends over a period of years with the same bull at the head of the family. This aids the survival of the young since their dependency period is so long. Maxwell (1967) is of the opinion that this dependency is related to cranial development, since the jaws must accommodate the crushing premolars before the animal can feed on its adult molluscan diet.

DISCUSSION

Evolutionary trends towards social life and alloparenting

As has been illustrated, the carnivores, in general, are adapted to a solitary mode of existence by reason of their main source of nutrition – that is, other animals. The capture of a prey animal smaller than a solitary hunter, the hunting technique of which depends on crypticity, is most effective when done alone. Only when the prey or food source differs from this most common type do changes in hunting strategies, social structures and, with them, changes in parental care systems come about. Table 1 summarizes the tendencies found in the various families of carnivores. A type species is given as representative of the majority of species in the family, followed by species showing a definite departure from this norm. Considering the total number of living carnivore species (around 245, not including the pinnepeds and almost twice this total number of subspecies), the number which have tended towards a social mode of life is relatively small. As illustrated, all these atypical species exhibit atypical feeding habits, either tending towards the taking of smaller prey items (especially invertebrates) perhaps combined with vegetarianism or towards pack-hunting prey larger than themselves. It is these changes in feeding behaviour which have laid the cornerstone for social life.

The trend towards becoming social appears to have arisen in two distinct ways within the order (Kleiman and Eisenberg, 1973). Either the male remains with the female during the time she is rearing the young (for example, *Suricata*, bobcat, most Canids, sealions and walrus) and this association continues on a permanent basis (as in certain canids and viverrids), or female young do not disperse on weaning and genetically related matriarchal groups are formed (for example, domestic cat, lion, hyena). Only in a few cases (highly social viverrids, wolf, dhole, hunting dog, for instance) are both systems operative. The pair-bond is permanent and young of both sexes remain with the parents after maturity. In the viverrids, this tendency is associated with protection against predation, but in the wolf and hunting dog, with more efficient hunting of large prey.

The most highly social systems known to date in the carnivores have arisen independently of one another in only three of the Families comprising this order, the viverrids, felids and canids (dwarf mongoose, lion and hunting dog). It is only in these highly social animals that parental care, in all its forms, is expressed by all members of the group; this including grooming, protecting, providing food and playing with the young. In addition, lactating females which are not the mother, if present, may even take over the purely maternal function of suckling. Here, therefore, the concept of familial care is clearly applicable.

Within these three species, a high degree of relatedness between group members exists and, from a sociobiological point of view, helper behaviour can be explained at the genetic level through the concepts of inclusive fitness and kin selection (Hamilton, 1964, 1971). Simply speaking, in the absence of its own young (which would carry a genetic complement of half the individual's genes; $r = 0.5$), advantages can be gained by the animal investing in the survival of close relatives, which could carry a

Table 1 Parental care systems and social structures in carnivores

Family	Prey and hunting methods	Basic social structure	Number of young Range	Number of young Mode	Lactation period	Maternal dependence period	'Helper' system
Mustelidae:							
Stoat, *Mustela ermina*	small-medium vertebrates	solitary	6-13	9	6 weeks	c. 6 months	none
Sea otter, *Enhydra lutris*	molluscs and echinoderms	♀♀ and ♂♂ groups	1	1	1 year	c. 1-2 years	none
Procyonidae:							
Racoon, *Procyon lotor*	omnivorous	solitary	3-7	4	16 weeks	c. 1 year	none
Coati, *Nasua narica*	omnivorous	♂♂ solitary, ♀♀ groups	3-7	4	16 weeks	c. 1 year	none
Ursidae:							
Brown bear, *Ursus arctos*	omnivorous	solitary	1-4	2	2 years	2-3 years	none
Polar bear, *Thalarctos*	piscivorous/carnivorous	solitary	1-4	2	21 months	2-3 years	none
Viverridae:							
Civet, *Civettictis civetta*	small vertebrates, reptiles, invertebrates	solitary	2-5	2	8 weeks	6 months	none
Suricate, *Suricata suricatta*	mainly invertebrates	family group	2-6	4	6-9 weeks	12 weeks (remain in group)	♂ protects young
Dwarf mongoose, *Helogale parvula*	mainly invertebrates	extended family group	2-6	4	6-7 weeks	12 weeks (remain in group)	siblings and father take over all maternal functions

Hyaenidae:

Striped hyena, *Hyaena vulgaris*	small vertebrates and carrion	solitary?	3-4	3	1 year	c. 1½ years	♂ may bring food to cubs
Spotted hyena *Crocuta crocuta*	large herbivores	pack or clan	2-3	2	16 months	c. 1½ years (remain in group)	♀♀ may babysit

Felidae:

European wild cat, *Felis sylvestris*	small vertebrates	solitary	1-8	4	8 weeks	c. 6 months	none
Domestic cat, *F. catus*	small vertebrates (associated with humans)	solitary, ♀♀ groups	1-8	4	8 weeks	c. 6 months ♂♂ disperse	♀♀ may protect and bring food to young
Lion, *Panthera leo*	large herbivores	♀♀ based family 1-4 ♂♂ associates	1-5	3	33 weeks	2-3 years	♀♀ take over all mothering in crèche system ♂♂ tolerant
Bobcat, *Lynx rufus*	small-medium vertebrates	solitary, ♂ associates during breeding	2-4	3	8 weeks	1 year	♂ helps feed young

Canidae:

Red fox, *Vulpes vulpes*	small vertebrates and invertebrates	seasonal pair bond	1-8	5	8 weeks	4-5 months	♂ helps feed young
Golden jackal, *Canis aureus*	small vertebrates and invertebrates	permanent pair bond	2-6	2	9 weeks	4-5 months	♂ and older ♀ siblings help feed young
Wolf, *Canis lupus*	small vertebrates and large herbivores	permanent mixed-sex group	1-8	4	6-8 weeks	4-5 months (disperse c. 2 years)	all group members feed, groom and protect young

continued

Table 1 continued

Family	Prey and hunting methods	Basic social structure	Number of young Range	Number of young Mode	Lactation period	Maternal dependence period	'Helper' system
Canidae: continued							
Hunting dog, *Lycaon pictus*	large herbivores	permanent mixed-sex group	1-16	10	6 weeks	4 months (do not disperse)	all group members take over mothering including lactation
Pinnepedia:							
Seals, e.g. elephant seal *Mirounga*; harbour seal, *Phoca vitulina*	piscivorous	harem or loose ♀♀ groups on land	1	1	4 weeks	4 weeks	none
Sealions and fur seals e.g. Cape fur seal, *Arctocephalus pusillus*; Galapagos sealion, *Zalophus californicus wollebaeki*	piscivorous	harem	1	1	6 months-1 year	1 year	none
	piscivorous	harem	1	1	6 months	1 year?	♂ protects pups
Walruses, *Odobenus* spp.	molluscs	family group	1	1	18 months	2½ years	♂ and group ♀♀ protect pup

Parental Care in Carnivores 141

quarter or an eighth of the individual's gene complement ($r = 0.25$ or $r = 0.125$) and thus increase its inclusive fitness by means of smaller increments. This theory may serve to explain the highly social systems found in these carnivores. It cannot, however, account for certain species having evolved such systems whilst closely related species have not. It is only by recourse to examination of the ecological factors that have shaped the social behaviour of the species concerned that an answer to this question can be obtained.

Ecological factors influencing group size and parental care techniques

It is probably no coincidence that the only carnivore species to have developed a truly social mode of life belong either to the gatherers or the harriers. The food sources available allow a relatively high population density of the species concerned within a given area, for the former because of its diversity, for the latter because of the quantity of food available at one time. No gatherer shares its prey and this trait is continued into the parental care system with very few exceptions (for example, striped hyena and red fox, both of which take larger food items or carrion, and it is these which are subsequently brought to the pups). Gatherers' food is not brought to the young but instead the young are taken to the food as soon as they can travel. As Ewer (1973) pointed out, this system is eminently sensible for an animal which is dependent on scattered small prey or vegetable matter, since such food takes a long time to collect. By the time the stomach contained sufficient food to warrant bringing it to the young it would be already partly digested. The bringing of individual small food items would only be efficient within the immediate vicinity of the den, thus depleting the area, and it is in exactly this area that the young would make their first foraging excursions.

The case of the harriers is somewhat different. Since their prey is large and highly mobile, most species have huge home ranges or territories which they patrol over long periods of time and most lead an almost nomadic existence. This is especially true of the large carnivores inhabiting steppe and Pan-Arctic tundra regions. Their movements are tied to those of their migrating herbivore prey, kills often being made at great distances from the dens. An exception to this is the spotted hyena population of the Ngrongoro Crater, whose prey animals no longer migrate, allowing the formation of smaller territories inhabited by a clan (Kruuk, 1972). Without the system of food regurgitation for the young found in the canid species involved, the wolf, dhole and hunting dog, it is difficult to imagine how the pups could be provided for, taking into consideration the short suckling times involved. The spotted hyena, however, has developed a different strategy of parental care to circumvent this problem by extending the suckling time to 16 months, at which point the young start joining the hunts themselves. A similar adaptation is found in the lion, which, although not a true harrier, is a group hunter faced with the same problem. Young lions are suckled for 33 weeks in comparison to 12 weeks for the cubs of its close and sympatric relative, the leopard, which remains a solitary, cryptic hunter.

Group size and prey size in the harriers also show a positive correlation. In most harrier species, and in the lion as well, group size seldom exceeds the number of animals capable of being satiated on a single kill. In terms of energy expenditure, it is more efficient for a group-hunting animal to kill a single prey item capable of feeding the entire group than several smaller prey which provide insufficient food, particularly given that the energy expended in the capture of a prey animal is practically the same whether the prey is large or small. The presence of a large amount of meat at one time ensures that, even in species where the cubs do not have precedence on a kill, sufficient food is usually available for them during the weaning phase and afterwards. Only in the hunting dog do numbers reach the point where some members do not reach satiety on a single kill. In this species, however, the habit of food-sharing between all pack members and the fact that young are afforded first rights to a kill ensure that they are all adequately fed. In the hunting dog, large group size is correlated with a factor other than food provision, that of food protection (Estes, 1967). Hunting dogs inhabit areas where other large predators live sympatrically: for example, the spotted hyena and the lion, both of which obtain a large proportion of their food by stealing. The augmented pack size found in the hunting dog, in comparison with the wolf (where such predators are absent), is probably an adaptation for efficient kill protection (Kleiman and Eisenberg, 1973).

There is therefore no single ecological factor which has encouraged carnivores to adopt a social mode of life. Size and availability of food, hunting techniques and predation all play a part in the shaping of the type of society evolved and, with this, the degree of parental care afforded the young.

Parent-young conflict and dispersal

For a species dependent on a food source that is hard to obtain, dependence on the mother as a provider of solid food cannot continue indefinitely. Most female carnivores inhabit relatively small territories and the retention of the young for a period beyond that in which they are dependent on her would result in rapid resource depletion. Although in all terrestrial carnivores the mother continues to kill for her young after the weaning period, this period is relatively short in the cryptic hunters of small prey, rarely exceeding six months, (by which time the young can fend for themselves and disperse gradually over a period of weeks). At the most, the young are permitted to accompany the mother until the next breeding season when she will actively drive them away. Once more, the main exceptions to this rule are found amongst the gatherers and harriers, where conflict between mother and young over available food sources is less acute and the young may remain for a period of years, or even permanently, within the family group. Here, however, one can no longer consider this a period of *maternal* dependence since the young are hunters in their own right, the difference being that the bond to the mother and siblings is not broken but maintained at a weaker level and bonds to other group members are formed or enhanced.

An important factor contributing to the time of the dispersal of the young which has not been mentioned so far, is the degree of aggression which exists between siblings. Almost without exception, carnivore young show extreme feeding-envy and serious fighting may take place in disputes over the possession of a food item. Zimen (1971) has shown for the wolf that the degree of aggression between siblings peaks at three months, when individual distance between them starts to increase and again at 21 months, the time at which dispersion or integration into the pack rank order takes place. The mechanism of inter-sibling aggression ensures that, in the majority of carnivores, dispersing siblings do not leave as a group, but singly, thus enhancing their chances of survival –especially in the case of solitary species. However there are some notable exceptions to this, curiously enough amongst the felids, where the solitary stalking of prey is most common. As has been mentioned, male lions leaving the group do so usually in pairs or trios (Bygott, Bertram and Hanby, 1979). This not only increases their chances of success in capturing prey without the assistance of females, but also their chances of displacing other males in the take-over of a female pride later. Another species in which males, usually siblings, remain as a group following separation from the mother is the cheetah. Here groups of males hunt together (Eaton, 1969; Schaller, 1968) on an almost permanent basis. Van der Werken (1968), however, is of the opinion that the main function of these all-male groups is to search for females. Both the lion and the cheetah, however, are hunters of relatively large prey and the association of siblings subsequent to independence can only be of advantage in these cases.

Maternal dependence and learning

In most species of carnivores, the period of maternal dependence is relatively short – as soon as the young are capable of feeding themselves efficiently, they disperse and take up an independent existence. In some species, where the finding, capture and handling of the adult food is difficult, however, the mother–young relationship may extend far beyond what is typical for the family concerned. Here again, there are two basic subdivisions which can be made within the carnivores with regard to a long period of maternal dependence: on the one hand, difficulty in capturing and handling food is associated with small litter size, the female investing a long period of time in the care of one or two young while; at the other extreme, this difficulty is associated with large litter size and subsequent pack formation.

The sea otter provides a good example for the first instance for, although the young of this species are the most precocial representatives of the *Mustelidae* at birth, they have by far the longest period of maternal dependency. The sea otter is an exception amongst the *Mustelidae* in that it feeds almost exclusively on molluscs and echinoderms, both of which have hard shells which must be broken before the meat inside can be reached. Although the location of such prey may not present any great problems, since the juvenile follows its mother on dives and can thus learn where prey can be found (Hall and Schaller, 1964), its detachment from the substrate and subsequent handling require special skills. Sandegren et al. (1973) suggest that the 'hammer and anvil' behaviour pattern shown in this species is probably innate since it may be used by pups in areas where the strong-jawed adults have no need of these tools to open their prey. Pups, however, practise this behaviour with empty shells provided by the mother, prior to using the method in feeding situations. The long period of maternal dependence here may be associated not only with skill in prey capture and handling but also with maturational factors, such as dentition and muscular development. It is a curious fact that the only other feeder on molluscs amongst the carnivores, the walrus, also has the longest period of maternal dependence amongst the *Pinnepedia*. It appears that adoption of this type of prey as a staple diet presents problems for the maturing young, the nature of which are not as yet adequately understood.

The pinnepeds, as a group, also produce only a single offspring. Maternal care, in the majority of seals at least, simply consists of providing the pup with sufficient fat reserves to survive the transition from a terrestrial to an aquatic mode of life on its own. There appears to be a correlation between the richness of the milk, suckling time and weight gain by the pup, and these, in turn, are related to the ambient temperature of the environment. Pinnepeds from the coldest environments have the shortest lactation times, the most concentrated milk, the fastest weight gains and the earliest independence for the pups; the fur seal genus can serve as an illustration. The Cape fur seal (*Arctocephalus pusillus*) which lives in relatively warm waters, has a milk fat content of 18.6 per cent and a suckling time of six months to one year while the Kerguelen fur seal (*A. tropicalis*) from the South Georgia region of Antarctica has the highest milk fat concentration for the genus (26.4 per cent) and the pups are weaned at 3½ months (Maxwell, 1967). In most pinnepeds, there appears to be no teaching of swimming or hunting skills by the mother and the young of many species are abandoned before they even enter the water. However, the pups seem to have some innate recognition of their future prey-types (Rasa, 1971), and even in species where the mother remains in association, prey-capture during and subsequent to weaning is a matter of trial and error.

The long period of maternal dependence found in many warm-water pinnepeds may be associated also with protection against predation. Sharks are especially prevalent pup predators before the young can swim well enough to avoid them (Heller, 1904), although these marine predators are more common in temperate than cold waters. Little is known of the behaviour of the pinnepeds once they leave the land so it is difficult to draw any firm conclusions regarding the length of the periods of maternal dependency in the species in which it occurs.

One group in which both a small litter size and a long period of maternal dependence are present are the omnivorous bears. Here, however, prey-capture and handling techniques are not likely to be factors which play a part in this apparent anomaly. The bear species on which data on suckling times and maternal dependency are available are all holarctic and all undergo a period of winter lethargy. For these species, the long period of maternal dependence could be associated with the extremely altricial state of the young at birth (Leitch et al., 1959) together with an absence of food in the environment at a time when the young are still growing rapidly and unable to lay down sufficient fat reserves to last them through the winter. By extending the suckling period into the cubs' second year of life, this problem can be overcome and, by their second winter, they are mature enough to form fat reserves of their own. Very little is known regarding maternal care in the tropical bear species but these appear to have evolved from the northern forms and have exploited this secondary habitat whilst retaining the ancestral parental care methods, although 'denning up' no longer occurs (Ewer, 1973).

The spotted hyena is a special case which falls between those of the pack hunters of large prey with their large litters on the one hand and the hunters of difficult-to-handle prey with a small litter size on the other. Although hyenas are pack hunters, their litters are small and the young suckle longer than is found in any of the other carnivore families, with the exception of the bears (Kruuk, 1972). The long period of maternal dependence here, however, is most probably associated with the cannibalism prevalent in this species. Since food is not brought to the den and the danger of being killed by adult hyenas in the clan exists until the cubs are almost fully grown, the most efficient means of circumventing these problems is to extend the suckling time and this is only practicable with a small litter.

The hunters of large ungulates, such as the lions, wolves and hunting dogs, are typified by relatively large litter size and a long period of dependence. This dependence is not usually on the mother herself but on the group as a whole. In the case of the two canid species, suckling times for the pups do not differ from those of other members of the family. The learning of specialized prey-capture techniques is probably not relevant to these species since no skilled hunting techniques or specialized killing bites are necessary, the prey being literally run down and torn to pieces. As Mech (1970) has shown, if the prey stops to defends itself, the hunt is usually broken off. It seems likely, therefore, that the long dependency period of these canids is associated with maturational factors. The endurance of the youngsters increases with age, enabling them to keep up with the hunting pack and the development of hunting skills themselves are probably of secondary importance. Even so, pups are usually the last to arrive at a kill (van Lawick and van Lawick-Goodall 1971) and there appears to be a gradual transition from accompanying a hunt to actually participating in it.

For the stalkers of large prey, however, such as the lions who do not extend their chases over long distances, the learning of hunting techniques by the young is of greater importance. Lions stalk their prey, as do the majority of felids, usually springing on it from ambush. In contrast to most other felids, however, they are co-operative hunters. Schaller and Lowther (1969) found that two lions hunting as a pair almost doubled their success rate over a single lion (52 per cent against 29 per cent success rate). Co-operation takes three main forms: either the pride fans out so that prey flushed by one animal can be caught by another, or one or two individuals circle round and drive the prey towards the rest of the pride lying in ambush, or prey may be driven into a dead end by the group, a tactic which infers knowledge of the topography. Schenkel (1966) observed two lionesses leading their six-month-old cubs on 'mock hunts', where stalking techniques could be practised. He was of the opinion that when the lionesses intended serious hunting, the cubs were left behind, but on such 'mock hunts' they were encouraged to join in. It is likely that during such training periods co-operative hunting is learned by imitation and experience. The only data on the effectiveness of imitative learning in felids available as yet is for the domestic cat. Chesler (1969) showed that a kitten learns to press a lever faster when it can copy its mother than when an unrelated but friendly cat is the 'teacher'. It is only in the large felids which hunt as a group that any evidence of 'teaching' of stalking methods by the mother is present (Schenkel, 1966, lion; Eaton, 1970, cheetah). In all other felids studied to date, the mother simply exposes the kittens to live prey during the critical period just prior to weaning, and the abilities to stalk and kill appear to be learned by trial and error.

With the exception of the lion (where some active teaching of the young regarding skills in prey capture appears to take place), all other cases of long maternal dependency periods seem to be related to factors in the ecology of the species concerned other than learning. Where prey-handling or capture is difficult, the mother does not appear to serve any active teaching role but simply provides a milieu in which the young can practise the necessary skills for themselves. There are two possible cases of prolonged maternal dependence probably functioning as a counter-active measure to predation (that is, the hyenas and warm-water pinnepeds) while, in the bears, it ensures the survival of the young through the first period of winter lethargy. Long maternal dependency thus appears to have evolved independently in the various families in response to special ecological problems. It is likely that it is a prime factor in the development of a social mode of life since all species in which this phenomenon is found, with the exception of the bears, later show either inter- or intra-sexual tolerance or both, at least during part of their life cycles.

CONCLUSIONS

For the majority of carnivores, owing to their source of nutrition, parental care is reduced to the level of maternal care only as a direct result of their food source. The social carnivores all show tendencies for their main food sources to differ from the typical medium-to-small prey that can be dispatched by a solitary hunter. Parental care in the most highly evolved of the social carnivores is the most advanced known for mammals, except man, and has reached a level of complexity rivalled only by that of the social insects (Wilson, 1975). In these cases all aspects of mothering, even lactation, can be taken over by other members of the group and strictly one should term this phenomenon familial care, since most of the group members involved are close relatives. Complex societies have developed because of the retention of the male–female bond beyond the breeding season, the non-dispersal of some, if not all of the siblings, and intra-sexual tolerance. These variations have primarily come about either as mechanisms to reduce the incidence of predation arising as a result of radiation into exposed biotopes or as a means by which prey-sources inaccessible for the solitary hunter can be efficiently annexed. Such exceptions, however, form only a relatively small percentage of the total number of carnivore species and the typical hunters remain, as they have probably existed since the Tertiary period, solitary, cryptic stalkers of prey, their survival depending on the production of a large number of young who rapidly become independent (usually prior to sexual maturity). Only the fittest of these, the most skilled hunters and those that escape predation themselves can survive and attempt to establish territories of their own in which to perpetuate the species.

REFERENCES

Adamson, J. 1969: *The Spotted Sphinx*. London: Collins & Harvill.
Allen, D. L. and Shapton, W. W. 1942: An ecological study of winter dens, with special reference to the eastern skunk. *Ecology*, 23, 59–68.
Asdell, S. A. 1964: *Patterns of Mammalian Reproduction*. 2nd edn, New York: Cornell University Press.
Baerends-van Roon, J. M. and Baerends, G. P. 1979: *The morphogenesis of the behaviour of the domestic cat*. Amsterdam: North-Holland Publishing Co.
Bygott, J. D., Bertram, B. C. R. and Hanby, J. P. 1979: Male lions in large coalitions gain reproductive advantages. *Nature*, 282, 840–1.
Carr, N. 1962: *Return to the wild*. London: Fontana.
Chesler, P. 1969: Maternal influence in learning by observation in kittens. *Science*, 166, 901–3.
Eaton, R. L. 1969: The cheetah. *Africana*, 3, 19–23.
—— 1970: Notes on the reproductive biology of the cheetah, *Acinonyx jubatus*. *International Zoological Yearbook*, 10, 86–9.
Eibl-Eibesfeldt, I. 1970: *Galapagos*. München: Piper Verlag.
Eisenberg, J. F. and Lockhart, M. 1972: An ecological reconnaissance of Wilpattu National Park. *Smithsonian Contribution to Knowledge*, (Zool), 101, 1–118.
Eloff, F. C. 1973: Ecology and behaviour of the Kalahari lion *Panthera leo vernayi* In R. Eaton (ed.) *International Conference on Ecology, Behaviour and Conservation of the World's Cats*.
Erlinge, S. 1968: Territoriality of the Otter *Lutra lutra* L. *Oikos*, 19, 81–98.
—— 1977: Spacing strategy in stoat *Mustela erminea* L. *Oikos*, 28, 32–42.
Estes, R. D. 1967: Predators and scavengers. Nat. Hist. N.Y. 76, 20–9, 38–47.
Estes, R. D. and Goddard, J. 1967: Prey selection and hunting behaviour in the African wild dog. *Journal of Wildlife Management*, 31, 52–70.
Ewer, R. F. 1963: The behaviour of the meerkat, *Suricata suricatta* (Schreber). *Zeitschrift für Tierpsychologie*. 20, 570–607.
—— 1973: *The Carnivores*. London: Weidenfeld and Nicolson.
Hall, K. R. L. and Schaller, G. B. 1964: Tool-using behaviour of the Californian sea otter. *Journal of Mammalogy*, 45, 287–98.
Hamilton, W. D. 1964: The genetical theory of social behaviour, I, II. *Journal of Theoretical Biology*, 7, 1–52.
—— 1971: Selection of selfish and altruistic behaviour in some extreme models. In *Man and beast: comparative social behaviour*, J. F. Eisenberg and W. S. Dillon (eds), Washington D.C.: Smithsonian Institute Press.
Hamilton, W. J. 1937: Winter activity of the skunk. *Ecology*, 18, 326–7.
Heller, E. 1904: Mammals of the Galapagos Archipelago, exclusive of the *Cetacea*. Papers of the Hopkins Stanford Galapagos Expedition, 1898–1899. *Proceedings of the Californian Academy of Science* (zoological), 3, 233–50.
Hershkovitz, P. 1969: The evolution of mammals on southern continents. VI. The recent mammals of the neotropical region: a zoogeographic and ecological review. *Quarterly Review of Biology*, 44, 1–70.
Imler, R. and Sarber, H. R. 1947: Harbor seals and sea lions in Alaska. US Department of the Interior, Fish and Wildlife Service. *Special Scientific Report*, 28, 1–22.

Kaufmann, J. H. 1962: Ecology and social behaviour of the coati, *Nasua narica*, on Barro Colorado Island, Panama. *University of California Publications in Zoology*, 60, 95-222.

Keynon, K. W. 1969: The sea otter in the eastern Pacific Ocean. US Department of the Interior, Bureau of sport, fisheries and wildlife publication, *North American Fauna*, 68, 1-352.

Keynon, K. W., Scheffer, V. B. and Chapman, D. G. 1954: A population study of the Alaska fur-seal herd. US Department of the Interior, Fish and Wildlife Service, *Spec. Sci. Rept. Wildl*, 12, 1-77.

Kleiman, D. G. 1972: Social behaviour in the Bush Dog (*Speothos venaticus*) and the Maned Wolf (*Chrysocyon brachyurus*): a study in contrast. *Journal of Mammalogy*, 53, 791-806.

Kleiman, D. G. and Eisenberg, J. F. 1973: Comparisons of canid and felid social systems from an evolutionary perspective. *Animal Behaviour*, 21, 637-59.

Kruuk, H. 1972: *The spotted hyena*. London: University of Chicago Press.

—— 1975: Functional aspects of social hunting by carnivores. In G. P. Baerends, C. Beer and A. Manning (eds), *Function and evolution in behaviour*, Oxford: Clarendon Press.

—— 1976: Feeding and social behaviour of the striped hyena (*Hyeana vulgaris* Desmarest). *East African Wildlife Journal*, 14, 91-111.

—— 1978: Foraging and spatial organisation in the European Badger, *Meles meles* L. *Behaviour Ecology and Sociobiology*, 4, 75-89.

Kruuk, H. and Turner, M. 1967: Comparative notes on predation by lion, leopard, cheetah and Wild Dog in the Serengeti area, East Africa. *Mammalia*, 13, 1-27.

Kühme, W. 1965: Communal food distribution and division of labour in African hunting dogs. *Nature*, 205, 443-4.

Lang, E. M. 1958: Zur Haltung des Strandwolfs (*Hyaena brunnea*). *Der Zoologische Garten Leipzig*, 24, 81-91.

Laundré, J. 1977: The daytime behaviour of domestic cats in a freeroaming population. *Animal Behaviour*, 25, 990-8.

van Lawick, H. 1974: *Solo: the story of an African wild dog*. Boston: Houghton Mifflin Co.

van Lawick, H. and van Lawick-Goodall, J. 1971: *Innocent killers*. Boston: Houghton Mifflin Co.

Leitch, I., Hytten, F. E. and Billewicz, W. Z. 1959: The maternal and neonatal weights of some mammalia. *Proceedings of the zoological Society, London*, 133, 11-28.

Leyhausen, P. 1965: Über die Funktion der Relativen Stimmungshierarchie (Dargestellt am Beispiel der phylogenetischen und ontogenetischen Entwicklung des Beutefangs von Raubtieren). *Zeitschrift für Tierpsychologie*, 22, 412-94.

Liburg, O. 1980: Spacing patterns in a population of rural free roaming domestic cats. *Oikos*, 35, 336-49.

—— 1981: Predation and social behaviour in a population of domestic cat. An evolutionary perspective. Dissertation, University of Lund, Sweden.

Lockie, J. D. 1966: Territory in small carnivores. *Symposium of the zoological Society*, London, 18, 143-65.

Maxwell, G. 1967: *Seals of the World*. London: Constable and Co. Ltd.

McDonald, D. W. 1976: Food caching by red foxes and some other carnivores. *Zeitschrift für Tierpsychologie*, 42, 170-85.

—— 1977: The behavioural ecology of the red fox. In C. Kaplan (ed.), *Rabies – the facts*, Oxford University Press.
McDonald, D. W. and P. J. Apps 1978: The social behaviour of a groups of semi-dependent farm cats *Felis catus*: a progress report. *Carnivore Genet. Newsletter*, 3, 256–68.
McPherson, A. H. 1969: The dynamics of Canadian arctic fox populations. *Canadian Wildlife Service Reports Series*, 8, Ottowa.
Mech, L. D. 1970: *The Wolf*. New York: Natural History Press.
Mech, L. D., Tester, J. R. and D. W. Warner 1966: Fall daytime resting habits of raccoons as determined by telemetry. *Journal of Mammalogy*, 47, 450–66.
Moehlman, P. D. 1979: Jackal helpers and pup survival. *Nature*,277, 382–3.
Muckenhirn, N. A. and Eisenberg, J. F. 1973: Spacing and predation by the Ceylon leopard (*Panthera pardus fusca*). In R. Eaton (ed.), *Proceedings of the 1st International Conference on the Ecology, Behaviour and Conservation of the World's Cats*.
Murie, A. 1944: The wolves of Mount McKinley. US *Fauna* Series, 5, 1–238.
Norris, T. 1969: Ceylon Sloth Bear. *Animals*, 12, 300–3.
Panaman, R. 1981: Behaviour and ecology of free-ranging female farm cats (*Felis catus* L.) *Zeitschrift für Tierpsychologie*, 56, 59–73.
Pedersen, A. 1957: *Der Eisbär: Verbreitung und Lebensweise*. Neue Brehm-Bücherei, Band. 201. Wittenberg-Lutherstadt: A. Ziemsen-Verlag.
—— 1962: *Polar Animals*. London: Harrap.
—— 1962b: *Das Walross*. Neue Brehm-Bücherei, Band. 306. Wittenberg-Lutherstadt: A. Ziemsen-Verlag.
Pienaar, U. de V. 1969: Predator–prey relationships amongst the larger mammals of the Kruger National Park. *Koedoe*, 12, 108–76.
Portmann, A. 1965: Über die Evolution der Tragzeit bei Säugetieren. *Revue Suisse de Zoologie*, 72, 658–66.
Rasa, O. A. E. 1971: Social interaction and object manipulation in weaned pups of the Northern Elephant Seal *Mirounga angustirostris*. *Zeitschrift für Tierpsychologie*, 29, 82–102.
—— 1973: Prey capture, feeding behaviour and its ontogeny in the African Dwarf mongoose (*Helogale undulata rufula*). *Zeitschrift für Tierpsychologie*, 32, 449–88.
—— 1977: The ethology and sociology of the Dwarf mongoose, *Helogale undulata rufula*. *Zeitschrift für Tierpsychologie*, 43, 337–407.
—— 1979: The effects of crowding on the social relationships and behaviour of the Dwarf Mongoose *Helogale undulata rufula*. Zeitschrift für Tierpsychologie, 49, 317–29.
—— 1981: Raptor recognition – an interspecific tradition? *Die Naturwissenschaften*, 68, 151–2.
—— 1983: Dwarf mongoose and hornbill mutualism in the Taru Desert, Kenya. *Behaviour, Ecology and Sociobiology*, 12, 181–90.
Rand, R. W. 1955: Reproduction in the female Cape Fur seal *Arctocephalus pusillus*(Schreber). *Proceeding of the Zoological Society*, London, 124, 717–40.
Robinette, W. L., Gashweiler, J. S. and Morris, O. W. 1969: Notes on cougar productivity and life history. *Journal of Mammalogy*, 42, 204–17.
Rood, J. P. 1974: Banded mongoose males guard young. *Nature*, 248, 176.
Sandegren, F. E. 1968: Sexual and maternal behaviour in the Steller Sea Lion

Eumetopias jubata in Alaska. M. Sc. Thesis, University of Alaska, May 1968, 1-135.

Sandegren, F. E., Chu, E. and Vandevere, J. 1973: Maternal behaviour in the Californian Sea Otter. *Journal of Mammalogy*, 54, 668-79.

Schaller, G. B. 1967: *The deer and the tiger*. Chicago: University of Chicago Press.

—— 1968: Hunting behaviour of the cheetah in the Serengeti National Park, Tanzania. *East African Wildlife Journal* 6, 95-100.

—— 1972: *The Serengeti lion*. Chicago: University of Chicago Press.

Schaller, G. B. and Lowther, G. R. 1969: The relevance of carnivore behaviour to the study of early hominids. *Southwestern Journal of Anthropology*, 25, 307-41.

Scheffer, V. B. 1958: *Seals, Sealions and Walruses*. California: Stanford University Press.

Schenkel, R. 1966: Play, exploration and territory in the wild lion. *Symposia of the Zoological Society of London*, 18, 11-22.

Schneider, D. G., Mech, D. L. and Tester J. R. 1971: Movements of female raccoons and their young as determined by radio-tracking. *Animal Behaviour Monograph*, 4, 1-43.

Schneirla, T. C., Rosenblatt, J. S. and Tobach, E. 1963: Maternal behaviour in the cat. In H. L. Rheingold (ed.) *Maternal behaviour in mammals*, New York: John Wiley and Sons.

Schönberger, D. 1965: Beobachtungen zur Fortpflanzungsbiologie des Wolfes. *Zeitschrift für Säugetierkunde*, 30, 171-8.

Sheldon, W. G. and Toll, W. G. 1964: Feeding habits of the river otter in a reservoir in central Massachusetts. *Journal of Mammalogy*, 45, 449-55.

Slijper, E. J. 1956: Some remarks on gestation and birth in *Cetacea* and other aquatic mammals. *Hvalrådets Skrifter*, 41, 1-62.

Steller, G. W. 1751: De Bestiis marinis. *Novi Comm. Acad. Sci. Petropolitanae*, 2, 289-398.

Storm, G. L. 1965: Movements and activities of foxes as determined by radio-tracking. *Journal of Wildlife Management*, 29, 1-13.

Thenius, E. 1969: Stammesgeschichte der Säugetiere (einschliesslich der Hominiden). *Handbuch der Zoologie*, VIII, 47 and 48.

Volf, J. 1963: Bemerkungen zur Fortpflanzungsbiologie der Eisbären, *Thalarctos maritimus* (Phipps) in Gefangenschaft. Zeitschrift für Säugetierkunde, 28, 163-6.

Wemmer, C. and Fleming, M. J. 1974: Ontogeny of playful contact in a social mongoose, the meerkat, *Suricata suricatta*. *American Zoologist*, 14, 415-26.

Werken van der, H. 1968: Cheetahs in captivity. *Der ZoologicscheGarten Leipzig*, 35, 156-61.

Wight, H. M. 1931: Reproduction in the eastern skunk (*Mephitis mephitis nigra*). *Journal of Mammalogy*, 12, 42-7.

Wilson, E. O. 1975: *Sociobiology, the new synthesis*. Cambridge, Massachusetts: Belknap Press of Harvard University Press.

Woolpy, J. P. 1963: The social organization of wolves. *Natural History*, 72, 46-55.

Wyman, J. 1967: The jackals of the Serengeti. *Animals*, 10, 79-83.

Yeager, L. E. and Woloch, J. P. 1962: Striped skunk with three legs. *Journal of Mammalogy*, 43, 420-1.

Zimen, E. 1971: Wölfe und Königspudel. Verlag, München: R. Piper und Co.

6 Parental Behaviour in Non-human Primates

J. D. Higley and S. J. Suomi

INTRODUCTION

Infants of most primate species, including *Homo sapiens*, present a certain paradox to the researchers who study their development. On the one hand, primate neonates seem unusually precocious in several domains, relative to neonates of most other mammalian species. For example, unlike rodents and carnivores, who are born with both eyes and both ears sealed, primate infants can see and hear within moments of birth and all their sensory systems develop to adult levels of sensitivity relatively early in ontogeny (Fobes and King, 1982). Virtually all neural cell proliferation is complete by the time of birth in advanced primate species, whereas the comparable figure for a newborn white rat is only about 25 per cent (Goldman, 1976; Rakic, 1985). While not as motorically precocious as some newborn ungulates, most primate infants can locomote independently within their first month of life; the contrast with neonatal marsupials is striking (see Gubernick and Klopfer, 1981).

On the other hand, in many other respects primate infants are profoundly altricial: they are nutritionally dependent upon their mothers for many months longer than are the young of most other mammalian species, and it takes years for their physical, cognitive, and social capabilities to mature fully (Harlow, 1959). As a consequence, most primate infants are inextricably tied to parents and other conspecifics not only for all of the basic needs shared by most other mammalian infants (including

---

This chapter is dedicated to the memory of Professor Harry F. Harlow, whose pioneering studies of captive-reared rhesus monkeys demonstrated beyond question that parental behaviour is much more than the product of the species' genetic history. Numerous individuals contributed to the preparation of this chapter, including Helen LeRoy, Sue Higley, Winston Hopkins, Rusty Dunner and Alita Ritzema. Special thanks are extended to Barbara Wright who was responsible for the preparation of the manuscript in its final form.

nourishment, thermoregulation, protection, and transportation in their first days and weeks of life), but also for additional social and emotional needs in the succeeding months and even years (Lindburg, 1971; Minami, 1975; Harlow and Suomi, 1970; Mason and Berkson, 1975). Thus, among most primates the period of neoteny and, correspondingly, the duration of species-normative parental care, is especially protracted.

Some authors have suggested that this extended period of parental involvement with offspring has evolved to increase behavioural flexibility and to decrease the relative degree of 'hardwiring' of behavioural patterns, permitting the remarkable environmental adaptability that is so characteristic of advanced primate species (Mason, 1979). Primate parental involvement goes beyond provision of basic biological requirements, and it is in these 'extra' provisions that the differences between primate and nonprimate species are especially evident. Extensive data collected on numerous primate species in both field and laboratory settings clearly indicate that primate parents must do more than feed and protect their offspring if they are to eventually contribute to their species' gene pool. Indeed, as primate infants mature and parental responsibilities for satisfying their basic biological needs diminish, especially after weaning, parental involvement regarding their infants' social and emotional needs becomes increasingly important (Harlow and Harlow, 1965). If these psychological needs are not met in adequate fashion for offspring throughout late infancy and childhood, then when the individuals reach maturity their higher-order behavioural capabilities may be largely compromised (see Suomi, 1982a, for a review of this issue).

This chapter examines parental behaviour across the primate order. It characterizes the various ways in which biological parents of different primate species interact with their infant offspring, and it examines how such interactions can differ from those with individuals who are neither parent nor offspring of the interactor. The chapter will begin by considering factors that might serve to elicit and maintain basic caretaking behaviour in biological parents (and, under some circumstances, other conspecifics). It will then focus on generalized caretaking activities characteristic of Old World monkey mothers, for whom the most complete data base exists within the primate order, as well as cross-species differences among this family of monkeys in certain parameters of maternal care. The Old World monkey maternal care data will then be compared and contrasted with those collected from prosimian, New World monkey, and ape mothers. A parallel cross-species review of paternal caretaking activities will follow. Finally, the phenomenon of 'alloparental care', that is, obvious caretaking (or caretaking-like behaviour performed on infants by individuals who are not their biological parents, will be examined.

INDUCTION AND MAINTENANCE OF PARENTAL BEHAVIOUR

In virtually all primate species newborn infants normally are active agents in eliciting and maintaining those basic aspects of parental care necessary for survival under any environmental circumstances. For example, neonates of many primate species stimulate maternal care by reflexively climbing up the mother's ventrum as they leave the birth canal (Mitchell, 1979). As will be discussed shortly, such active stimulation, however effective in the long term it may be for infant survival and subsequent social, cognitive, and emotional development (see Klaus and Kennell, 1982), is only the beginning from the infant's standpoint.

These facts not withstanding, surprisingly little is known about the induction of labor and subsequent maintenance of parental behaviour in non-human primates. This area could best be summed up as hypotheses in need of data. Even the hormonal and physiological variables remain generally uninvestigated. In 1965 and earlier, Harlow called for a systematic study of the physiological and hormonal variables affecting maternal induction (Harlow and Harlow, 1965); yet, in a more recent review of hormones and maternal behavior, Rosenblatt and Siegel (1981) pointed out that they were unable to review the hormonal mediation of maternal behaviour in primates because there were virtually no data to cite. This situation has not changed dramatically in the years since then.

The evidence suggesting a possible hormonal role in parental behaviour, excluding lactation *per se*, is largely correlational. Gross and colleagues (1977) found a correlation between serum cortisol, prolactin, and the quality of maternal care in pigtail macaque females. Rosenblum (1972) reported that female squirrel monkeys near parturition were more interested in and more likely to retrieve infants than were squirrel monkey females in the early stages of pregnancy. On the other hand, Gibber (1981) failed to find any enhancement of infant-directed responses in pregnant rhesus monkeys as little as two days before parturition. Cross and Harlow (1963) also observed that there was little interest in an infant by primiparous rhesus monkey females during the entire 12-month period before their first delivery. However, there was a dramatic increase in interest shortly after birth. Thus, even the correlational evidence for hormonal mediation of maternal behaviour appears to be mixed.

Some investigators have suggested possible neurological controls for the basic infant caretaking behaviour patterns shown by monkey mothers (Mitchell, 1979). Most of the relevant data have come from lesion studies. For example, Franzen and Myers (1973) found a decrease in infant care by macaque mothers following lesions in their anterior association cortex. Other researchers have reported deficiencies in maternal care following amygdalectomy (Kling and Steklis, 1976). However, virtually all the studies of neurological controls using ablation techniques have been largely preliminary in nature and at best provide incomplete explanations for how such controls might operate.

Other factors may play a role in the release and maintenance of infant caretaking behaviour, especially those factors associated with the infant. For example, many authors have suggested that primate neonates possess unique stimulus features which serve to release species specific caregiving responses (Bowlby, 1969; Lewis and Rosenblum, 1974; Bell and Harper, 1977). One of the stimulus features most frequently implicated by ethologists is coat and/or skin coloration. Neonates among many species of primates (as well as some nonprimate infants) have coats with distinctive coloration that eventually change to the adult shade. Some ethologists have hypothesized that this unique infantile coloration has evolved to elicit special care by adults and to evoke maternal solicitude (Booth, 1962; Jay, 1962; Poirier, 1968; Hrdy, 1976; Alley, 1980). Such an assertion is usually based on the observed temporal correlation between the special treatment that an infant receives from conspecifics and the maintenance of its infantile coloration. As long as an infant possesses its infantile coat, it is treated with solicitude, protection, and often a surprising level of tolerance by adults in its social group. However, as its distinctive coat colour fades, so does its special treatment (Poirier, 1968, 1970; Struhsaker, 1971; Hall and DeVore, 1965; Burton, 1972; Jay, 1963; Alley, 1980). For example, among patas monkeys, sub-adult 'aunting' behaviour (maternal-like care directed toward an infant by those other than the mother or father) is usually practised only on infants still retaining their neonatal pelage (Gartlan, 1969; Struhsaker, 1971; Lancaster, 1971), while maternal rejection associated with the weaning process increases dramatically during the time when the infant's coat begins to change from its natal colour (Alley, 1980). Furthermore, it has been reported in many nomadic primate species that while troop members typically will leave behind a dead or unconscious adult, adolescent, or juvenile as they move throughout their habitat, infants who die before their natal coats change colour often continue to be carried by their mothers for several days after death (Alley, 1980; Hrdy, 1976). Booth (1962) reported that cercopithecine monkeys became agitated in the presence of a moving, neonatal-coloured stuffed skin.

There has been at least one systematic experimental study of the attention-eliciting capabilities of infantile facial/fur coloration. Higley and colleagues gave both multiparous and primiparous rhesus monkey females the opportunity to choose between juvenile monkeys whose facial skin had been cosmetically tinted either a greenish colour, a reddish infant-like colour, or left with their normal skin colour. The investigators found that both categories of female rhesus monkeys spent more time with juveniles dyed with the reddish neonatal skin colour than with juveniles who had normal or green-dyed skin tone (Higley, Hopkins, Suomi, Hirsch and Marra, 1983). This study notwithstanding, the general phenomenon and significance of infantile coloration as an elicitor of attention and, presumably, caretaking behaviour still remains largely uninvestigated under controlled experimental conditions (see Alley, 1980, for an excellent review of this subject).

Another potential mechanism posited to be a maternal elicitor by some researchers involves infantile vocalizations (see Bell and Harper, 1977; Noirot, 1972; and Newman, 1985, for reviews of this issue). Infant monkey cries can be spectographically differentiated from adult vocalizations; furthermore the infant vocal frequency in many primate species is in the auditory range where a mother's sensory threshold is the lowest (Struhsaker, 1967a). While few systematic studies have set out specifically to investigate infant vocalizations as possible releasers of caretaking or otherwise solicitous behaviour from parents, other studies are suggestive (see Newman, 1985, for a comprehensive review of the literature regarding 'isolation peeps').

A number of authors have noted that macaque mothers exposed to either their own or to an unfamiliar infant's distress cries demonstrate increased disturbance and anxiety; in contrast, males seem relatively unaffected (Simons, Bobbitt and Jensen, 1968; Simons and Bielert, 1973). Kaplan and colleagues found similar results in squirrel monkey mothers, but only when they were tested with their own infant (Kaplan et al., 1978). In a recent elegant playback study Symmes and Biben (1985) demonstrated individual recognition (indeed, intense interest) by squirrel monkey mothers in response to the isolation peeps of their own seven-month-old offspring, compared with those peeps of familiar but genetically unrelated conspecific infants. Van Lawick-Goodall (1967) reported that young infant cries were more likely to produce retrieval than were an older infant's cries. Rogers and Davenport (1970) suggested that infant distress cries can produce an unpleasant state which females may quickly learn to 'turn off' by retrieving the infant. In support of this hypothesis is the finding that when an infant vervet monkey's distress calls are recorded and played back via a speaker, other females in the group will look at the mother whose infant is vocalizing (Cheney and Seyfarth, 1980). Finally, Harlow and colleagues long ago noted that when rhesus mothers are remiss in retrieving their vocalizing infant, they may be threatened by the other nearby females (Harlow, Harlow and Hansen, 1963). These findings suggest a role for infant vocalizations in the induction or elicitation of caretaking behaviour in mothers and others, but direct experimental demonstration of proximal mechanisms is largely lacking.

A different mechanism suggested for the initial induction of primate maternal behaviour is placentaphagia. Ethologists have long hypothesized that because of its widespread occurrence across the class *mammalia*, the post-natal eating of the placenta by mothers may play a role in the induction of maternal care (Hartman, 1928; Cairns, 1972). Placentaphagia occurs throughout the primate order (Brandt and Mitchell, 1971; Goodall and Athumani, 1980). However, its actual role or function remains largely unknown. Brandt and Mitchell (1973) reported that among laboratory-housed rhesus monkey females, 85 per cent of the mothers surveyed consumed the placenta. The relative incidence of placental consumption appears to be much lower in great ape mothers (Rogers and Davenport, 1970; Brandt and Mitchell, 1971; Goodall and Athumani, 1980). Even in species in which placental consumption is common, its occurrence does not necessarily assure the initiation of maternal care (Hopf, 1967; Brandt and Mitchell, 1971; Sekulic, 1982), nor does nonconsumption predict a subsequent lack of maternal care (Rowell, Hinde and Spencer-Booth, 1964; Meier, 1965; van Lawick-Goodall, 1967; Swartz and Rosenblum, 1981).

These facts notwithstanding, a number of hypotheses have been offered to explain why placentaphagia should be expected to enhance the appearance of appropriate maternal behaviour. For example, the placenta is rich in sex hormones (Thau and Lanman, 1975), and the consumption of these hormones could, conceivably, elicit or otherwise facilitate maternal behaviour. On balance, however, this hypothesis remains somewhat doubtful since all the presumably relevant hormones would most likely be rapidly broken down into the respective amino acids, precluding any long-term influences.

Ewer (1968) suggested a very different reason why placentaphagia may aid in maternal induction. Immediately after birth non-human primate mothers who consume the placenta and umbilical cord typically follow this act by licking their newborn infant, seemingly as part of a larger sequence. Infant-licking is a nearly universal mammalian maternal response that occurs shortly after birth (Bo, 1971; Mitchell and Brandt, 1971; Fisher, 1972; Brandt and Mitchell, 1973). Among simian primates it is accompanied by frequent pauses to inspect the infant visually, with special attention usually directed toward the infant's genitalia – especially in male infants. Some researchers have suggested that this early social interaction plays a special role in facilitating maternal care by serving as a means by which the mother can 'label' the infant as her own. In support of this hypothesis, Lundblad and Hogden (1980) found that swabbing a rhesus monkey infant with vaginal secretions before returning it to its mother significantly increased the probability that the infant would receive maternal care in laboratory settings.

Poirier (1977) has suggested an alternative role for maternal placentaphagia. The placenta is high in protein and so consuming it may allow the mother increased, uninterrupted time with her infant to recover from the delivery process and interact before she is forced to return to foraging for food. Alternatively, placentaphagia may not have any direct link with maternal behaviour induction. It may instead serve no special purpose other than possibly hiding evidence of a birth from potential predators. Clearly, the hypotheses concerning this phenomenon exceed the actual definitive data at present.

Another possible mechanism of maternal behaviour induction may involve stimulation from an infant clinging to its mother and other physical contact. Harlow and Harlow (1965) suggested that infant clinging may be a primary stimulus in elicitation and maintenance of maternal behaviour. A number of studies have indicated that clinging may act as a possible maternal regulator. Harlow found that a rhesus mother who had her infant removed shortly after birth adopted a kitten and even allowed it to nurse. However, when the kitten could no longer maintain ventral contact and after it repeatedly fell to the floor of the cage, the mother's interest began to wane. After several days of retrieving the kitten, the mother lost interest and thereafter largely ignored it (Harlow and Harlow, 1965). Rhesus monkey infants which can cling are more readily adopted by females than those which cannot. In attempts to cross-foster infants to new mothers, those cases of occasional failures in inducing adoptive care occurred when the adopted infant failed to cling to or respond to retrieval gestures by the foster mother (Higley and Champoux, personal observation). Snowdon and his colleagues (1982), studying New World monogamous marmosets, found that following attempts by the mother to elicit clinging, infants who failed to cling immediately after birth were soon abandoned. Similar findings have been reported in prosimian species. Harlow also noted the phenomenon in rhesus monkeys:

A single case in which attempted adoption failed was equally informative ... The data suggested that visual and auditory variables of an infant monkey elicited maternal affectional responses but that these never culminated in adoption because the infant made no effort to contact or cling to the mother. That the infant, not the mother, was to blame became apparent when this same mother adopted a congenitally blind baby. (Harlow and Harlow, 1965, p. 307)

A final mechanism that could be posited as an elicitor of maternal behaviour involves social interaction. From birth the infant is primed to be socially responsive to conspecifics. Initially the infant uses its primitive grasping, clinging, climbing and sucking reflexes to maintain protracted, intimate contact. However, during the neonatal period infants quickly develop a number of social reflexes. For example, during their first few days of life macaque monkeys will visually follow large salient stimuli, such as a face (Boothe, Kiorpes, Regal and Lee, 1982), and auditory stimuli, such as lipsmacks (Schneider, Marra, Suomi and Higley, 1986). In many species, the infant spends long intervals staring into its mother's face (Hines, 1942) and responding to her eye contact

(Chevalier-Skolnikoff, 1973: Mendelson, Haith and Goldman-Rakic, 1982; Mendelson, 1982). Conversely, the mother spends much of the first few days grooming, licking and adjusting to her infant's positions. She, like her infant, spends relatively long periods looking at the other's face, and she may at times lipsmack to restore eye contact (Mitchell and Brandt, 1975). Similar and more frequent mother–infant sequences are reported in the great apes (Hoff, Nadler and Maple, 1981; van Lawick-Goodall, 1967). Such dyadic, synchronous sequences have been postulated to play a major role in human mother–infant bonding (Stern, 1974; Robson, 1967). Although it is difficult to test such a supposition, it may be that this plays an important role in inducing and subsequently maintaining maternal behaviour. It is interesting that during the first 30 days, as rhesus monkey infants become increasingly socially responsive, maternal grooming shows a concomitant increase (Harlow, Harlow and Hansen, 1963).

In summary, there are a number of potential candidates for crucial roles in the process of inducing primate maternal behaviour. It may be that a number of mechanisms function together to elicit and maintain caretaking behaviour. At this point, the evidence for any one of them is fragmentary and at best weak. We suggest this area is ripe for further study.

MATERNAL CARE ACTIVITIES: CROSS-SPECIES SIMILARITIES AND DIFFERENCES

Primate mothers typically display a large repertoire of caretaking activities that serve to enhance their infant's likelihood of survival. These activities vary with the infant's age and, to a lesser extent, with the sex of the baby. A mother's activities neatly mirror her infant's needs and the environment. As the infant matures, the mother's responses towards her infant change also. Nevertheless, there are also wide species differences in both the kind and quantity of maternal care.

Old World monkeys: prototypical maternal caretaking activities

Without question, the most extensive data bases on maternal care and the development of mother–infant relationships in non-human primates can be found among a number of Old World monkey species. This is true for data obtained in the field as well as for data gathered under laboratory conditions. Such extensive data bases have permitted investigators not only to make meaningful cross-species comparisons in maternal caretaking practices and tendencies, but also to make within-species comparisons between mothers rearing their infants in various laboratory environments and those in the field, as well as between individual mothers sharing the same physical and social environment.

Neonatal caretaking. Virtually all Old World monkey mothers spend the overwhelming majority of their offspring's first days and weeks of life attending to its immediate physical needs, in a manner characterized by its striking intimacy and closeness. Physical contact between mother and infant in most Old World species is almost constant throughout most of the neonatal period (Harlow, 1963; DeVore, 1963; Rosenblum, 1971a). Being mammals, of course, primate mothers suckle their infants. During its first few days nursing is the infant's most frequent social behaviour other than the nearly continuous ventral contact. In fact, among Old World monkeys, nipple contact is maintained over 80 per cent of the time during the first month (Harlow et al., 1963; Bolwig, 1980). This fact of biology predisposes each Old World monkey mother to be the central figure in her infant's initial social world. When the infant later takes its first awkward steps, she continually monitors its movement so that she can respond quickly to retrieve or protect her infant from any perceived threat.

Old World monkey mothers are their infants' primary locomotor vehicle during this early period. They are also more restrictive and protective during this period than they will be at any other period in their offsprings' lives (Rosenblum, 1971b). When normally nomadic mothers are not travelling, visual and physical inspections of neonates are frequent (DeVore, 1963; Rijksen, 1978). Such inspections are marked by grooming, licking, and combined auditory and facial signals, for example lipsmacking. Since most Old World monkey mothers tend to respond positively to all neonates, these intimate, close-quarter interactions probably allow the mothers to discriminate their own specific infant from all others in their social group (Hansen, 1966; Harlow, 1963; Rosenblum, 1972; see also Harlow and Harlow, 1965; DeVore, 1963; Rhine and Hendy-Neely, 1978), most likely via multiple sensory modalities.

Caretaking during infancy. As Old World monkey infants gain voluntary control over their reflexes and acquire increased motor co-ordination in their third and fourth weeks of life, they begin to locomote away from their mothers. As the frequency, duration and distance of these forays increase, their relationships with their mothers pass into a new phase that Harlow aptly termed the state of security and trust (Harlow and Harlow, 1965). Mothers now become psychologically more than physically important for their infants, essentially serving as a secure base to explore from and return to (Harlow, 1958; Harlow and Harlow, 1965; Hansen, 1966). Mason (1973) has suggested that these 'yo-yo' excursions are probably the result of arousal modulation on the part of the infants. Mothers seem to impart a sense of reliability in being able to reduce excessive arousal to a comfortable level, not only through protection and retrieval, but also by providing reliable access to contact clinging which quickly reduces over-arousal (McCulloch, 1939; Harlow, 1958; Mason, 1967; Candland and Mason, 1968). Harlow's now classic work with surrogate mothers has identified this variable as paramount in the infant's psychological development (Harlow, 1958; Harlow and Suomi, 1970). Through contact-maintaining clinging, Old World monkey infants learn to self-regulate their relative arousal (Mason, 1967, 1973) and, as a result, they develop over time a sense of security in the general presence of their mothers even when not in actual physical contact with them.

Mothers also provide frequent and varied kinesthetic stimulation to their infants. This variable has been shown to be important also in emotional development, in that laboratory-born infants who were reared on stationary surrogate mothers developed bizarre self-rocking patterns (Mason and Berkson, 1975). Surrogate mothers that provided proprioceptive-kinesthetic stimulation (as opposed to surrogate mothers that were immobile) produced infants who were more emotionally stable, more curious and more likely to explore their environment (Berkson, 1967; Mason and Berkson, 1975). As biological mothers transport and carry their infants in field environments they provide the necessary kinesthetic stimulation that is important for normal development.

Old World monkey mothers also provide contingently responsive stimulation to their infants. They respond to their infants' cries and other vocal signals, affiliative gestures, and inappropriate behaviours. Mothers can differentially reinforce, punish, or ignore each of their infant's social response (for example, Ruppenthal et al., 1974). Although this type of social stimulation has received considerable attention in the human developmental literature (Korner and Thoman, 1972; Pedersen and Ter Vrugt, 1973; Seligman, 1975), the value of contingent maternal activity has been largely ignored in non-human primates (Mason, 1979; Suomi, 1981); yet this contingent social stimulation seems to be of critical importance to an infant's social and emotional development. For example, infants reared in isolation demonstrate the rudiments of many social skills, but without an opportunity to practise these behaviours, many social responding skills become exaggerated or entirely absent (Harlow and Harlow, 1971; Sackett, 1982). When infants are reared without adults in a peer-only group, they can direct behaviours towards their peers but the peers are much less likely to respond contingently. As a consequence, these infants tend to grow up emotionally unstable and labile (Suomi, 1977a). Surrogate mothers that provide contingent movement (Mason, 1979) and puppets which provide contingent interaction (Suomi, 1981) have been shown to produce less emotional infants who grow up to be better adjusted juveniles, adolescents and, eventually, adults.

Weaning and maternal rejection. As the onset of weaning approaches, Old World monkey infants spend increasing amounts of time away from their mothers, which in turn show less solicitude and more frequent rejections. This new stage period in the mother–infant relationship is known as the period of ambivalence (Harlow and Harlow, 1965). The ensuing changes in maternal behaviour are thought to be crucially important for the infants' subsequent social development. At this point mothers clearly play an active role in reducing the frequency and duration of physical interactions with their infants with the emergence and expansion of punitive and rejecting behaviour in Old World monkey mothers.

In the 1960s and 1970s a major debate developed over the role of maternal rejection and punishment in the development of infant independence in these species. Those who argued that mothers are active in promoting infant independence pointed out that the mother increases her punishments and rejections significantly just before weaning and conversely, she decreases her restraints and retrievals as the infant's locomotor skills develop (Hansen, 1966; Hall and DeVore, 1965; Hinde and Spencer-Booth, 1967a; Hall, 1968; Jensen et al., 1968, 1969, 1971, 1973; Hinde, 1975; Hinde and Atkinson, 1970; Hinde and White, 1974; Altmann, 1980; Berman, 1980a; Bolwig 1980; Ransom, 1981). Furthermore, both rearing by surrogate mothers and rearing infants in peer-only groups (both situations lack incidence of rejection and punishment) produce a protracted period of dependency in infants (Harlow and Harlow, 1965). On the other hand, both Rosenblum (1971b) and Kaufman (1974) argued that behaviours such as punishment and rejection actually tend to inhibit independence; that is, infants who are frequently punished by their mother increase their proximity and contact with her.

Two kinds of evidence were used to support this hypothesis: first, Kaufman and Rosenblum (1969) discovered that pigtail macaque mothers direct much higher levels of punishment to their infants than do their closely related counterparts, bonnet macaques. Yet pigtail macaques show a significantly longer period of dependency than do bonnet macaques. Second, rhesus monkey infants were reared on surrogate mothers which were able to dispense punishment, with the result that the infants actually increased their levels of clinging in response to punishment (Rosenblum and Harlow, 1963; Harlow and Harlow, 1971). However, Kaplan (1981) tested squirrel monkeys at an age closer to weaning, and in this New World monkey species at this age it was reported that punishment had the effect of decreasing contact and clinging in infants tested. Thus, it may be that the age or the species tested are important in evaluating both the short- and long-term effects of punishment on social-emotional and cognitive development in young primates. Finally, Suomi (1976) found evidence that under some circumstances maternal punishment may be largely irrelevant for the development of infant independence.

Part of the controversy about the effects of punishment may have been due in large part to a general lack of consideration for contingent versus noncontingent punishment. Recent studies have shown that infants which are reared in peer-only groups, but which are given training with contingent, positive experiences are more confident and more likely to explore their environment (Champoux, 1983). Unlike the punishing surrogates, mothers dispense punishment and rejection contingent on a wide range of behaviours. Furthermore, in large social groups there are a number of other interesting individuals that an infant can interact with or, as it grows older, that it can seek out for psychological comfort and social support.

The debate notwithstanding, several points are generally accepted in current discussions on the caretaking contribution of maternal rejection and punishment:

1 The infant's age determines in part the relative degree and intensity of mother–infant interactions. The younger the infant, the more likely a mother is to restrict and restrain her infant (Hinde and Spencer-Booth, 1967a; Hinde, 1975; Berman, 1980a; Bolwig, 1980). Conversely, older infants spend more time away from their mother and leave more often to explore their environment and interact with other infants than do very young infants, all other factors being equal (Hinde and Spencer-Booth, 1967b; Harlow, 1969; Suomi and Harlow, 1975; White and Hinde, 1975; Suomi, 1976; Berman, 1980a; Bolwig, 1980).

2 The more varied and interesting the environment, the more often an infant will leave its mother and the longer it will stay away (Jensen, Bobbitt and Gordon, 1973; White and Hinde, 1975; Suomi, 1976; Berman, 1980a; Bolwig, 1980; Young and Hankins, 1979). Today, we would add that this will be true as long as the environment is not too interesting or too varied (see Suomi, 1983).

3 There is agreement that these prolonged periods of non-mother–infant interaction contribute to the waning mother–infant bond (Harlow, 1963; Hinde, 1975; Rosenblum, 1971b; Suomi, 1976).

Other maternal activities. Mothers also play a role in their infant's ontogenic learning history. Among Hani and Kawamura's well-known sweet potato washing Japanese macaques, both sweet potato washing and candy eating was quickly transmitted via the mother to the infant (Kawamura, 1958). Learning from watching mother seems to be a widely generalized primate characteristic. Observational learning has been reported for termite fishing among chimpanzees (van Lawick-Goodall, 1967), and for certain tasks there are even reports of prosimian infants learning the correct foods to eat by watching what their mother eats and picking up what she drops (Hall, 1965; Kawamura, 1958). In advanced primate species, the mother may take a more active role by actually sharing her food with her infant (Fox, 1972; Kuroda, 1980; Schessler and Nash, 1977; van Lawick-Goodall, 1967). Infants are also quickly conditioned by their mothers' fear responses (Baldwin, 1969; Mineka, Davidson, Cook and Keir, 1984); for example, Mineka and colleagues found that offspring who were not afraid of snakes quickly developed a snake phobia when they saw their mothers exhibiting fearful behaviour in the presence of snakes. Thus, the mother, either in a passive or in some cases active manner, can play an important role in her infant's learning history.

A major activity of Old World monkey mothers is social grooming. Old World monkey mothers seem to relish grooming their infants. Missakian (1974) noted that of 27,000 grooming bouts recorded in a free-ranging rhesus monkey colony, 15,000 were between mother and offspring. Ventral contact aside, grooming is the most frequent mother–infant social interaction (Harlow and Harlow, 1965; Oki and Maeda, 1973; Rhine and Hendy-Neely, 1978). The mother is the principal individual who grooms any one infant (Lindburg, 1973; Berman, 1980), and furthermore, she grooms her infant much more often than it grooms her (Sade, 1965; Lindburg, 1973; McKenna, 1978; Suomi, 1979b). Thus, among Old World monkey species, maternal grooming not only provides the infant with whatever benefit or pleasurable experience that grooming gives, it also allows the infant to learn first-hand how to participate in this potentially reciprocal social activity. In contrast, grooming appears to occur far less often among mother–infant pairs in New World monkeys and more primitive primate species.

Species differences in maternal behaviour among Old World monkey species

Among most Old World monkey species, the infant's mother is almost without exception the individual responsible for carrying the infant. The method of transportation varies with both species and the age of the infant. Among Old World monkey species, the norm for terrestrial species is for the mother to initially carry her infant on her ventrum and later for the infant to graduate to its mother's dorsal surface (DeVore, 1963; Kaufman, 1966; Struhsaker, 1967b; Bertrand, 1969; Lindburg, 1971; Ransom and Rowell, 1972; Murray and Murdoch, 1977; Altman, 1980). Mitchell (1979) reported that dorsal carrying may not be seen in laboratory environments and we have noted this also. Among the studies of arboreal species reviewed, all reported instances of infant carrying were ventral in nature (Jay, 1965; Horwich and Mansky, 1975; Struhsaker, 1975; Hrdy, 1977; Roonwal and Mohnot, 1977). There are additional reports of ventral-only carrying among semi-terrestrial species. The mode of infant carrying thus appears to vary with the type of environment inhabited by the species in question (Gartlan, 1969).

Other aspects of maternal behaviour have been shown to vary somewhat among Old World species, however, this variation is often more in quantity than kind. One noteworthy variation in Old World monkeys is in how restrictive mothers are in allowing others in their troop to interact with their infant. For example, among most macaque and baboon species, a mother will usually not allow other members to hold or even touch her infant, and as her infant matures she dictates the other monkeys the infant may interact with (Rosenblum, 1971a; Rowell, 1967). On the other side of the continuum are the langur and colobus species. Mothers among these species are characterized as allowing permissive access to their infant by others. For example, an infant langur may not even be dry before the mother allows others to handle and carry it (Jay, 1963; McKenna, 1979) and in fact during its first day the neonate may be carried by several different individuals. (This is not to say the mother is uncaring – at the first sign of distress she quickly retrieves her infant (see Jay, 1963; Hrdy, 1977).) Horwich and Manski (1975) even observed that among colobus monkeys, other females in the troop may spend more time caring for the infant than the infant's own mother. Thus, there are considerable species differences in how restrictive the mother is with her infant.

Even within closely related species, restrictiveness may vary. Rosenblum (1971b) found that pigtail macaque mothers are restrictive, as are rhesus monkey mothers. In contrast, bonnet macaques, members of the same genus *Macaca*, allow other females to groom, carry their infant and interact relatively freely with other members of their group.

These patterns of maternal care in Old World species seem to be influenced by both the availability of blood relatives and by various moment-to-moment changes in the environment. For example, female baboons will restrict access to their infant by other unrelated females, but they will allow the infant's older siblings to care and interact with it. Similar, less restrictive, patterns of infant sharing by mothers with kin have been observed in other Old World monkey species (Spencer-Booth, 1968; Breuggeman, 1973; Altmann, 1980; Johnson et al., 1980). Other aspects of the environment can influence maternal styles: baboon mothers show far more restrictive maternal patterns when living in laboratory environments than when in the field (Rowell et al., 1968; Ransom and Rowell, 1972; Nash, 1978). Furthermore, Chalmers (1973) reported that arboreal monkeys tend to show less restrictiveness than terrestrial species.

Rowell et al. (1964) and Hrdy (1976) both indicate that part of this restrictive/permissive dichotomy may be related to dominance hierarchies. Species which show a rigid, well-defined dominance hierarchy are much more likely to have a restrictive pattern of maternal behaviour. This may be in part because females who are low-ranking have difficulty retrieving their infants from higher ranking females (Rowell et al., 1964; Rowell, 1967; Hrdy, 1976; Altmann, 1980).

Mothers also impart rank to their infant. High-ranking females have high-ranking infants (Kawai, 1965; Koyama, 1967; Sade, 1967; Chikazawa, Gordon, Bean and Bernstein, 1979; Raney and Rodman, 1981; Bramblett, Bramblett, Bishop and Coelho, 1982). Studies indicate several ways in which this might occur: first, mothers may take an active role in their infant's rank acquisition by supporting and defending their infant when it is confronted by a lower ranking female's infant (Kawai, 1965; Cheney, 1977; Berman, 1980b). Altmann (1980) indicates that infants may learn their position by watching their mother interact with other group members and observing which she defers to and which defer to her. (Walters' (1980) findings support this interpretation.) Among yellow baboons (*Papio cynocephalis*) mothers do not intervene in the infant's skirmishes with other adult females, yet when these infants become adolescents and young adults, it is the females below their mother's rank that they target for domination. Another contributing factor to rank acquisition is the nature and number of the infant's social companions. Infants with high ranking mothers chose social companions with high ranking mothers (Cheney, 1977; Caine and Mitchell, 1979).

It seems likely that in rhesus monkeys rank is learned, since the juvenile monkey will retain its rank if the mother loses her status (Sade, 1972; Walters, 1980). This rank may remain even if a juvenile and all its age mates are removed from the troop and put in a new environment as a peer group (Loy and Loy, 1974). However, this is not necessarily true for all Old World monkey species, for not all species show a strong female rank order, and among these species maternal rank plays little part in the infant's social development (Nicholson, 1977; Dolhinow, 1972; Hrdy, 1977).

Maternal care in prosimians. Although no longer considered by most to be a member of the primate order (see Luckett, 1980), the tree shrew provides an illuminating contrast to the maternal behaviour of 'true' primates. Among common tree shrews (*Tupia glis*), maternal care is limited to nest building and visits to nurse young once every 48 hours. When tree shrew infants are distressed or disturbed, their mothers seldom respond or retrieve them (Martin, 1966; Sorenson, 1970).

True prosimian primate mothers exhibit infant caretaking repertoires that are, in many ways, similar to those of Old World monkey mothers. They feed, groom, protect and defend their infants, and in feral environments most respond to their infants' cries (see Klopfer and Boskoff, 1979; and Tattersall, 1982, for recent reviews). Klopfer (1972) also reported that like Old World monkey females, prosimian mothers provide a secure base for early exploration by their infants.

One major difference between prosimian neonates and those of other species of non-human primates is in the rate of growth. In many prosimian species, the infant is weaned by two months of age, while in virtually all prosimian species the infants reach sexual maturity by one year of age (Doyle, 1979). This more rapid rate of development is undoubtedly one factor contributing to the differences that do appear between prosimian maternal care and that displayed by mothers from more advanced primate species. Among prosimian species themselves, there is a wide range in the quantity and quality of maternal care (see Klopfer and Boskoff, 1979).

For example, the mothers of several prosimian species routinely will leave their infant unattended for relatively long periods of time while they forage. This is especially true among the solitary and nocturnal species, in which mothers typically build nests for their offspring. Each night the infants remain in the nest while the mothers search for food (Martin, 1972; Klopfer and Dugard, 1976; Charles-Dominique, 1977). A second unusual (for primates) pattern of maternal behaviour seen in some prosimian species has been termed 'parking'. Here mothers leave their infants literally hanging motionless from a branch of a tree while they go elsewhere, sometimes for prolonged periods of time. Among one species of potto, mothers display an almost ritualistic behaviour sequence in carrying out this 'parking' behaviour. Typically, a mother first intensely grooms and licks her infant (this tends to create a high level of excitement and movement in the infant). At first the infant moves around on its mother, then later it grasps a branch. After a period of continued grooming and licking by the mother the infant becomes motionless and the mother then sets out for a night of foraging. She returns near dawn to retrieve her infant (Charles-Dominique, 1977). Infant 'parking' has been observed in a number of different nocturnal prosimian species, including tarsiers (*Tarsius bancanus*, Niemitz, 1974), pottos and golden pottos (*Perodicticus potto* and *Arctocebus calabarensis*, Charles-Dominique, 1977) and bushbabies (*Galago senegalensis*, Charles-Dominique, 1977). Some species simply deposit their infant in a tree-fork or hole in a tree when foraging (Petter, 1965). Such 'tree-sitting' is largely possible because the infants are highly camouflaged and because they tend to remain relatively motionless during the periods of maternal absence (Charles-Dominique, 1977; Doyle, 1979). Furthermore, these species are largely nocturnal and, as a consequence, the infants are probably less easily seen than if 'parking' were to occur during daylight hours. In most species who leave their infant at night, the mothers typically scent-mark their infant for identification (Niemitz, 1974). This may preclude a mother from adopting an unrelated infant that she passes and mistakes as her own.

Infant carrying by prosimian mothers is also somewhat more varied in characteristic style than in Old World monkeys. Like the Old World species, the majority of prosimian mothers initially carry their infants ventrally, and only later do the infants graduate to their mothers' dorsal surface. As mentioned previously, most prosimian infants mature very rapidly and are locomoting independently a few weeks after birth (Tattersall, 1982). On the other hand, in many prosimian species the newborn infants are unable to cling to their mother at birth or at most do so awkwardly. In these species, especially those primarily arboreal, it is common practice for the mothers to carry their infants in their mouth (Petter-Roussequx, 1964; Martin, 1972; Doyle, 1979; Klopfer and Boskoff, 1979; Niemitz, 1979). Another peculiar form of infant carrying style is transversal carrying, seen in some prosimian species (see Klopfer and Boskoff, 1979). In this form of carrying the infant clings transversely, instead of in a head-first fashion. Both kinds of carrying graduate into dorsal carrying (Sauer, 1967; Klopfer and Boskoff, 1979; Doyle, 1979; Tattersall, 1982). Often, in times of danger, the later form of carrying may revert to the earlier form (Sauer, 1967).

There is also some evidence that among some species of prosimians, females cannot tell their own infant from another female's infant. Swayamprabha and Kadam (1980) found that loris mothers would respond to any loris infant as if it were their own. Another notable difference from Old World primates is that mothers in many prosimian species play with their infants. Extensive play bouts have been recorded between mother and infant for bushbabies (Charles-Dominique, 1977; Ehrlich, 1974). Mother–infant play has also been recorded in sifakas (*Propithecus verreauxi*, Richard and Heimbuch, 1975) and tarsiers (Niemitz, 1979). In contrast, mother–infant play occurs relatively infrequently in most Old World monkey species (see Suomi, 1979b).

Maternal care in New World monkeys. All New World species are arboreal and among these forest dwellers two basic maternal care patterns have emerged. One pattern is characteristic of the monogamous species; the other pattern is characteristic of the monkeys living in large social groups. Among the monogamous species (for example, tamarins, marmosets, and titis) in which fathers carry out extensive caretaking activities, the mothers play a relatively limited caretaking role. Unlike virtually all other simian species, twin births are the norm in many of these monogamous species (see Snowdon and Suomi, 1982). Mothers usually assist in carrying the infants the first few days, but in succeeding weeks and months this responsibility shifts to the male until finally mothers' interactions with the infants become largely limited to those times of nursing (Epple, 1975; Box, 1975). While this pattern seems to be the norm, there is considerable variability among these monogamous species: in some species mothers give their infants to the male soon after birth (Rothe, 1975; Vogt, Carlson and Menzel, 1978); in other species the mother may be the primary caretaker for several weeks (Box, 1977a; Hoage, 1977, 1982) or even through weaning (Izawa, 1978). (The reasons for these species differences are largely unknown at present.)

Maternal care among the monogamous New World monkey species is different from that of Old World monkey mother, not only in overall involvement, but also in certain specific caretaking activities. Monogamous New World mothers rarely groom their infants (Box, 1975; Izawa, 1978). Furthermore, when mothers do carry their infants, they carry them almost exclusively in a dorsal fashion (Box, 1975). Most likely as a result of mother's relative non-involvement and the male's principal involvement in child care, she is not used by the infant as a secure base when other adults are around. Instead the male or another family member is preferred as a secure base for exploration (Heltne, Wojecik and Pook, 1981).

Among the other group of New World monkeys, the most thoroughly studied, both in the lab and in the field, is the squirrel monkey. Like Old World monkeys, squirrel monkeys live in large social groups (Thorington, 1968; Baldwin, 1969). However, mothers living in these large social groups (many of which exclude adult males) differ from Old World monkey females markedly. Squirrel monkey mothers tend to be relatively passive, unconcerned and at times seemingly oblivious to their infant (Rumbaugh, 1965; Rosenblum, 1968; Baldwin, 1969). During the first month infants spend over 85 per cent of their time unaided on their mother's back (Rosenblum, 1968). Mothers spend virtually no time directly interacting with their infant in more active fashion (Baldwin, 1969), in striking contrast to Old World monkey mothers who actively carry the infant on their ventrum, cradle it and interact socially for several hours each day.

Despite this apparent early non-concern, Rumbaugh (1965) found that squirrel monkey mothers will take a more active role in their infant's care as the infants mature and develop their locomotion skills. As the infants discover that they can climb off their mothers to explore the immediate world, the mothers become more restrictive and actively attempt to prevent their infant's departure. Often a mother will retrieve the infant when it escapes her clutches. In other instances she might use an equally effective strategy to prevent the infant from leaving, shifting her body to stand bipedally, so that the infant must hold fast, and then by moving her arm so that the infant cannot climb down (Rosenblum, 1968). Restrictive behaviour by the squirrel monkey mother increases sharply in the beginning of the second month and then drops off sharply by the third month. The chronology is similar to that seen in many Old World monkey species, but the specific behavioural strategy is different. Old World monkey mothers actively hold their infant back as they wander too far by physically restraining them; these New World monkey mothers prevent their infant from physically getting off their body when possible.

Squirrel monkey mothers are considered to be generally permissive, in that they may let other females in the troop carry their infant for protracted periods. However, the infants are quickly retrieved by their respective mothers at the first signs of distress. In addition, squirrel monkey mothers are used as secure bases from which to explore (Baldwin, 1969).

Among the squirrel monkey's close relatives, the howler monkeys' (*Aloutta palliata*), characteristic maternal caretaking patterns are very much the same. The mother seems at times oblivious to her infant and takes a very passive role in its care. Baldwin and Baldwin (1973) reported that in their field study 'mothers were never seen to groom, fondle, or direct overtly affectionate behaviours toward their tiny infants'. Neville (1972) reported similar findings for red howlers. However, unlike squirrel monkeys, red howler monkey mothers carry their infant ventrally as often as dorsally, but the infant is still primarily responsible for maintaining physical contact. Also, like the squirrel monkeys, Neville (1972) reported high levels of 'aunting' in the red howler monkey groups. However, there were no reports of aunting among the howler monkey groups in southwestern Panama studied by the Baldwins. In fact, other females, although tolerant of other infants, remained generally disinterested (Baldwin and Baldwin, 1973). Clearly, there are differences among New World monkeys in the relative incidence of allomaternal care, as will be discussed in a later section.

Maternal behaviour in apes. The great apes (orang-utans, gorillas and chimpanzees) demonstrate remarkably similar maternal care to the Old World monkeys. With a few interesting exceptions, what few differences there are tend to be largely in quantity rather than in kind, most of which can be related to the greatly protracted period of infancy. Like many Old World monkeys, ape mothers initially carry their infants on their ventrum, but later carry them dorsally. But oran-utans and 'lesser ape' gibbons are exceptions. Those mothers carry their infants ventrally or in a side-ventral fashion (MacKinnon, 1974; Maple, 1980; Ellefson, 1974). Like Old World monkey mothers, ape mothers also act as a secure base, and they similarly nurse, groom and protect their infants (Schaller, 1963; Reynolds and Reynolds, 1965; van Lawick-Goodall, 1967; Nadler, 1974; Rijksen, 1978; Fossey, 1979; Miller and Nadler, 1981).

As mentioned previously, the period of infancy is much longer in great apes; consequently the early period of intense maternal solicitude lasts longer than in Old World monkeys. Maternal care is maintained with relatively intense vigour for up to three or four years in orang-utans (Horr, 1977), about three years in gorillas (Fossey, 1979), about four years in chimpanzees (van Lawick-Goodall, 1967); the comparable period is intermediate between monkeys and great apes for the lesser apes (gibbons and siamangs) (Berkson, 1967). On the other hand, great apes are noted for their increased cognitive capability over that of monkeys and lesser apes. In an ultimate sense, Mason (1979) indicates that a protracted period of dependency is one price a species must pay for increased behavioural flexibility and cognitive capability. At any rate, ape mothers differ in kind from Old World monkey species in two fundamental respects: first, they take a more active and 'purposeful' role in their infant's social and cognitive development; secondly, they engage in extensive social play interactions with their infants.

From birth through the early months of life, the prototypical great ape mother is very active in interacting with her infant. She frequently inspects and physically manipulates her infant (van Lawick-Goodall, 1967; Nicholson, 1977; Hoff, Nadler and Maple, 1981). There are many reports of mothers physically moving their infant's face to re-establish eye contact, of prolonged gazes, of hugs, nuzzling and even 'kissing' by great ape mothers (van Lawick-Goodall, 1967; Horr, 1977; MacKinnon, 1974; Nicholson, 1977; Fossey, 1979; Maple, 1980), although the incidence of such behaviour is far less than that which is characteristic of human mothers (see Papousek et al., 1986). Ape mothers are also likely to use affiliative gestures such as pats and touches or to put their arm around their infant. Often these behaviours are done without any apparent external purpose (van Lawick-Goodall, 1967; Fossey, 1979; Maple, 1980). These affiliative interactions are reported to occur much more frequently in great apes than they do in Old World monkeys.

Another difference between ape and monkey maternal behaviour concerns the sharing of solid food with infant offspring. Food-sharing has been reported for a number of ape species including chimpanzees (van Lawick-Goodall, 1967; Silk, 1978), orang-utans (MacKinnon, 1974), gibbons (Schessler and Nash, 1977) and siamangs (Fox, 1972). Eventually, many ape infants develop begging gestures to obtain desired food. Because the probability of a successful begging gesture increases initially and then drops off after weaning, Silk (1978) suggests that this behaviour may be an active part of the weaning process.

Some of the most interesting great ape maternal behaviours appear, at least from some observers' view, to be of a didactic nature, as if the mother is purposely teaching her infant. Locomotion training is the most frequent activity mentioned as a possible example of active training. A similar type of activity has been noted in some Old World monkey species also. Bolwig (1980) described how a pigtail macaque female places her infant on the ground, backs away two or three steps and then places her head down at the infant's level and displays a lipsmack or pout face until the infant walks towards her. Analogous phenomena have been reported for other Old World species (Ransom and Rowell, 1972; Bolwig, 1980). Interestingly, Burton (1972) reports identical locomotion 'training' by adult males in Barbary macaques.

However, locomotion training seems to occur much more frequently in ape mothers. Chimpanzees, gorillas and orang-utans frequently take their infants hands and 'force' walk them –often to an infant's consternation and screams (Schaller, 1963; Nicholson, 1977; Fossey, 1979; Miller and Nadler, 1981). Nicholson (1977) describes the following sequence for chimpanzees.

By the time Delta was a month old, Gigi [Delta's biological mother] was manipulating her in a more purposeful manner, seemingly testing and encouraging the infant's developing motor skills. During one observation period, Gigi repeatedly placed Delta's foot against an enclosure fence, bending the infant's toes until they gripped the wire. When Delta was 6 weeks old, Gigi initiated 'walking lessons.' She would sit with Delta on the ground between her legs, the infant clinging to Gigi's body with her hands only; then Gigi would shuffle backwards, forcing Delta to take a few steps forward. This, too, caused Delta to whimper, but within a few weeks the lessons were tolerated, if not willingly engaged in. This form of tuition continued throughout the study period. Gigi later refined her technique, walking backwards bipedally while holding Delta's hands; this enabled Gigi to walk Delta somewhat farther and at a faster pace. When Gigi first began to break contact deliberately with Delta (the infant was then 16 weeks old), she would walk up to the enclosure fence, detach Delta, and place Delta's hands on the fencing. She would then back away a few meters, leaving the infant clinging to the fence, with her feet usually on the ground. Gigi watched Delta carefully during the separation and prevented all other animals from touching her. Sometimes Gigi turned and rubbed her perineum on Delta's back. A variation of the separation procedure was to put Delta in a shallow hole in the ground, again backing a few meters away and watching her.' (pp. 541-2)

Orang-utans also engage in locomotion training. Aulman (as reported in Maple, 1980) describes the following mother–infant interaction:

At the age of three months the baby does not move at all on his own, although the mother has started educating it towards climbing from its tenth day. She does this by taking the baby with one hand round the waist, and with the other, places its hands and feet round the bars of the cage. So far, the baby is very clumsy and does not grip well round anything except the fur of its mother, she also tries in another way to incite the baby to move on its own. She places its belly down on the floor of the cage. Then settling herself on a high shelf, she observes with great interest the baby's effort to walk towards her – whining miserably while it does so. If the baby makes no progress, she comes down and gives it a finger to grip. Then she pulls it gently along the floor. (p. 153)

Similar sessions that take on the appearance of a tutorial have been reported for brachiation and climbing (Maple, 1980).

Finally, unlike Old World monkey mothers, great ape mothers play with their infants. Suomi (1979b) found that among rhesus monkeys, mother–infant play is rare. On the other hand, among the great apes, all investigations in which extensive mother–infant observations have been obtained to date have reported wide-spread mother–infant play (for example chimpanzees, van Lawick-Goodall, 1967; Nicholson, 1977; Miller and Nadler, 1981; orang-utans, MacKinnon, 1974; Rijksen, 1978; Miller and Nadler, 1981; Maple, 1982; gorillas, Fossey, 1979; Maple and Hoff,

1982). In fact, among great apes the mother is often an infant's major play partner during its first two years of life (Miller and Nadler, 1981). From an evolutionary view, the high incidence of maternal play may reflect the fact that there are usually far fewer play partners that are the infant's age in ape species than in most Old and New World monkey species.

PATERNAL CARE

Males also display considerable caretaking behaviour toward infants in many diverse primate species. During the past decade there has been a surge in the study of male–infant interactions and care, in part due to Mitchell's influential reviews in 1969 and 1972 which called attention to this neglected area (Mitchell, 1969; Mitchell and Brandt, 1972). Since Mitchell's reviews several other recent reviews have appeared (Hrdy, 1976; Redican, 1976; Parke and Suomi, 1981; Fedigan, 1982; Snowdon and Suomi, 1982). However, many additional studies have subsequently been conducted in this area and, consequently, wherever possible we shall attempt to up-date these reviews.

A major problem noted by those who study primate paternal care involves identification of the biological father of any infant under study. For self-evident reasons, identifying the infant's biological mother is usually very easy. In sharp contrast, paternal identification is often impossible without using such complicated and costly genetic analyses as blood group markers. In some species and/or social groups, genetic consanguinity is so inextricably interwoven that even blood typing cannot identify the biological father with any certainty (Shively and Smith, 1983). Even among species where fathers can be routinely identified through genetic analysis, prolonged lab interbreeding can make paternity moot.

Since in most cases biological relatedness is difficult to prove, paternal behaviour is more often designated as 'male care'. Two kinds of male care have been delineated in the literature (see Snowdon and Suomi, 1982): substitute care and *complementary care.* Substitutive care refers to infant-directed behaviours that are 'normally' exhibited by mothers: for example, mothers carry, retrieve, groom, feed, punish, promote independence, hug and calm distressed infants. When these activities or roles are performed or provided by a male, they are termed substitutive care. With the exception of nursing, males of certain species have been shown to perform all of the above varieties of substitutive care. For example, among Barbary macaques (*Macaca sylvanus*), stumptail macaques (*macaca arctoides*), most callithricidae species and great apes, males carry, groom, retrieve and comfort infants (Deag, 1980; Estrada and Sandoval, 1977; Epple and Katz, 1983; Tilford and Nadler, 1978). Complimentary care is represented by behaviours that are not usually engaged in by females in most species, and these behaviours are thought of as more-or-less male specific. For example, males in most species are more aggressive than are females (Goy, 1966) and so they are more likely to participate in group protection and territorial defence.

Suomi (1977b) indicates that when rhesus macaque males living in the lab exhibit defence behaviours, it is most frequently in defence of an infant. In fact, a review of the literature indicates that among many Old World non-monogamous and great ape species, direct defence of the infant from other individuals within the infant's social group and protection from predators is a very common form of infant care by adult males (Itani, 1959; Jay, 1963; Carpenter, 1965; Hall and DeVore, 1965; Alexander, 1970; Struhsaker and Gartlan, 1970; Poirier, 1970; Lindburg, 1971; Ransom and Ransom, 1971; Wooldridge, 1971; Neville, 1972; Rhine and Owens, 1972; Rowell, 1974; Bernstein, 1975, 1976; Hrdy, 1976; Wolters, 1977; Altmann, 1980; Taub, 1980a; Bolin, 1981; Green, 1981). Rowell (1974) suggests that protection and defence is a very primitive male role which has been selected for during evolutionary history. She also suggests that infant defence should be more male-specific among those species where sexual dimorphism is extant. Given her prediction, it is no surprise that this form of complementary male care is less variable among different species than are most types of substitutive care. Complementary care is seen across the order (although among some species, especially New World monogamous species, females may be as likely as males to engage in infant and group defence (for example, see Bernstein, 1975; Suomi, 1977b).

Substitutive care, on the other hand, is highly variable across the primate order and this variability is not taxonomically systematic. Even within species, the probability of substitutive male care varies within specific groups, across social situations and between individual males. As a general rule, however, virtually all males are less interested in and less likely to care for an infant than are females. A major exception appears in the New World monogamous species, as will be described next. 'Male care' will subsequently refer to substitutive male care unless noted otherwise.

Species differences in primate male care

Substitutive male care varies in incidence across the primate order from the virtually nonexistent to the more extensive maternal care performed by females of the latter primate species. Closest to the nonexistent end of the spectrum are the males of most prosimian species. The modal form of adult male caretaking of infants in these prosimian species is no care, attributable in part to the solitary nature and general absence of group-living adults in these species see Tattersall, 1982).

While prosimian males tend to be uninvolved in infant caretaking activities, they are nevertheless generally benign in their treatment of infants. The only report to date of infant-directed male aggression is in captivity under crowded conditions (Buettner-Janusch, 1964). However, in most cases, both in the feral environment and in captivity, males ignore infants (for example, see Jolly, 1966; Doyle, Pelletier and Bekker, 1967; Doyle, Andersson and Bearder, 1971; Sussman, 1977; Harrington, 1978). Prosimian males appear to be generally tolerant of infants and may even allow them to climb on and over their bodies (Doyle et al., 1971). While the typical prosimian male may not show any substitutive care for his infant offspring, in some species he may play with the infant as the infant grows older (Sussman, 1977; Pages, 1980), and as the infant reaches the more mature juvenile stage, it may sleep with the adult male (Pollock, 1980). In at least one species, male care seems limited not because of lack of interest, rather because the male's contacts are fewer (*Phaner furcifer*, Charles-Dominique and Petter, 1980).

An exception to this prototypical prosimian paternal pattern is among the sifakas (*Propithecus verreauxi*). Males among this species have been observed to show various forms of substitutive parental care, including carrying, grooming and playing with the infant (Jolly, 1966; Eaglen and Boskoff, 1978). Jolly (1972) reported that the levels of care among male sifakas may exceed that of many Old World monkeys. Richard (1976) noted, however, that in one family group of sifakas, the male made no contact with infants; in fact, when the group's sub-adult male attempted to contact or groom any infant, the mother slapped and bit the sub-adult for its attempts. Eaglen and Boskoff (1978), also studying sifakas, reported that the adult male's very early attempts to interact with the infant were met with punitive responses from the mother. Among a number of the lemur species the adult males' lack of interest in infants may be related to how restrictive the female is and how she treats the male's attempts to interact with the infant (Jolly, 1966; Petter and Klopfer, 1970). Further evidence for this comes from Klopfer's (1972) findings that when the mother is removed, male care of infants increases dramatically. In fact, Jolly (quoted in Sussman, 1977) observed a ringtail lemur (*Lemur catta*) male adopt an abandoned infant. Thus, while male care may seldom be found in prosimians, it may be because the male is restricted from doing so either by the social circumstance or by the female, rather than from a lack of interest and/or caretaking competence.

Monogamous New World monkey species

At the other end of the caretaking spectrum are the New World monogamous species: extensive male care among these species is the norm, not the exception. In fact, our review of the literature on parental care among New World monogamous species found only one species in which the male does not make a significant contribution to infant care – *Saguinus nigricollis* (Izawa, 1978). In all other monogamous species males play an important role in rearing their offspring (for example, *Callimico goeldi*, Pook, 1978; Heltne, Turner and Wolhandler, 1973; Lorenz, 1972; *Leontopithecus roselia*, Hoage, 1977, 1982; *Callithrix jacchus*, Box, 1975, 1977a,b; Locke-Haydon and Chalmers, 1983; Ingram, 1977; 1978a,b; Rylands, 1981; Epple, 1978; *Saguinus fuscicollis*, Epple 1975; Epple and Katz, 1983; *Saguinus o. oedipus*, Wolters, 1978; Hampton, Hampton and Landwehr, 1966; Cleveland and Snowdon, 1984; *Callicebus moloch*, Fragaszy, Schnarz and Shimosaka, 1982; *Aotus trivikagtus*, Dixson and Fleming, 1981; Dixson, 1983).

Little is known about how parental care is induced in these New World monogamous males. Some researchers have recorded altered levels of the hormone prolactin in males who are caring for an infant (Dixson and George, 1982; Dixson, 1983), whilst Sassenrath, Mason, Fitzgerald and Kenney (1980) found significant decreases in testosterone in males who were caring for infants. However, these studies were correlational in design and thus failed to demonstrate cause and effect.

More clear is the role that prior experience in caring for infants plays in determining whether any particular male will exhibit competent, species-normative infant caretaking behaviour as an adult. In fact, adult males raised in captivity with no prior experience with an infant seem to find infant contact aversive (Fragaszy et al., 1982). Infants are more likely to be left alone and unprotected when both the male and the female are inexperienced than when only the female is inexperienced (Epple, 1975; Hoage, 1977; Epple and Katz, 1983). Ingram (1978b) noted that the probability of a male showing care for infants who have been abandoned by their mother is directly related to whether the male has had prior experience with infants. In fact, among captive brownheaded tamarins (*Saguinus fuscicollis*), Epple and Katz (1983) found that almost all primiparous parents with limited social experience killed their infants. Interestingly, prior experience seems less important for the male than for the female, in that infants are less likely to be abused if the male is inexperienced than if the female is inexperienced, and inexperienced males are more likely to carry an infant than are inexperienced females (Epple and Katz, 1983).

Although adult males in these monogamous species do show significant infant care, and in some species they are the primary caretakers, infant care in these species is usually a family project. The norm is that siblings, other group members and the mother all share the infant caretaking duties with the adult male. In fact, in some marmoset and tamarin species, siblings and other members of the group make as much or more of a contribution to infant care than the male (for example, see

Epple, 1975; Hoage, 1977; Ingram, 1977; Pook, 1977; Vogt et al., 1978; Locke-Haydon and Chalmers, 1983; Cleveland and Snowdon, 1984). How much the female contributes seems related to whether siblings or other group members are present – if they are, her contribution diminishes. In contrast, the male's contribution to infant care seems to remain the same, regardless of the presence or absence of substitute caretakers (Epple, 1975; Cleveland and Snowdon, 1984).

In the *Saguinus* species adult males share food with infants (Izawa, 1978; Wolters, 1978; Cleveland and Snowdon, 1984), they groom infants (Rothe, 1975; Pook, 1978; Woodcock, 1983; Locke-Haydon and Chalmers, 1983), and they promote the infants' emerging independence (Dixon, 1983; Fragaszy et al., 1982; Locke-Haydon and Chalmers, 1983). There are even reports of some males, following their mate's parturition, licking off birth fluids to clean the infant when the female is too tired to do so (Epple and Katz, 1983; Stevenson, 1976). Males in most New World monogamous species are often used as secure bases to explore from and to alleviate distress (Ingram, 1977; Dixon, 1983; Cleveland and Snowdon, 1984). As in most other primate species the adult males are also active in protecting infants from other group members (Wolters, 1978; Epple and Katz, 1983) and from potential predators (Wolters, 1978).

The major source of variability in male caretaking activities among these species seems to be the timing of onset. In some species, males initiate infant care before the infant is dry (Stevenson, 1976; Fragaszy et al., 1982; Epple and Katz, 1983; Locke-Haydon and Chalmers, 1983), while in others, male care may not start for days or weeks (Hoage, 1977, 1982; Dixon and Fleming, 1981). Even within the same species considerable variation may occur in the timing of male initiation of infant caretaking (Pook, 1978; Hoage, 1977). Pook (1978) indicates that the point when male care first starts is directly related to when the female begins to first push the infant off herself. He found that the more time the infant spends off its mother, the more time it spends on the father.

This suggests a general principle for New World monkeys that is similar to the prosimian species in that the male contribution to infant care is a function of female restrictiveness. If females are restrictive, male care decreases; if they are not, male care increases. Stevenson (1976) compared groups of marmoset mothers along a restrictive, permissive, and rejecting scale and found that there was less male interest and care for the infant when the mother was restrictive than when the mother was permissive or rejecting. Fragaszy et al. (1982) suggests that this increase in male care may in part be infant induced – when the infant is near the mother she is more likely to use rejection and show high intolerance of it. Fathers, on the other hand, are more tolerant and accepting in these species and so, as the infant is rejected, it actively seeks out a more receptive source, that is, its father. Thus, it appears that males among

New World monogamous species show high levels of infant care, and this care is related to the males' prior experience and the females' permissive/restrictive nature.

Male care in other primate species

Non-monogamous New World monkeys, Old World monkeys and apes show a high degree of variability in the amount of species-normative interaction with infants and actual infant care. Among most species, the modal patterns of care typically take one of two forms: either a benign tolerance with otherwise general disinterest, or a pattern of limited but clearly significant parental care, usually complementary to that provided by the mother. Nevertheless, there is considerable diversity, not only within each of these different forms, but also between closely related species, and within identical species in different settings. Even within the same group some males just seem to 'like' babies, while others merely seem disinterested.

Males of many species show passive but benign disinterest, generally ignoring any infant contacts or solicitations. Included among these species are squirrel monkeys (*Saimiri sciureus*, Ploog, 1969; Vaitl, 1977), rhesus monkeys (*Macaca mulatta*, Southwick, Beg and Siddigi, 1965; Lindburg, 1971; Taylor et al., 1978), patas monkeys (*Erythrocebus patas*, Hall, 1966), Sykes monkeys (*Cercopithecus albogulans*, Rowell, 1974), vervets (*Cercopithecus aethiops*, Gartlan and Brain, 1968; Rowell, 1974 [others have found that vervets show low but significant rates, for example, Johnson et al., 1980]), talapoin monkeys (*Miopithecus talapoin*, Hill, 1966; Rowell, 1974), lion-tail macaques (*Macaca silenus*, Kumar and Kurup, 1981), chimpanzees (*Pan troglodytes*, van Lawick-Goodall, 1967; Horvat and Kraemer, 1977), orang-utans (Maple, 1980), gorillas (*Gorilla gorilla beringei*, Fossey, 1979), red colobus (*Colobus badius*, Struhsaker, 1975), langurs (*Presbytis entellus*, Jay, 1963; Hrdy, 1977; *Presbytis johnii*, Poirier, 1968), capuchin monkeys (*Cebus apella*, Izawa, 1980) and howler monkeys (*Alouatta palliata, Alouatta seniculus*, Baldwin and Baldwin, 1973). Some species initially show no interest in neonates, but demonstrate significant care as the infants mature and leave their mothers (for example, gibbons (*Hylobates lar*) and siamangs (*Symphalangus syndactylus*), Ellefson, 1974; Fox, 1972, and various colobus monkeys (Chalmers, 1968; Horwich and Manski, 1975).

Among these species males show a special tolerance for infants that they display toward no other class of conspecifics. Activities that would elicit rage if perpetrated by adults, adolescents, or even juveniles, may be ignored if they are performed by an infant (for example, see Itani, 1959; Jay, 1963; Bernstein, 1975; Ploog, 1969). Some of these activities include playing very close to or even on top of the adult male (Jay, 1963;

Alexander, 1970; Fox, 1972; Fossey, 1979), interrupting coitus (Gartlan and Brain, 1968) and even directly threatening the adult male (Bernstein, 1975; Ploog, 1969).

In a large number of species, males not only tolerate infants and their antics but also actively seek out interaction with infants. Most males in these species show significant levels of parental care and solicitude. Often this care includes carrying the infant, retrieving it, or even kissing, licking, and chattering to it. These males may occasionally be used as secure bases for infants and juveniles to explore from or to alleviate fear, and they may also groom infants and play with them. Species engaging in these activities include Barbary (*Macaca sylvana*, Lahiri and Southwick, 1966; Deag and Crook, 1971; Taub, 1980b), stumptail macaques (*Macaca arctoides*, Gouzoulez, 1975; Estrada and Sandoval, 1977; Hendy-Neely and Rhine, 1977), crab-eating macaques (*Macaca fascicularis*, Brandt, Irons and Mitchell, 1970), and Japanese macaques (*Macaca fuscata*, Itani, 1959). A number of baboon and closely related species are also included in this list (*Papio hamadryas*, Kummer, 1967; *Papio anubis*, Ransom and Ransom, 1971; Ransom, 1981; *Papio cynocephalus*, Rowell, 1974; Seyfarth, 1978; Altmann, 1980; *Theropithecus gelada*, Bernstein, 1975; Mori, 1979; *Cercocebus atys*, Chalmers, 1968; Bernstein, 1975). Also included are various colobus monkeys (Wooldridge, 1974; Horwich and Manski, 1975), howler monkeys (*Alouatta palliata*, Bolin, 1981), capuchins (*Cebus apella*, Izawa, 1980) and, under certain circumstances, adult silver-back male gorillas (Harcourt, 1978; Fossey, 1979).

The fact that many males participate in parental care within these species has been somewhat obscured by the fact that most studies of infant caretaking by non-human primates have been performed on four species: rhesus macaques, squirrel monkeys, chimpanzees and Hanuman langurs. These species are notorious for their relative lack of interest or outright antagonism toward infants. For example, in 900 hours of social interaction, Lindburg (1971) saw only two incidents of rhesus males carrying infants. Brandt and colleagues (1970) found when they compared rhesus monkey males with males of three other macaque species, rhesus males were the least interested in infants. In contrast, Gouzoules (1975) reported that within the stumptail macaque troop he observed, male care was almost as frequent as female care. Taub (1980b) reported that all the Barbary macaque males that he studied at some point demonstrated infant care and that the only instance of infant mortality occurred in an infant that failed to receive any male care. Among baboon species, males occasionally adopt and subsequently raise weaned infants whose mothers have died (Fossey, 1979; Ransom, 1981). However, in all these species male care seldom equals or exceeds female care.

Factors affecting adult male caretaking activities

There is substantial variability of male care within the same species and even within the same troop or social group. Itani (1959) found that although two groups of Japanese macaques might live in similar surroundings, one group might show substantial male care while males in the other group might show none. Furthermore in any one group, males that were highly interested in infants in one year might show no interest the next. Itani speculated that this might represent a cultural phenomenon rather than a genetic one. Similar 'cultural' differences have been found for vervets (Rowell, 1974; Gartlan and Brain, 1968; Johnson et al., 1980) and stumptail macaques (Bertrand, 1969; Gouzoules, 1975).

Not only is there widespread variation in male care across different groups and locales, but also there are dramatic individual differences within specific troops. Some males seem 'baby hungry', while others seem to be disinterested (for example, Itani, 1959; Ransom and Ransom, 1971; Wooldridge, 1971; Gouzoules, 1975; Bernstein, 1975, 1976; Taylor et al., 1978; Altmann, 1980; Bolin, 1981). Even among rhesus monkey males, which as a species are at the extreme of indifference, certain males have occasionally been noted to spend substantial time caring for an infant and even adopting it (Taylor et al., 1978).

The basis for individual differences among males in infant-directed activity is unknown at present. However, some clues do exist. Itani (1959) was able to demonstrate a correlation between certain personality traits, such as lack of aggression, and male care. Taylor et al. (1978) also identified a personality factor in the rhesus monkey who most frequently interacted with infants.

Another possible reason for the paucity of male care among many species is the relative lack of access to infants. As mentioned in an earlier section, among many species mothers are quite restrictive with their infants; they generally deny access to the infants by other individuals, including adult males. In other species the infant may not be available to the male because when the infant is off the mother, other females or siblings carry it (DuMond, 1968; Spencer-Booth, 1968; Lancaster, 1971; Caine and Mitchell, 1980), and often when the male approaches the infant, the mother quickly retrieves it. Hunt and colleagues (1978) found a negative correlation between overall protective behaviour by the mother and sub-adult male–infant care. Furthermore, there was a negative correlation between those classes of individuals towards which most aggression was shown when approaching the infant and the infant's level of interaction with those classes.

Other examples of males increasing their care when infants become 'available' come from cases in which mothers were 'careless' or actually died. Ransom and Ransom (1971) found that males increase their contacts with infants if mothers ignore or fail to retrieve infants at appropriate times. Fossey (1979) found that even previously disinterested adult silver-back gorilla males actively adopt and care for orphans that are abandoned. These findings support the theory that the lack of adult male care in some species may be related to how restrictive the females are of that species. However, it should be noted that other investigators have found that kinship is a better predictor than restrictiveness in who cares for an infant. Nevertheless, it may be that females are less restrictive with kin than with individuals who are not blood relatives (see Johnson et al., 1980). All in all, there is considerable support for the hypothesis that male care, or the lack of it, at least for many species, may be related to the level of restrictiveness by mothers (see Redican, 1976, for a detailed discussion of this hypothesis).

Several positive tests of this hypothesized correlation between male care and female restrictiveness have emerged. Vaitl (1977) observed male–infant interactions in squirrel monkey males, a species noted for its lack of male care, both with all females present, and with all of the females removed. When the females were removed, male care rapidly was transformed from nonexistent to huddling with each infant, sleeping with it and playing with it.

Redican's dissertation findings are also illuminating (Redican, 1976; Redican and Mitchell, 1973). After postulating that the lack of male care in rhesus monkeys was due to restrictiveness, not to a lack of ability or innate motivation, Redican decided to remove the mothers and let single caged males rear the infants. Not only were the males tolerant of the infants, but also, in time, they made good 'surrogate' mothers. These males exhibited high levels of carrying and grooming, even though the carrying was, at least initially, awkward and less frequent than their single caged, control group mother–infant dyads. However, the males demonstrated higher levels of play and aggression than did the control mothers. This is consistent with Suomi's (1977b) findings with nuclear family housed rhesus monkeys, where the quality of male care improved over time. It may be that the males' interactions with infants are significantly inhibited because they have no opportunities to interact and learn how to care for an infant, or to experience the associated pleasures that mothers seem to receive.

Other investigators have similarly found that male care increases when mothers cannot inhibit males from interacting with their infant. For example, Gibber (1981) – much to her surprise – found that when a neonatal infant and an adolescent male or female rhesus monkey were paired and placed as a dyad in a single cage, both sexes were equally interested in the neonatal infants. Furthermore, except for retrieval and social contact there were no sex difference in adolescent male or female treatment of the infant. However, if females had given birth and cared for an infant they were more likely to retrieve and care for an infant than were nulliparous females. Consistent with the hypothesis that male care is related to female restrictiveness, when males and females were paired together, even the nulliparous females that were paired with a male were quick to retrieve the infant and preclude the male from making contact. Others have also found minimal sex differences in the treatment of infants when females cannot intervene. Using a videotape of an infant in distress, Lande et al. (1985) found few sex differences in positive behaviours directed towards the infant. Thus, male care may be limited because females generally monopolize or restrict access to the infant.

However, female restrictiveness as the basis for the lack of male care may not apply to all species. Hrdy (as quoted in Redican, 1976) notes that langur mothers are not restrictive of their infant; resident males just do not seem interested in the infants.

As an alternative explanation of the between- and within-species variability in male care, some researchers have suggested that the degree of male care may be related to how often a male associates with a certain female. Among some Old World species, a male who forms a close and prolonged relationship with a female is more likely to interact and care for that female's infant than those who do not form close relationships (Altmann, 1980; Johnson et al., 1980). These findings, however, could also be explained as a mother showing less restrictiveness to individuals she 'trusts', or to the male merely being more available.

Increased male care for infants of mothers with a male consort suggests a role for male paternal certainty in infant care. Sociobiologists argue that the more certain a male is of his progeny, the more he should invest in their care. For example, the extremes in male care are generally found among those species that are most likely to have paternal certainty and uncertainty. The monogamous New World monkeys generally show the most male care; they are the most likely species to be certain that any infants born belong to them. On the other hand, multimale groups such as rhesus macaques show little infant care; they are the species that have the highest uncertainty as to paternity. Furthermore, Smith (1981) found that the infants with whom a rhesus male most frequently interacts are the infants most likely to have been sired by him. Suomi (1977b) also found that when rhesus males' paternity was artificially made certain by restricting mothers and fathers in a nuclear family caging arrangement, fathers directed more positive responses to their own offspring – and the fathers preferred their own infants over nonrelated ones (Suomi et al., 1973).

While a review of the literature indicates that possibly the male certainty hypothesis can account for a large part of the variance in explaining the degree of male care characteristic of a species, there are clearly many exceptions (for an in-depth review of this issue the interested reader is referred to Snowdon and Suomi, 1982). For example, gibbon and siamang males are monogamous, yet they ignore neonates.

Several authors have suggested that male care is a selfish activity in some species and has strong benefits for those males that engage in it. One frequently postulated benefit involves the phenomenon of *agonistic buffering*. Agonistic buffering has been reported in a number of species (for example, Barbary macaques, Deag and Crook, 1971; stumptail macaques, Gouzoules, 1975; Japanese macaques, Itani, 1959; baboons, Kummer, 1967; Ransom and Ransom, 1971; Packer, 1980). Agonistic buffering has been most thoroughly studied in Barbary macaques. Essentially, the phenomenon consists of a male picking up an infant, carrying it towards another male and holding it as the initiating male interacts with the other male. Both males might then groom the infant. Possession of the infant, in this case, is believed to enhance the carrier's status and inhibit aggression directed towards the carrier. Deag and Crook (1971) found that when lower ranking males were socially threatened, they would run to an infant, pick it up, carry it towards the more dominant male and present it to him. They also found that by carrying an infant, these low-ranking males clearly improved their status.

However, recent studies have called this interpretation into question, at least for Barbary macaques. Not all investigators found that agonistic buffering inhibits all aggression or that infants are never injured in the process, nor does aggression necessarily precede these episodes (Packer, 1980; Taub, 1980a; Smith and Peffer-Smith, 1982). In fact, Deag (1980) subsequently modified his stand, suggesting that the primary benefit may be increased social status and social facilitation. However, others have failed to replicate the findings supporting this interpretation (Smith and Peffer-Smith, 1982).

A number of authors have pointed out that agonistic buffering encounters differ in quality from species to species. In olive baboons the infant is not handed to the other male, nor does the infant seem to make a difference to the social facilitation; however, among gelada baboons, males carrying infants are less likely to receive aggression (Mori, 1979; Packer, 1980). Among gelada baboons, immigrant males may use agonistic buffering to form relationships and coalitions with adults in their new group (Packer, 1980). Finally, some authors have pointed out that the infants can benefit from these episodes also. They have the opportunity to acquire powerful allies, learn adult male roles and frequently be groomed (Packer, 1980).

ALLOPARENTAL CARE

Mothers and fathers are not the only conspecifics who show concern for primate infants. In many species siblings, unrelated sub-adults, older females and, in some species, even unrelated adult males display care-like behaviour toward infants. When such behaviour occurs in individuals other than the mother or father the term 'alloparental care' has been utilized throughout the primate literature. Alloparental care has been defined by Wilson as 'assistance by individuals other than the parents in the care of offspring' (Wilson, 1975, p. 578).

Alloparental care has been noted in both female and male non-human primates (for examples, see Hrdy, 1976; Quiatt, 1979; Snowdon and Suomi, 1982). However, among most species of primates, alloparental care is more likely to be displayed by females than by males (Altmann, 1980; Breuggeman, 1973; DuMond, 1968; Horwich and Manski, 1975; Hunt et al., 1978; Johnson et al., 1980; Lancaster, 1971; Poirier, 1968; Rosenblum, 1968; Scollay and DeBold, 1980; Spencer-Booth, 1968; Struhsaker, 1971; Wooldridge, 1971). New World monogamous species are an exception to this rule (see Snowdon and Suomi, 1982). For most species alloparental care is most frequently performed by sub-adult females (Horwich and Manski, 1975; Poirier, 1968; Struhsaker, 1971). When unrelated or related females, especially sub-adult females, engage in alloparental care, the term 'aunting' is typically used (Baldwin, 1969; Hrdy, 1977; Rowell, Hinde and Spencer-Booth, 1964; Spencer-Booth, 1968).

While the basis for this sex difference is unknown, there are some reasons to believe that the tendency to aunt by females may represent, at least in part, a biological predisposition. Females seem to be more interested in neonates than males from very early in life (Chamove, Harlow and Mitchell, 1967). Indeed some females may even show aunting to neonates while they are still infants themselves (Lancaster, 1971; Scollay and DeBold, 1980). Such sex differences in infant interest have also been observed in young preadolescent rhesus monkeys that have never seen or had experience with infants (Chamove *et al.*, 1967; Higley, Hopkins, Suomi, Hirsch and Marra, 1983). The young females' greater interest occurs across different environments, including feral (Altmann, 1980; Breuggeman, 1973; Hendy-Neely and Rhine, 1977; Lancaster, 1971; Neville, 1972; Ransom and Rowell, 1972) and laboratory or captive settings (Chamove et al., 1967; Horwich and Manski, 1975; Hunt et al., 1978; Scollay and DeBold, 1980). Females continue to show high levels of interest and attempt interactions despite rebuffs from the infant's mother (Breuggeman, 1973; Chism, 1978; Horwich and Manski 1975; Hunt et al., 1978; Poirier, 1968; van Lawick-Goodall, 1967) and high levels of resistance and protest from the infant being aunted (Chism, 1978; Gartlan, 1969; Horwich and Manski, 1975). Indeed in species where mothers are restrictive with their infant, such interest in infants may lead juvenile females to use subterfuge and distraction to get near or touch another female's infant (Chism, 1978; Gartlan, 1969; Hinde and Spencer-Booth, 1967b; Lancaster, 1971; Rowell, Din and Omar, 1968; Spencer-Booth, 1968).

The interest in infants by females is generally strongest during the infant's early neonatal period (Altmann, 1980; Horwich and Manski, 1975; Hrdy, 1977; Johnson et al., 1980; Rowell et al., 1964; Scollay and DeBold, 1980), but it usually declines as the infant matures (Horwich and Manski, 1975; Kurland, 1977). It is noteworthy that this decline in aunting by other females as the infant matures parallels the age-related decline in caretaking shown by the mother (Rosenblum, 1968; Scollay and DeBold, 1980).

The reason for the decline in interest in infants as they mature is largely unknown and to date has received surprisingly little direct research attention. One possible reason for this decline in interest may be due to changes in the infant's appearance. For example, in species where infants are born with characteristically coloured natal coats, the decline in interest in infants shown by nonmaternal females parallels the change in coat colour (Altmann, 1980; DeVore, 1963; Jay, 1963; Lancaster, 1971). However, it should be noted that not all species show qualitative coat changes as they develop. Some species, such as rhesus macaques, show coat changes, but those coat changes generally occur during the first weeks and do not parallel the temporal sequence for the preferential treatment and care the infant receives (Higley, Hopkins, Suomi, Hirsch and Marra, 1983). Thus the nature of such mechanisms can only be speculated about at this time.

Other infant factors also seem to make a difference in how attractive infants are to the females in a group. Infants born early in the year (Gartlan, 1969; Lancaster, 1971), to more dominant mothers (Lee, 1983; White and Hinde, 1975), and to previously nulliparous females (Lee, 1983), all seem to be more interesting to the aunts in a group than other infants. Certain features of potential aunts also seem to influence the appearance and intensity of alloparental care. Gartlan (1969) and Altmann (1980) both noted that older females generally show less interest in other neonates after their own infants are born. In addition, a number of studies have reported that some females in a social group have a personality style that seems to promote increased aunting. In other words, some females just seem to be more interested in infants, while other females just seem uninterested in infants (Altmann, 1980; Breuggeman, 1973; Lancaster, 1971). Furthermore, some females act calmly and in a relaxed manner when handling infants, while other females seem tense and anxious (Altmann, 1980; Jay, 1965).

Infants are usually not passive partners in the aunting process. They may resist being taken from their mother and may locomote back to their mother at the first opportunity (Chism, 1978; Horwich and Manski, 1975; Wooldridge, 1971). In other cases, they may transfer to another female freely (Baldwin, 1969; Rosenblum, 1969).

Species differences

While alloparental care is widespread across the primate order, there are major species differences in how often this phenomenon occurs. For example, it is especially prominent in the *Colobinae* subfamily of Old World monkeys (Horwich and Manski, 1975; Hrdy, 1976; Scollay and DeBold, 1980). There are reports that among some *Colobinae* groups, aunts may care for an infant more often than the infant's own biological mother (Horwich and Manski, 1975), especially during certain periods of the day (Scollay and DeBold, 1980). However, in Cercopithecinae Old World monkeys, aunting is generally far less common (for examples see McKenna, 1979; Rosenblum, 1971b; Simonds, 1965).

One plausible explanation for the differences seen between species in the frequency of aunting is that in some species mothers are very restrictive with their infants, not only to adult males but also to other females. For example, among Cercopithecines, mothers seem to be more likely to punish other females for attempts to interact with their infant, to withdraw from females interested in their infant and retrieve their infant when it strays from their grasp than are *Colobinae* mothers (for examples compare Horwich and Manski, 1975; Hrdy, 1976, 1977; Jay, 1963; Scollay and DeBold, 1980; Wooldridge, 1971, with Altmann, 1980; Lindburg, 1971; Rhine and Hendy-Neely, 1978; Rosenblum, 1971b; Simonds, 1965; Southwick, Beg and Siddiqi, 1965). Furthermore, among cercopithecines there are individual differences in the level of restrictiveness shown by mothers in a troop and infants belonging to mothers who are more restrictive typically receive less alloparental care than infants who belong to mothers who are less restrictive (Altmann, 1980; Gartlan, 1969; Lancaster, 1971; Kurland, 1977). In addition, among these more restrictive species, as the infant begins to leave its mother's grasp and becomes more accessible, aunting increases significantly (Rhine and Hendy-Neely, 1978). Thus one reason why aunting occurs more frequently in *Colobinae* species may be because these females generally have increased access to infants.

Nevertheless, the basis for species differences in mothers restricting access to their infants is largely unknown. Some investigators have noted that this permissive-restrictive dichotomy may be in part the result of low-ranking mothers' inability to regain their infant after a more dominant female obtains it (Rowell et al., 1964). Such a hypothesis would predict that subordinate mothers would be more restrictive of their infant, and that species without a rigid dominance hierarchy would be more likely to show alloparental care. Both of these predictions have received considerable empirical support (for example, Altmann, 1980; Horwich and Manski, 1975; Hrdy, 1976, 1977; Jay, 1965; Rowell et al., 1964; Spencer-Booth, 1968; Wooldridge, 1971). However, not all studies support this rule. For example, Zucker and Kaplan (1981) found that relative dominance did not indicate which females demonstrated aunting behaviour; nevertheless, it did predict which mothers were unable to regain their infants when the infants had been taken by another female. Moreover, there is evidence that some low-ranking mothers may restrict their infant from more dominant females for reasons other than an inability to re-obtain their infant. Dominant females may simply be rougher with a subordinate's infants, and those mothers, as a consequence, may protect their infants from rough treatment by keeping them close. Kurland (1977) found that dominant Japanese macaque females handle infants more roughly than subordinate females, and when he controlled for handling awkwardness, dominance as an indication of which females mothers allowed to aunt their infants disappeared.

A second difference between species in the phenomenon of alloparental behaviour involves the age of the aunts. In many species, as mentioned earlier, the aunts are usually subadults (for example, Japanese macaques – Kurland, 1977; bonnet macaques – Caine and Mitchell, 1980; vervets – Gartlan, 1969; Lancaster, 1971; Lee, 1983; rhesus monkeys – Brueggeman, 1973; Rowell et al., 1964; Spencer-Booth, 1968; baboons – DeVore, 1963; Ransom and Rowell, 1972; chimpanzees – van Lawick-Goodall, 1967). Among these species, the probability of aunting occurring increases dramatically during the prepubertal period and peaks in the season prior to the time that these females have their first infant (Brueggeman, 1973; Caine and Mitchell, 1980; Lee, 1983; Scollay and DeBold, 1980; Spencer-Booth, 1968). While in some species older females may also show alloparental care (for examples see DeVore, 1963; Gartlan, 1969; Jay, 1965; Kurland, 1977; Wooldridge, 1971), such behaviour in older females generally declines with age, and, conversely, the probability of female aggression to infants increases. However, in squirrel monkeys and patas monkeys the majority of aunting is limited to older, more experienced females (Baldwin, 1969; Chism, 1978; Hunt et al., 1978). The reason for this species difference is largely unknown; however, some have suggested that in these species experienced adult females handle infants in a more competent fashion (Jay, 1965; Baldwin, 1969) and as a consequence the mothers are less likely to restrict access to their infant with adult aunts than with maternally inexperienced sub-adults (Baldwin, 1969; Hunt et al., 1978). However, this suggestion does not explain why, in other species, females allow subadult females to aunt their infant since in these species juvenile females are also more likely to be awkward in handling infants. Indeed, in many species older experienced mothers may even show more rough handling than younger females (Hrdy, 1977) and punish infants who come too close (Hansen, 1966).

Explanations for the appearance of alloparental care

At present there are a number of hypotheses that attempt to explain both the proximate and ultimate causes of alloparental behaviour. Several recent reviews present competing arguments for different evolutionary roles for aunting (Hrdy, 1976, 1977; McKenna, 1979; Quaitt, 1979). These explanations include the learning of mothering skills (Baldwin, 1969; Hrdy, 1976; Lancaster, 1971), kin selection (McKenna, 1979), 'baby sitting' (Lancaster, 1971; Poirier, 1970) and replacement in case of maternal death (Horwich and Manski, 1975; Lancaster, 1971). Other less well-developed hypotheses include more rapid socialization for infants and increased troop social cohesion (Baldwin, 1969; Horwich and Manski, 1975; Jolly, 1972). One might note that many of these explanations are not mutually exclusive.

While an in-depth review of each of these explanations is beyond the scope of this chapter, it is apparent that explanations which adequately explain aunting for one species are inadequate or even wrong for another species (Hrdy, 1977; McKenna, 1979; Quiatt, 1979). We might also note that very few studies to date have experimentally manipulated any of the variables hypothesized to be important to the etiology of aunting. Amost all studies in the literature have been experiments of nature or field investigations that correlate observations with postulated hypotheses. Thus, we note that any postulated causes for the aunting phenomenon at this point seem preliminary at best. The interested reader is referred to the above reviews for more detailed discussions of possible causal factors and adaptive significance.

Other investigators have suggested that alloparental behaviour may not have been directly selected for by natural selection. Instead, it is argued, aunting may be a corollary result of nature selecting for females to show care for their own infants. Thus, it may be that aunting behaviour is no more than a consequence of the selective pressure for females to develop and maintain maternal behaviour. As Quiatt (1979) has suggested,

Allomaternal behavior may be no more than a fortuitous outcome of selection for a behavioral orientation toward young conspecifics that is essential to normal maternal performance in species in which early stages of growth and development are markedly prolonged (hardly a startling suggestion); but it appears to be an outcome, however fortuitous, that, once having occurred, serves as a conservative selective feature that works efficiently to maintain effective levels of maternal performance within populations. (Quiatt, 1979, p. 316)

CONCLUDING DISCUSSION

In this chapter we have extensively reviewed the primary literature concerning parental behaviour across the primate order. It has been argued from the outset that among most primates the species-normative patterns of caretaking encompass much more than direct satisfaction of the infant's basic biological needs and that caretaking activities typically persist long past the time of nutritional weaning. Yet, although there are obviously some features of normal caretaking behaviour shared by virtually all primate species (for example, nursing and initial provision of thermoregulatory support by the mother) there is also substantial variability in normative caretaking activities between different primate species, as well as within each species. We believe that from this huge literature, filled with both cross-species commonalities and striking species and individual differences, a number of basic principles concerning primate parental care can be gleaned. These principles come from different perspectives.

From a developmental perspective, a major lesson from the primate parental literature is that the demands on parents change dramatically during the period of their offsprings' physical and psychological dependency. For primates, this period is prolonged compared with most other mammals, affording greater opportunity for offspring to undergo ontogenic change in their capabilities as well as their needs, and for parents and interested others to devise effective ways to meet such needs even while adjusting to their offsprings' expanding capabilities. Thus, the changing needs of developing infants necessitate changes in caretaking activities and involvement on the part of parents and others. Certain factors that are crucial for the survival of neonates may well be irrelevant for juveniles, while parental support of juveniles in contests with peers has no counterpart in normal neonatal care. Moreover, specific caretaking responsibilities and the degree of involvement with infants may shift from one parent to another, or to others in the social unit, as those infants are growing up. In the vast majority of primate species, mothers bear the brunt of most early caretaking duties, but as their infants get older others in the group may 'take over' from the mothers in many ways, especially with respect to socialization processes.

Other principles concerning parental care in primates derive from a comparative perspective. Clearly there are major species differences in parental behaviour across the primate order, even as there are some features of parental behaviour that are common to all species. It is the actual pattern of cross-species similarities and differences that is of primary interest from a comparative perspective. At the risk of oversimplification, it can be argued that similarities between species are generally strongest for aspects of maternal caretaking, especially in the earliest days of the infant's life, whereas the greatest differences between species are apparent in the caretaking activities of individuals other than the mother, especially when the infant is older (see Suomi, 1979b). Thus, most differences in maternal care tend to be of a basically quantitative nature (for example, more/less restrictiveness, high/low rates of rejection during weaning, and so on), with relatively few qualitative differences (for example, ventral or dorsal mode of carrying infants). In contrast, the nature of caretaking by adult males is enormously variable across the primate order, not only in the degree of involvement (for example, marmosets compared with rhesus monkeys), but also whether the care is basically substitutive or complementary in nature (Parke and Suomi, 1981), and during which periods of infant development it tends to be the most prominent. Similar argument can be made with respect to alloparental behaviour by unrelated juvenile and adult females.

The specific patterns of cross-species similarities and differences in infant care provide fertile grounds for generating hypotheses concerning selective factors that influence the appearance of such caretaking activities. At the very least, one might expect such selective factors to be different for the various potential caretakers, for different ages of infants and for different physical and social settings in which particular patterns of caretaking activities have evolved. Why, for example, should rhesus and pigtail macaque mothers be so restrictive in granting access to their infants when bonnet macaque mothers pass their infants to other adult females? Why is paternal involvement in infant care so much more predominant in monogamous New World monkeys than in those New World species in which life-long pair bonds are not maintained between the sexes? Consideration of the patterns of cross-species similarities and differences in primate care can also be used to test hypotheses examining the relationship between patterns of parental care and species-normative social structure and/or habitat selection.

Finally, the present review should make it obvious that proximate factors affecting parental care are far from being fully understood at this time. Most of the studies to date, especially those conducted in feral environments, are basically descriptive in nature. Very little is known about the actual mechanisms underlying the induction, facilitation and maintenance of parental behaviour. Both the developmental and the comparative perspectives suggest that such mechanisms are not likely to be identical for all caretakers, or across all species, or even for different stages of infant development. Clearly, much work remains to be done before our understanding of parental behaviour across the primate order matches our knowledge at the descriptive level.

REFERENCES

Alexander, B. K. 1970: Parental behavior of adult male Japanese monkeys. *Behaviour*, 36, 270–84.

Alley, T. R. 1980: Infantile colouration as an elicitor of caretaking behavior in old world primates. *Primates*, 21, 416–29.

Altmann, J. 1980: *Baboon Mothers and Infants*. Cambridge, Massachusetts: Harvard University Press.

Baldwin, J. and Baldwin, J. 1973: Interactions between adult female and infant howling monkeys (*Alouatta palliata*). *Folia Primatologica*, 20, 27–71.

Baldwin, J. 1969: The ontogeny of social behavior of squirrel monkeys (*Saimiri sciureus*) in a seminatural environment. *Folia Primatologica*, 11, 35–79.
Bell, R. and Harper, L. 1977: *Child Effects on Adults*. New York: Halsted Press.
Berkson, G. 1967: Development of an infant in a captive gibbon group. *The Journal of Genetic Psychology*, 108, 311–25.
Berman, C. 1980a: Mother–infant relationships among free-ranging rhesus monkeys Cayo Santiago: A comparison with captive pairs. *Animal Behaviour*, 28, 860–73.
—— 1980b: Early agonistic experience and rank acquisition among free-ranging infant rhesus monkeys. *International Journal of Primatology*, 1, 153–70.
Bernstein, I. S. 1975: Activity patterns in a gelada monkey group. *Folia Primatologica*, 23, 50–71.
—— 1976: Activity patterns in a sooty mangabey group. *Folia Primatologica*, 26, 185–206.
Bertrand, M. 1969: *The Behavioral Repertoire of the Stumptail Macaque*. New York: S. Karger.
Bo, W. 1971: Parturition. In E. Hafez (ed.), *Comparative Reproduction of Non-human Primates*, Springfield: C. Thomas, 302–14.
Bolin, I. 1981: Male parental behavior in black howler monkeys (*Alouatta Palliata pigra*) in Belize and Guatemala. *Primates*, 22, 349–60.
Bolwig, N. 1980: Early social development and emancipation of *Macaca nemestrina* and species of *Papio*. *Primates*, 21, 357–75.
Booth, C. 1962: Some observations on behavior of cercopithecus monkeys *Annals of the New York Academy of Sciences*, 102, 477–87.
Boothe, R., Kiorpes, L., Regal, D. and Lee, C. 1982: Development of visual responsiveness in *Macaca nemestrina* monkeys. *Developmental Psychology*, 18, 665–70.
Bowlby, J. 1969: Attachment and loss. *Attachment* (vol. 1). New York: Basic Books.
Box, H. 1975: A social development study of young monkeys (*Callithrix jacchus*) within a captive family groups. *Primates*, 16, 419–35.
—— 1977a: Quantitative data on the carrying of young captive monkeys (*Callithrix jacchus*) by other members of their family groups. *Primates*, 18, 475–84.
—— 1977b. Social interactions in family groups marmosets (*Callithrix jacchus*). In D. G. Kleiman (ed.), *The Biology and Conservation of the Callithrichidae*, Washington, DC: Smithsonian Institution Press, 239–49.
Bramblett, C., Bramblett, S., Bishop, D. and Coelho, A. Jr. 1982: Longitudinal stability in adult status hierarchies among vervet monkeys (*Cercopithecus aethiops*). *American Journal of Primatology*, 2, 43–51.
Brandt, E. Irons, R. and Mitchell, G. 1970: Paternalistic behavior in four species of macaques. *Brain Behaviour and Evolution*, 3, 415–20.
Brandt, E. and Mitchell, G. 1971: Parturition in primates: Behavior related to birth. In L. Rosenblum (ed.), *Primate Behavior Developments in Field and Laboratory Research*, New York: Academic Press, 177–223.
—— 1973: Labor and delivery behavior in rhesus monkeys (*Macaca mulatta*). *American Journal of Physiological Anthropology*, 38, 519–22.
Breuggeman, J. 1973: Parental care in a group of free-ranging rhesus monkeys (Macaca mulatta). *Folia Primatologica*, 20, 178–210.

Bucher, K., Myers, R. E. and Southwick, C. 1970: Anterior temporal cortex and maternal behavior in monkeys. *Neurology*, 20, 415.

Buettner-Janusch, J. 1964: The breeding of galagos in captivity and some notes on their behavior. *Folia Primatologica*, 2, 93–110.

Burton, F. 1972: The integration of biology and behavior in the socialization of *Macaca sylvana* of Gibraltar. In F. Poirier (ed.), *Primate Socialization*, New York: Random House.

Caine, N. and Mitchell, G. 1979: The relationship between maternal rank and companion choice in immature macaques (*Macaca mulatta* and *Macaca radiata*). *Primates*, 20, 583–90.

—— 1980: Species differences in the interest shown in infants by juvenile female macaques (*Macaca radiata* and *Macaca mulatta*). *International Journal of Primatology*, 1, 323–32.

Cairns, R. 1972: Attachment and dependency: A psychobiological and social learning synthesis. In J. L. Gewirtz (ed.), *Attachment and Dependency*, Washington, DC: W. H. Winston and Sons.

Candland, D. and Mason, W. 1968: Infant monkey heartrate: Habituation and the effects of social substitutes. *Developmental Psychology*, 1, 254–6.

Carpenter, C. R. 1965: The howlers of Barro Colorado Island. In I. DeVore (ed.), *Primate Behavior: Field studies of monkeys and apes*, New York: Holt, Rinehart and Winston.

Chalmers, N. B. 1968: The social behavior of free living mangabeys in Uganda. *Folia Primatologica*, 8, 263–81.

—— 1973: Arboreal and terrestrial monkeys. In R. Michael and J. Crook, (eds), *Comparative Ethology and Behavior of Primates*, New York: Academic Press.

Chamove, A., Harlow, H. and Mitchell, G. 1967: Sex differences in the infant-directed behavior of preadolescent rhesus monkeys. *Child Development*, 38, 329–35.

Champoux, M. 1983: The effects of controllability on behavior, development, and responses to stress in infant rhesus monkeys (*Macaca mulatta*). Unpublished master's thesis, University of Wisconsin.

Charles-Dominique, P. 1977: *Ecology and Behaviour of Nocturnal Primates: Prosimians of Equatorial West Africa*. New York: Columbia University Press.

Charles-Dominique, P. and Petter, J. J. 1980: Ecology and social life of *Phaner furcifer*. In P. Charles-Dominique, H. M. Cooper, A. Hladik, C. M. Hladik, E. Pages, G. F. Pariente, A. Petter-Rousseaux, A. Schilling and J. J. Petter (eds), *Nocturnal Malagasy Primates: Ecology, physiology and behavior*, New York: Academic Press.

Cheney, D. 1977: The acquisition of rank and the development of reciprocal alliances among free-ranging immature baboons. *Behavioral Ecology and Sociobiology*, 2, 303–18.

Cheney, D. and Seyfarth, R. 1980: Vocal recognition in free-ranging vervet monkeys. *Animal Behaviour*, 28, 362–7.

Chevalier-Skolnikoff, S. 1973: Visual and tactile communication in *Macaca arctoides* and its ontogenetic development. *American Journal of Physical Anthropology*, 38, 515–18.

Chikazawa, D., Gordon, T., Bean, C. and Bernstein, I. 1979: Mother–daughter dominance reversal in rhesus monkeys (*Macaca mulatta*). *Primates*, 20, 301–5.

Chism, J. 1978: Relationships between patas infants and group members other than the mother. In D. J. Chivers and J. Herbert (eds), *Recent Advances in Primatology*, New York: Academic Press.

Cleveland, J. and Snowdon, C. T. 1984: Social development during the first twenty weeks in the Cotton-top Tamarin (*Saguinus o. oedipus*). *Animal Behaviour*, 32, 432-44.

Cross, H. and Harlow, H. 1963: Observation of infant monkeys by female monkeys. *Perceptual and Motor Skills*, 16, 11-15.

Davis, D., Fouts, R. and Hannum, M. 1981: The maternal behavior of a home-reared, language using chimpanzee. *Primates*, 22, 570-73.

Deag, J. 1980. Interactions between males and unweaned Barbary macaques: Testing the agonistic buffering hypothesis. *Behaviour*, 75, 54-81.

Deag, J. and Cook, J. 1971. Social behavior and 'agonistic buffering' in the wild Barbary macaque (*Macaca sylvana*). *Folia Primatologica*, 15, 183-200.

DeVore, I. 1963: Mother-infant relations in free-ranging baboons. In H. Rheingold (ed.), *Maternal behavior in Mammals*, New York: John Wiley and Sons.

Dixson, A. 1983: The Owl monkey (*Aotus trivirgatus*). In J. Hearn (eds.), *Reproduction in New World Primates*, Boston: MTP Press Limited.

Dixson, A. and Fleming, D. 1981: Parental behaviour and infant development in Owl monkeys (*Aotus trivirgatus griseimembra*). Journal of Zoology, 194, 25-39.

Dixson, A. F. and George, L. 1982: Prolactin and parental behavior in a male new world primate. *Nature Lond.*, 299, 551-3.

Dolhinow, P. 1972: *Primate Patterns*. New York: Holt, Rinehart and Winston.

Doyle, G. A. 1974: Behavior of prosimians. In A. M. Schrier and F. Stollnitz (eds), *Behavior of Nonhuman Primates*, vol. 5, New York: Academic Press.

Doyle, G. A. 1979: Development of behavior in prosimians with special reference to the lesser bushbaby, *Galago senegalensis moholi*. In G. A. Doyle and R. D. Martin (eds), *The Study of Prosimian Behavior*, New York: Academic Press.

Doyle, G., Pelletier, A. and Bekker, T. 1967: Courtship, mating and parturition in the lesser bushbaby (*Galago senegalensis moholi*) under semi-natural conditions. *Folia Primatologica*, 7, 169-97.

Doyle, G., Andersson, A. and Bearder, S. 1971: Maternal behavior in the lesser bushbaby (*Galago senegalensis moholi*) under semi-natural conditions. *Folia Primatologica*, 11, 215-38.

DuMond, F. 1968: The squirrel monkey in a seminatural environment. In L. Rosenblum and R. Cooper (eds), *The Squirrel Monkey*, London: Academic Press.

Eaglen, R. and Boskoff, K. 1978: The birth and early development of a captive sifaka, *Propithecus verreauxi coquereli*. *Folia Primatologica*, 30, 206-19.

Ehrlich, A. 1974: Infant development in two prosimian species: Greater galago and slow loris. *Developmental Psychology*, 7, 439-54.

Ellefson, J. 1974: A natural history of white-handed gibbons in the Malayan Penninsula. In D. M. Rumbaugh (ed.), *Gibbon and Siamang*, London: S. Karger.

Epple, G. 1975: Parental behavior in *Saguinus fuscicollis* ssp. (Callitrichidae). *Folia Primatologica*, 24, 221-38.

—— 1978: Reproductive and social behavior of marmosets with special reference to captive breeding. *Primates in Medicine*, 10, 50-62.

Epple, G. and Katz, Y. 1983: The saddle back tamarin and other tamarins. In J. Hearn (ed.), *Reproduction in New World Primates*, Boston: MTP Press Limited.
Estrada, A. and Sandoval, J. 1977: Social relations in a free-ranging troop of stumptail macaques (*Macaca arctoides*): Male-care behaviour 1. *Primates*, 18, 793-813.
Ewer, R. 1968: *Ethology of Mammals*. New York: Plenum Press.
Fedigan, L. 1982: *Primate Paradigms: Sex roles and social bonds*. Canada: Eden Press Inc.
Fisher, L. 1972: The birth of a lowland gorilla at Lincoln Park Zoo, Chicago *International Zoo Yearbook*, 12, 106-8.
Fobes, J. and King, J. E. (eds) 1982: *Primate Behavior*. New York: Academic Press.
Fossey, D. 1979: Development of the mountain gorilla (*Gorilla gorilla beringei*): The first thirty-six months. In D. A. Hamburg and E. B. McCown (eds), *The Great Apes*, Menlo Park, California: Benjamin/Cummings.
Fox, G. 1972: Some comparisons between siamang and gibbon behavior. *Folia Primatologica*, 18, 122-39.
Fragaszy, D., Schwarz, S. and Shimosaka, D. 1982: Longitudinal observations of care and development of infant Titi monkeys (*Callicebus Moloch*). *American Journal of Primatology*, 2, 191-200.
Franzen, E. and Myers, R. 1973: Neural control of social behavior: Prefrontal and anterior temporal cortex. *Neuropsychologia*, 11, 141-57.
Gartlan, J. 1969: Sexual and maternal behavior of the vervet monkey, *Cercopithecus aethiops*. *Journal of Reproductive Fertility – Supplement*, 6, 137-50.
Gartlan, J. S. and Brain, C. K. 1968: Ecology and social variability in *Cercopithecus aethiops and C. mitis*. In P. C. Jay (ed.), *Primates: Studies in adaptation and variability*, New York: Holt, Rinehart and Winston.
Gibber, J. 1981: Infant-directed behaviors in male and female rhesus monkeys. Doctoral dissertation, University of Wisconsin, Madison.
Goldman, P. S. 1976: Maturation of the mammalian nervous system and the ontogeny of behavior. *Advances in the Study of Behavior*, 2, 1-90.
Goodall, G. and Athumani, J. 1980: An observed birth in a free-living chimpanzee (*Pan troglodytes schweinfurthii*) in Gombe National Park, Tanzania. *Primates*, 21, 545-9.
Gouzoules, H. 1975: Maternal rank and early social interaction of infant stumptail macaques (*Macaca arctoides*). *Primates*, 16, 405-18.
Goy, R. 1966: Role of androgens in the establishment and regulation of behavioural sex differences in mammals. *Journal of Animal Science*, 25, 21-35.
Green, K. 1981: Preliminary observations on the ecology and behavior of the capped langur, *Presbytis pileatus*, in the Nadhupur Forest of Bangladesh. *International Journal of Primatology*, 2, 131-54.
Gross, R., Schiller, H. and Bowden, D. 1977: Post partum serum cortisol and prolactin and their relationship with maternal and affective behaviors in the *Macaca nemestrina*. Paper presented at the American Society of Primatologists Meeting, Seattle, Washington.
Gubernick, D. J. and Klopfer, P. H. 1981: *Parental Care in Mammals*. New York: Plenum Press.
Hall, K. R. L. 1965: Behavior and ecology of the wild patas monkey, *Erythrocebus patas* in Uganda. *Journal of Zoology*, 148, 15-87.

—— 1968: Behavior and ecology of the wild patas monkey (*Erythrocebus patus*) in Uganda. In P. C. Jay (ed.), *Primates: Studies in adaptation and variability*, New York: Holt, Rinehart and Winston.

Hall, K. and DeVore, I. 1965: Baboon social behavior. In I. DeVore (ed.), *Primate Behavior: Field studies of monkeys and apes*, New York: Holt, Rinehart and Winston.

Hamptom, J., Hamptom, S. and Landwehr, B. 1966: Observations on a successful breeding colony of the marmoset, *Oedipomidas oedipus*. *Folia Primatologica*, 4, 265-87.

Hansen, E. 1966: The development of maternal and infant behavior in the rhesus monkey. *Behaviour*, 27, 107-49.

Harcourt, A. 1978: Activity periods and patterns of social interaction: A neglected problem. *Behaviour*, 66, 121-35.

Harlow, H. F. 1958: The nature of love. *American Psychologist*, 13, 673-85.

—— 1959: The development of learning in the rhesus monkey. *American Scientist*, 47, 459-79.

—— 1963: The maternal affectional system. In B. M. Foss (ed.), *Determinants of Infant Behavior* (vol. 2), London: Methuen Ltd.

—— 1969: Age-mate or peer affectional system. In D. Lehrman, R. Hinde and E. Shaw (eds), *Advances in the Study of Behavior* (vol. 2), New York: Academic Press.

Harlow, H. F. and Harlow, M. K. 1965: The affectionals systems. In A. M. Schrier, H. F. Harlow and F. Stollnitz (eds), *Behavior of nonhuman primates*, II, New York: Academic Press.

—— 1971: Psychopathology in monkeys. In H. D. Kimmel (ed.), *Experimental psychopathology*, New York: Academic Press.

Harlow, H. F., Harlow, M. K. and Hansen, E. W. 1963: The maternal affectional-system of rhesus monkeys. In H. L. Rheingold (ed.), *Maternal Behavior in Mammals*, New York: Wiley.

Harlow, H. F. and Suomi, S. J. 1970: The nature of love – simplified. *American Psychologist*, 25, 161-8.

Harrington, J. 1978: Development of behavior in *Lemur macaco* in the first nineteen weeks. *Folio Primatologica*, 29, 107-28.

Hartman, C. 1928: The period of gestation in the monkey, macacus rhesus, first description of parturition in monkeys, size, and behavior of the young. *Journal of Mammalia*, 9, 181-94.

Heltne, P. G., Turner, D. C., Wolhandler, J. 1973: Maternal and paternal periods in the development of infant *Callimico Goeldi*. *American Journal of Physical Anthropology*, 38, 555-60.

Heltne, P., Wojcik, J. and Pook, A. 1981: Goeldi's monkey, genus Callimico. In A. Coimbra-Filho and R. Mittermeier (eds), *Ecology and Behavior of Neotropical Primates*, Rio de Janeiro: Acadmia Brasileira de Ciencias.

Hendy-Neely, H. and Rhine, R. J. 1977: Social development of stumptail macaques (*Macaca arctoides*): Momentary touching and other interactions with adult males during the infants' first 60 days of life. *Primates*, 18, 589-600.

Higley, J. D., Hopkins, W. D., Suomi, S. J., Hirsch, R. and Marra, L. 1983: Infantile coloration as a possible maternal elicitor in rhesus monkeys. Paper presented at the meetings of the American Society of Primatologists, East Lansing, MI.

Hill, W. C. O. 1966: Laboratory breeding, behavioural development and relations of the talapoin (*Miopithecus talapoin*). *Extrait De Mammalia*, 30, 353-70.

Hinde, R. 1975: Mother's and infant's roles: Distinguishing the questions to be asked. *Ciba Foundation Symposium*, 33. Amsterdam: Elsevier.

Hinde, R. and Atkinson, S. 1970: Assessing the roles of social partners in maintaining mutual proximity, as exemplified by mother-infant relations in rhesus monkeys. *Animal Behaviour*, 18, 169-76.

Hinde, R. and Spencer-Booth, Y. 1967a: The behavior of socially living rhesus monkeys in their first two and a half years. *Animal Behaviour*, 15, 169-96.

—— 1967b: The effect of social companions on mother-infant relations in rhesus monkeys. In D. Morris (ed.), *Primate Ethology*, New York: Aldine Publishing Co.

Hinde, R. and White, L. 1974: Dynamics of a relationship: Rhesus mother-infant ventro-ventral contact. *Journal of Comparative and Physiological Psychology*, 86, 8-23.

Hines, M. 1942: The development and regression of reflexes, postures, and progression in the young macaque. *Contributions to Embryology*, Washington, DC: Carnegie Institute, 30, 153-209.

Hoage, R, 1977: Parental care in *Leontopithecus rosalia rosalia*: Sex and age differences in carrying behavior and the role of prior experience. In D. Kleiman (ed.), *Biology and Conservation of the Callitrichidae*, Washington, DC: Smithsonian Institution Press.

—— 1982: Social and physical maturation in captive lion tamarins, *Leontopithecus rosalia rosalia* (Primates: Callitrichidae) *Smithsonian Contributions of Zoology*, V, 354.

Hoff, M., Nadler, R. and Maple, T. 1981: Development of infant independence in a captive group of lowland gorillas. *Developmental Psychobiology*, 14, 251-65.

Hopf, S. 1967: Notes on pregnancy, delivery and infant survival in captivity. *Primates*, 8, 323-32.

Horr, D. 1977: Orang-utan maturation: Growing up in a female world. In S. Chevalier-Skolnikoff and F. Poirier (eds), *Primate Bio-social Development: Biological, social and ecological determinants*, New York: Garland Publishing.

Horvat, J. R. and Kraemer, H. C. 1981: Infant socialization and maternal influence in chimpanzees. *Folia Primatologica*, 36, 99-110.

Horvat, J. R. and Kraemer, H. C. 1982: Behavioral changes during weaning in captive chimpanzees. *Primates*, 23, 488-99.

Horwich, R. and Manski, D. 1975: Maternal care and infant transfer of two species of colobus monkeys. *Primates*, 16, 49-73.

Hrdy, S. 1976: Care and exploitation of nonhuman primate infants by conspecifics other than mother. In J. S. Rosenblatt, R. A. Hinde, E. Shaw, and C. Beer (eds), *Advances in the Study of Behavior*, New York: Academic Press.

—— 1977: The langurs of Abu: Female and male strategies of reproduction. Cambridge, Massachusetts: Harvard University Press.

Hunt, S., Gamache, K and Lockard, J. 1978: Babysitting behavior by age/sex classification in squirrel monkeys (*Saimiri scuireus*). *Primates*, 19, 179-86.

Ingram, J. 1977: Parent-infant interaction in the common marmoset (*Callithrix*

jacchus). In D. G. Kleiman (ed.), *The biology and conservation of the Callitrichidae*, Washington, DC: Smithsonian Institution Press.
—— 1978a: Infant socialization within common marmoset family groups. In H. Rothe, H. J. Wolters and J. P. Hearn (eds), *Biology and Behavior of Marmosets*, Gottingen, West Germany: Eigenrelag Hartmut Rothe.
—— 1978b. Preliminary comparisons of parental care of wild-caught and captive born common marmosets. In H. Rothe, H. J. Wolters and J. P. Hearn (eds), *Biology and Behavior of Marmosets*, Gottingen, West Germany: Eigenrelag Hartmut Rothe.
Itani, J. 1959: Paternal care in the wild Japanese monkey (*Macaca fuscata fuscata*). *Primates*, 2, 61–285.
Izawa, K. 1978: A field study of the ecology and behavior of the black mantle tamarin (*Saguinus nigricollis*). *Primates*, 19, 241–74.
—— 1980: Social behavior of the wild black-capped capuchin (*Cebus apella*). *Primates*, 21, 443–67.
Jay, P. 1962: Aspects of maternal behavior among langurs. *Annals of the New York Academy of Sciences*, 102, 468–76.
—— 1963: Mother–infant relations in langurs. In H. Rheingold (ed.), *Maternal Behavior in Mammals*, New York: John Wiley and Sons.
—— 1965: The common langur of North India. In I. DeVore (ed.), *Primate Behavior: Field studies of monkeys and apes*, New York: Holt, Rinehart and Winston.
Jensen, G. D., Bobbitt, R. A. and Gordon, B. N. 1968: Effects of environment on the relationship between mother and infant pigtailed monkeys (*Macaca nemestrina*). *Journal of Comparative and Physiological Psychology*, 66, 259–63.
—— 1969: Patterns and sequences of hitting behavior in mother and infant monkeys (*Macaca nemestrina*). *Journal of Psychiatric Research*, 7, 55–67.
—— 1971: Mother and infant roles in the development of independence of *Macaca nemestrina*. In C. Carpenter (ed.), *Behavioral Regulators of Behavior in Primates*, Lewisburg, Pennsylvania: Bucknell University Press.
—— 1973: Mother's and infant's roles in the development of independence of *Macaca nemestrina*. *Primates*, 14, 79–88.
Johnson, C., Koemer, C., Estrin, M. and Duoos, D. 1980: Alloparental care and kinship in captive social groups of vervet monkeys (*Cercopithecus aethiops sabaeus*). *Primates*, 21, 406–15.
Jolly, A. 1972: *The Evolution of Primate Behavior*. New York: Macmillan.
—— 1966: *Lemur Behavior: A Madagascar field study*. Chicago: University of Chicago Press.
Kaplan, J. 1973: Responses of mother squirrel monkeys to dead infants. *Primates*, 14, 89–91.
—— 1981: Effects of surrogate-administered punishment on surrogate contact in infant squirrel monkeys. *Developmental Psychobiology*, 14, 523–32.
Kaplan, J., Winship-Ball, A. and Sim, L. 1978: Maternal discrimination of infant vocalizations in squirrel monkeys. *Primates*, 19, 187–93.
Kaufmann, J. 1966: Behavior of infant rhesus monkeys and their mothers in a free-ranging band. *Zoologica: New York Zoological Society*, 51, 11–27.

Kaufman, I. and Rosenblum, L. 1969: The waning of the mother–infant bond in two species of macaques. In B. Foss (ed.), *Determinants of Infant Behavior*, vol. 4, London: Methuen.
Kaufman, I. 1974: Mother–infant relations in monkeys and humans: A reply to Professor Hinde. In N. White (ed.), *Ethology and Psychiatry*, Toronto: University of Toronto Press.
Kawai, M. 1965: On the system of social ranks in a natural troop of Japanese monkeys. *Primates*, 1, 111–30.
Kawamura, S. 1958: The process of sub-culture propagation among Japanese macaques. *Primates*, 2, 43–54.
Klaus, M. H. and Kennel, J. H. 1982: *Parent–infant Bonding*. St. Louis: Mosby.
Kling, A. and Steklis, H. 1976: A neural substrate for affiliative behavior in nonhuman primates. *Brain, Behavior and Evolution*, 13, 216–38.
Klopfer, P. H. 1970: Discrimination of young in galagos. *Folia Primatologica*, 13, 137–45.
—— 1972: Patterns of maternal care in lemurs: II. Effects of group size and early separation. *Zeitschrift für Tierpsychologie*, 30, 277–96.
Klopfer, P. and Boskoff, K. 1979: Maternal behavior in prosimians. In G. Doyle and R. Martin (eds), *The study of prosimian behavior*, New York: Academic Press.
Klopfer, P. and Dugard, J. 1976: Patterns of maternal care in lemurs. III. *Lemur variegatus. Zietschrift für Tierpsychologie*, 40, 210–20.
Korner, A. and Thoman, E. 1972: The efficacy of contact and vestibular-proprioceptive stimulation in soothing neonates. *Child Development*, 43, 443–53.
Koyama, N. 1967: On dominance rank and kinship of a wild Japanese monkey troop in Arashiyama. *Primates*, 8, 189–216.
Kumar, A. and Kurup, G. U. 1981: Infant development in the lion-tailed macaque, *Macaca silenus* (Linnaeus): The first eight weeks. *Primates*, 22, 512–22.
Kummer, H. 1967: Tripartite relations in Hamadryas baboons. In S. Altman (ed.) *Social Communication among Primates*, Chicago: University of Chicago Press.
Kurland, J. 1977: Kin selection in the Japanese monkey. In F. S. Szalay (ed.), *Contributions to Primatology*, New York: S. Karger.
Kuroda, S. 1980: Social behavior the pygmy chimpanzees. *Primates*, 21, 181–97.
Lahiri, R. and Southwick, C. 1966: Parental care in *Macaca sylvana. Folia Primatologica*, 4, 257–64.
Lancaster, J. 1971: Play-mothering: The relations between juvenile females and young infants among free-ranging vervet monkeys (*Cercopithecus aethiops*). *Folia Primatologica*, 15, 161–82.
Lande, J. S., Higley, J. D., Snowdon, C. T., Goy, R. W. and Suomi, S. J. 1985: Elicitors of parental care in rhesus monkeys. Paper presented at the Annual Meeting of the American Society of Primatologists, Niagara Falls, New York.
Lee, P. 1983: Caretaking of infants and mother–infant relationships. In R. Hinde (ed.), *Primate Social Relationships*, Sunderland, Massachusetts: Sinauer Associates, Inc.
Lewis, M. and Rosenblum, L. 1974: *The Origins of Behavior: The effect of the infant on its caregiver*. New York: John Wiley and Sons.

Lindburg, D. 1971: The rhesus monkey in North India: An ecological and behavioral study. In L. A. Rosenblum (ed.), *Primate Behavior: Developments in field and laboratory research*, New York: Academic Press.

—— 1973: Grooming behavior as a regulator of social interactions in rhesus monkeys. In C. Carpenter (ed.), *Behavioral Regulators of Behavior in Primates*, Lewisburg, Pennsylvania: Bucknell University Press.

Locke-Haydon, J. and Chalmers, N. 1983: The development of infant–caregiver relationships in captive common marmosets (Callithrix jacchus). *International Journal of Primatology*, 4, 63–81.

Lorenz, R. 1972: Management and reproduction of the Goeldi's monkey *Callimico goeldii* (Thomas, 1904) Callimiconidae, primates. In D. D. Bridgewater (ed.), *Saving the Lion Marmoset*, West St. Paul, Minnesota: Viking Services.

Loy, J. and Loy, K. 1974: Behavior of an all-juvenile group of rhesus monkeys. *American Journal of Physical Anthropology*, 40, 83–96.

Luckett, W. 1980: *Comparative Biology and Evolutionary Relationships of Tree Shrews*. New York: Plenum Press.

Lundblad, E. and Hogden, G. 1980: Induction of maternal–infant bonding in rhesus and cynomolgus monkeys after cesarean delivery. *Laboratory Animal Science*, 30, 913.

MacKinnon, J. 1974: The behaviour and ecology of wild orang-utans (*Pongo pygmaeus*). *Animal Behaviour*, 22, 3–74.

Maple, T. L. 1980: *Orang-utan Behavior*. New York: Van Nostrand Reinhold Co.

—— 1982: Orang-utan behavior and its management in captivity. In L. E. M. DeBoer (ed.), *The Orang-utan: Its biology and conservation*, Boston: Dr W. Junk Publishers.

Maple, T. L. and Hoff, M. P. 1982: *Gorilla Behavior*. New York: Van Nostrand Reinhold Co.

Martin, R. D. 1966: Tree shrews: Unique reproductive mechanism of systematic importance. *Science*, 152, 1402–4.

—— 1972: A preliminary field-study of the lesser mouse lemur (*Microcebus murinus*, J. F. Miller). *Zeitschrift für Tierpsychologie*, 9, 43–89.

Mason, W. A. 1967: Motivational aspects of social responsiveness in young chimpanzees. In H. W. Stevenson, E. H. Hess and H. L. Rheingold (eds), *Early Behavior: Comparative and developmental approaches*, New York: John Wiley and Sons.

—— 1973: Regulatory functions of arousal in primate psychosocial development. In C. Carpenter (ed.), *Behavioural Regulators of Behavior in Primates*, Lewisburg, PA: Bucknell University Press.

—— 1979: Ontogeny of social behavior. In P. Marler and G. Vandenbergh (eds), *Handbook of Behavioral Neurobiology: vol. 3. Social behavior and communication*, New York: Plenum Press.

Mason, W. A. and Berkson, G. 1975: Effects of maternal mobility on the development of rocking and other behaviors in rhesus monkeys: A study with artificial mothers. *Developmental Psychobiology*, 8, 197–211.

McCulloch, T. 1939: The role of clasping activity in adaptive behavior of the infant chimpanzee: III. The mechanism of reinforcement. *The Journal of Psychology*, 7, 305–16.

McKenna, J. J. 1978: Biosocial functions of grooming behavior among the common Indian langur monkey (*Presbytis entellus*). *American Journal of Physical Anthropology*, 48, 503-10.
—— 1979: The evolution of allomothering behavior among colobine monkeys: Function and opportunism evolution. *American Anthropologist*, 81, 818-40.
Meier, G. W. 1965: Maternal behaviour of feral- and laboratory-reared monkeys following the surgical delivery of their infants. *Nature*, 206, 492-3.
Mendelson, M. 1982: Clinical examination of visual and social responses in infant rhesus monkeys. *Developmental Psychology*, 18(5), 658-64.
Mendelson, M., Haith, M. and Goldman-Rakic, P. 1982: Face scanning and responsiveness to social cues in infant rhesus monkeys. *Developmental Psychology*, 18, 222-8.
Miller, L. C. and Nadler, R. D. 1981: Mother–infant relations and infant development in captive chimpanzees and orang-utans. *International Journal of Primatology*, 2, 247-61.
Mineka, S., Davidson, M., Cook, M. and Keir, R. 1984: Observational conditioning of snake fear in rhesus monkeys. *Journal of Abnormal Psychology*, 93, 355-72.
Missakian, E. A. 1974: Mother–offspring grooming relations in rhesus monkeys. *Archives of Sexual Behavior*, 3, 135-41.
Mitchell, G. D. 1969: Paternalistic behavior in primates. *Psychological Bulletin*, 71, 399-417.
—— 1979: *Behavioral Sex Differences in Nonhuman Primates*. New York: Van Nostrand Reinhold Co.
Mitchell, G. and Brandt, E. M. 1972: Paternal behavior in primates. In F. E. Poirier (ed.), *Primate Socialization*, New York: Random House.
—— 1975: Behavior of the female rhesus monkey during birth. In G. H. Bourne (ed.), *The Rhesus Monkey*, New York: Academic Press.
Mori, U. 1979: Individual relationships within a unit. *Contributions to Primatology*, 16, 93-124.
Murray, R. D. and Murdoch, K. M. 1977: Mother–infant dyad behavior in the Oregon troop of Japanese macaques. *Primates*, 18(4), 815-24.
Myers, R. E., Swett, C. and Millar, M. 1973: Loss of social group affinity following prefrontal lesions in free-ranging macaques. *Brain Research*, 64, 257-69.
Nadler, R. D. 1974: Periparturitional behavior of a primiparous lowland gorilla. *Primates*, 15, 55-73.
Nash, L. T. 1978: The development of the mother–infant relationship in wild baboons (*Papio anubis*). *Animal Behaviour*, 26, 746-59.
Neville, M. K. 1972: Social relations within troops of red howler monkeys (*Alouatta seniculus*). *Folia Primatologica*, 18, 47-77.
Newman, J. D. 1985: Squirrel monkey communication. In L. Rosenblum and C. Coe (eds): *Handbook of Squirrel Monkey Research*, New York: Plenum Press.
Nicholson, N. A. 1977: A comparison of early behavioral development in wild and captive chimpanzees. In S. Chevalier-Skolnikoff and F. E. Poirier (eds), *Primate Bio-social development: Biological, social, and ecological determinants*, New York: Garland Publishing.
Niemitz, C. 1974: A contribution to the postnatal behavioral development of *Tarsius bancanus*, Horsfield, 1821, studied in two cases. *Folia Primatologica*,

21, 250-76.
—— 1979: Outline of the behavior of *Tarsius bancanus*. In G. Doyle and R. Martin (eds), *The Study of Prosimian Behavior*, New York: Academic Press.
Noirot, E. 1972: The onset of maternal behavior in rats, hamsters and mice: A selective review. *Advances in the Study of Behavior*, 4, 107-46.
Oki, J. and Maeda, Y. 1973: Grooming as a regulator of behavior in Japanese macaques. In C. R. Carpenter (ed.), *Behavioral Regulators of Behavior in Primates*, Lewisburg, Pennsylvania: Bucknell University Press.
Packer, C. 1980: Male care and exploitation of infants in Papio anubis. *Animal Behaviour*, 28, 512-20.
Pages, E. 1980: Ethoecology of Microcebus coquereli during the dry season. In P. Charles-Dominique, H. M. Cooper, A. Hladik, C. M. Hladik, E. Pages, G. F. Pariente, A. Petter-Rousseaux, A. Schilling and J. J. Petter (eds), *Nocturnal Malagasy Primates: Ecology, physiology, and behavior*, New York: Academic Press.
Parke, R. and Suomi, S. 1981: Adult male-infant relationships: Human and non-human primate evidence. In K. Immelmann, G. Barlow, L. Petrinovich and M. Main (eds), *Behavioral Development: The Bielefeld Interdisciplinary Project*, New York: Cambridge University Press.
Pederson, D. and TerVrugt, D. 1973: The influence of amplitude and frequency of vestibular stimulation of the activity of 2-month old infants. *Child Development*. 44, 122-8.
Peter, P. H. and Klopfer, M. S. 1970: Patterns of maternal care in lemurs: I. Normative description. *Zietschrift für Tierpsychologie*, 27, 984-96.
Petter, J. 1965: The lemurs of Madagascar. In I. DeVore (ed.), *Primate behavior, Field Studies of Monkeys and Apes*, New York: Holt, Rinehart and Winston.
Petter, J. J. and Peyrieras, A. 1975: Preliminary notes on the behavior and ecology of Hapalemur griseus. In I. Tattersall and R. W. Sussman (Eds), *Lemur biology*, New York: Plenum Press.
Petter-Rousseaux, A. 1964: Reproductive physiology and behavior of the lemuroidea. In J. Buettner-Janusch (ed.), *The evolutionary and genetic biology of primates*, vol. II, New York: Academic Press.
Ploog, D. 1969: Early communication processes in squirrel monkeys. In R. J. Robinson (ed.), *Brain and early behavior: Development in the fetus and infant*, New York: Academic Press.
Poirier, F. 1968: The Nilgiri langur (*Presbytis johnii*) mother-infant dyad. *Primates*, 9, 45-68.
—— 1970: The nilgiri langur of South India. In L. A. Rosenblum (ed.), *Primate Behavior*, vol. 1, New York: Academic Press.
—— 1977: Introduction. In S. Chevalier-Skolnikoff and F. Poirier (eds), *Primate Bio-social Development: Biological, social and ecological determinants*, New York: Garland Publishing, Inc.
Pollock, J. I. 1980: Field observations on Indri indri: A preliminary report. In I. Tattersall and R. W. Sussman (eds), *Lemur Biology*, New York: Plenum Press.
Pook, A. G. 1978: A comparison between the reproduction and parental behavior of the goeldi's monkey (*Callimico Goeldii*) and of the true marmosets

(*Callitrichidae*). In H. Rothe, H. J. Wolters and J. P. Hearn (eds), *Biology and Behavior of Marmosets*, West Germany: Eigenvelag Hartmut Rothe.

Quiatt, D. 1979: Aunts and mothers: Adaptive implications of allomaternal behavior of nonhuman primates. *American Psychologist*, 81, 310-19.

Rakic, P. 1985: Limits of neurogenesis in primates. *Science*, 227, 1054-6.

Raney, D. F. and Rodman, P. S. 1981: Dominance and the social behavior of adult female. *Primates*, 22, 368-78.

Ransom, T. 1981: *Beach Troop of the Gombe*. Lewisburg, Penn.: Bucknell University Press.

Ransom, T. W. and Ransom, B. S. 1971: Adult male-infant relations among baboons (*Papio anubis*). *Folia Primatologica*, 16, 179-95.

Ransom, T. W. and Rowell, T. E. 1972: Early social development of feral baboons. In F. Poirier (ed.), *Primate Socialization*, New York: Random House.

Redican, W. 1976: Adult male-infant interactions in nonhuman primates. In M. Lamb (ed.), *The Role of the Father in Child Development*, New York: John Wiley and Sons.

Redican, W. and Mitchell, G. 1973: A longitudinal study of paternal behavior in adult male rhesus monkeys: I. Observations on the first dyad. *Developmental Psychology*, 8, 135-6.

Reynolds, V. and Reynolds, F. 1965: Chimpanzees of the Budongo forest. In I. DeVore (ed.), *Primate Behavior: Field studies of monkeys and apes*, New York: Holt, Rinehart and Winston.

Rhine, R. J. and Hendy-Neely, H. 1978: Social development of stumptail macaques (*Macaca arctoides*): Momentary touching, play and other interactions. *Primates*, 19, 115-23.

Rhine, R. J. and Owens, N. W. 1972: The order of movement of adult male and black infant baboons (*Papio anubis*) entering and leaving a potentially dangerous clearing. *Folia Primatologica*, 18, 276-83.

Richard, A. F. 1976: Preliminary observations on the birth and development of *Propithecus verreauxi*: To the age of six months. *Primates*, 17, 357-66.

Richard, A. F. and Heimbuch, R. 1975: An analysis of the social behavior of three groups of Propithecus verreauxi. In I. Tattersall and R. W. Sussman (eds), *Lemur Biology*, New York: Plenum Press.

Rijksen, H. D. 1978: A field study on Sumatran orang-utans (*Pongo pygmaeus abelii* lesson 1827). *Ecology, Behavior, and Conservation*, 3, 420.

Robson, K. 1967: The role of eye-to-eye contact in maternal-infant attachment. *Journal of Child Psychology and Psychiatry*, 8, 13-25.

Rogers, C. M. and Davenport, R. K. 1970: Chimpanzee maternal behavior. *The Chimpanzee*, (3), New York: Karger.

Roonwal, M. and Mohnot, S. 1977: *Primates of South Asia*. Cambridge, Massachusetts: Harvard University Press.

Rosenblatt, J. and Siegel, H. 1981: Factors governing the onset and maintenance of maternal behavior among nonprimate mammals. In D. Gubernick and P. Klopfer (eds), *Parental Care in Mammals*, New York: Plenum Press.

Rosenblum, L. A. 1968: Mother-infant relations and early behavioral development in the squirrel monkey. In L. A. Rosenblum and R. W. Cooper (eds), *The Squirrel Monkey*, New York: Academic Press.

—— 1971a: Ontogeny of mother–infant relations in macaques. In H. Moltz (ed.), *Ontogeny of Vertebrate Behavior*, New York: Academic Press.
—— 1971b: Infant attachment in monkeys. In H. R. Schaffer (ed.), *The Origins of Social Relations*, New York: Academic Press.
—— 1972: Sex and age differences in response to infant squirrel monkeys. *Brain, Behavioral Evolution*, 5, 30–40.
Rosenblum, L. A. and Harlow, H. 1963: Approach-avoidance conflict in the mother–surrogate situation. *Psychological Reports*, 12, 83–5.
Rothe, H. 1975: Influence of newborn marmosets (*Callithrix jacchus*) behaviour on expression and efficiency of maternal and parental care. *Symposia of the Fifth Congress of the International Primatological Society*, Japan Science Press, Japan.
Rowell, T. E. 1967: A quantitive comparison of the behavior of a wild and a caged baboon group. *Journal of Animal Behavior*, 15, 499–509.
—— 1974: Contrasting adult male roles in different species of nonhuman primates. *Archives of Sexual Behavior*, 3, 143–9.
Rowell, T. E., Din, N. A. and Omar, A. 1968: The social development of baboons in their first three months. *Journal of Zoology*, 155, 461–83.
Rowell, T., Hinde, R. and Spencer-Booth, Y. 1964: 'Aunt'–infant interaction in captive rhesus monkeys. *Animal Behaviour*, 12, 219–26.
Rumbaugh, D. 1965: Maternal care in relation to infant behavior in the squirrel monkey. *Psychological Reports*, 16, 171–6.
Ruppenthal, G. Harlow, M., Eisele, C., Harlow, H. and Suomi, S. 1974: Social development of infant monkeys reared in a nuclear family environment. *Child Development*, 45, 670–82.
Rylands, A. B. 1981: Preliminary field observations on the marmoset, *Callithrix humeralifer intermedius* (Hershkovitz, 1977) at Dardanelos, Rio Aripuana, Mato Grosso. *Primates*, 22, 46–59.
Sackett, G. 1970: Innate mechanisms, rearing conditions and a theory of early experience effects in primates. In M. Jones (ed.), *Effects of Early Experiences*, Coral Gables, Florida: University of Miami Press.
—— 1982: Can single processes explain effects of postnatal influences on primate development. In R. N. Ende and R. J. Harmon (eds), *The Development of Attachment and Affiliative Systems*, New York: Plenum Press.
Sade, D. S. 1965: Some aspects of parent–offspring and sibling relations in a group of rhesus monkeys, with a discussion of grooming. *American Journal of Physical Anthropology*, 23, 1–18.
—— 1967: Determinants of dominance in a group of free-ranging rhesus monkeys. In S. A. Altmann (ed.), *Social Communication among Primates*, Chicago: University of Chicago Press.
—— 1972: A longitudinal study of social behavior of rhesus. In R. Tuttle (ed.), *The Functional and Evolutionary Biology of Primates*, Chicago: Aldine and Atherton.
Sassenrath, E. N., Mason, W. A., Fitzgerald, R. C. and Kenney, M. D. 1980: Comparative endocrine correlates of reproductive status in *Callithrix* (titi) and *Saimiri* (squirrel) monkey. *Antropologie Contemporanea*, 3, 265.
Sauer, E. G. F. 1967: Mother–infant relationships in galagos and the oral child-transport among primates. *Folia Primatologica*, 7, 127–49.

Schaller, G. B. 1963: *The Mountain Gorilla: Ecology and behavior.* Chicago: University of Chicago Press.
Schessler, T. and Nash, L. T. 1977: Short communications: Food sharing among captive gibbons (Hylobates lar). *Primates*, 18, 677–89.
Schneider, M. D., Marra, L. M., Suomi, S. J. and Higley, J. D. 1986: Manuscript in preparation.
Scollay, P. A. and DeBold, P. 1980: Allomothering in a captive colony of hanuman langurs (*Presbytis entellus*). *Ethology and Sociobiology*, 1, 291–9.
Seay, B. 1966: Maternal behavior in primiparous and multiparous rhesus monkeys. *Folia Primatologica*, 4, 146–68.
Seay, B. M., Alexander, B. K. and Harlow, H. F. 1964: Maternal behavior of socially deprived rhesus monkeys. *Journal of Abnormal and Social Psychology*, 69, 345–54.
Sekulic, R. 1982: Birth in free-ranging howler monkeys (*Alouttinae seniculus*). *Primates*, 23, 580–2.
Seligman, M. E. P. 1975: *Helplessness.* San Francisco: W. H. Freeman Press.
Seyfarth, R. M. 1978: Social relationships among adult male and female baboons. II. Behaviour throughout the female reproductive cycle. *Behaviour*, 64, 227–47.
Shively, C. and Smith, D. 1983: Reproduction and social status in macaques. Paper presented at the 6th annual meeting of the American Society of Primatologists, East Lansing, Michigan.
Silk, J. B. 1978: Patterns of food sharing among mother and infant chimpanzees at Gombe National Park, Tanzania. *Folia Primatologica*, 29, 129–41.
Simonds, P. E. 1965: The bonnet macaque in south India. In I. DeVore (ed.), *Primate Behavior: Field studies of monkeys and apes*, New York: Holt, Rinehart and Winston.
Simons, R. C., and Bielert, C. F. 1973: An experimental study of vocal communication between mother and infant monkeys. *American Journal of Physiological Anthropology*, 38, 455–62.
Simons, R. C., Bobbitt, R. A. and Jensen, G. D. 1968: Mother monkeys' (*Macaca memestrina*) responses to infant vocalizations. *Perceptual and Motor Skills*, 27, 3–10.
Smith, D. G. 1981: A test of randomness of paternity of members of maternal sibships in six captive groups of rhesus monkeys (*Macaca mulatta*). *International Journal of Primatology*, 3, 461–9.
Smith, E. O. and Peffer-Smith, P. D. 1982: Triadic interaction in captive Barbary macaques (*Macaca sylvana, linnaeus*, 1758): 'Agonistic buffering'. *American Journal of Primatology*, 2, 99–107.
Snowdon, C. T. and Suomi, S. J. 1982: Paternal behavior in primates. In H. E. Fitzgerald, J. A. Mullins and P. Gage (eds), *Child nurturance*, 3, New York: Plenum Press.
Sorenson, M. W. 1970: Behavior of tree shrews. In L. A. Rosenblum (ed.), *Primate Behavior*, vol. 1, New York: Academic Press.
Southwick, C., Beg, M. and Siddigi, M. 1965: Rhesus monkeys in North India. In I. DeVore (ed.), *Primate Behavior: Field studies of monkeys and apes*, New York: Holt, Rinehart and Winston.
Spencer-Booth, Y. 1968: The behavior of group companions towards rhesus monkey infants. *Animal Behaviour*, 16, 541–57.

Stern, D. N. 1974: Mother and infant at play: The dyadic interaction involving facial, vocal, and gaze behaviors. In M. Lewis and L. A. Rosenblum (eds), *The Effect of the Infant on its Caregiver*, New York: John Wiley and Sons.

Stevenson, M. 1976: Birth and perinatal behavior in family groups of the common marmoset (*Callithrix jacchus jacchus*) compared to other primates. *Journal of Human Evolution*, 5, 365–81.

Struhsaker, T. T. 1967a: Auditory communication among vervet monkeys (*cercopithecus aethiops*). In S. A. Altman (ed.), *Social Communication among Primates*, Chicago: University of Chicago Press.

—— 1967b: Behavior of vervet monkeys (*Cercopithecus aethiops*). *University of California Publications in Zoology*, 82, 1–64.

—— 1975: *The Red Colobus Monkey* (1st edn). Chicago: University of Chicago Press.

—— 1971: Social behavior of mothers and infant vervet monkeys. *Animal Behavior*, 19, 233–50.

Struhsaker, T. T. and Gartlan, J. S. 1970: Observations on the behavior and ecology of the patas monkey (*Erythrocebus patas*) in the Waza Preserve, Cameroon. *Journal of Zoology*, 161, 49–63.

Suomi, S. 1976: Mechanisms underlying social development: A reexamination of mother–infant interactions in monkeys. In A. D. Pick (ed.), *Minnesota Symposia on Child Psychology*, vol. 10, Minneapolis: University of Minnesota Press.

—— 1977a: Development of attachment and other social behaviors in rhesus monkeys. In T. Alloway, P. Pliner and L. Krames (eds), *Advances in the Study of Communication and Affect*, vol. 3, New York: Plenum Press.

—— 1977b: Adult male–infant interactions among monkeys living in nuclear families. *Child Development*, 48, 1255–70.

—— 1979a: The perception of contingency and social development. In E. Thoman (ed.), *Origins of the Infant's Social Responsiveness*, Hillsdale, New Jersey: Lawrence Erlbaum Associates.

—— 1979b: Differential development of various social relationships by rhesus monkey infants. In M. Lewis and L. Rosenblum (eds), *Genesis of Behavior*, vol. 2, *The Child and its Family*, New York: Plenum Press.

—— 1981: The perception of contingency and social development. In L. Sherrod and M. Lamb (eds), *Infant Social Cognition: Empirical and theoretical considerations*. Hillsdale, New Jersey: Lawrence Erlbaum Associates.

—— 1982a: Abnormal behavior and primate models of psychopathology. In J. Fobes and J. King (eds), *Primate Behavior*, New York: Academic Press.

—— 1982b: Genetic, maternal, and environmental influences on social development in rhesus monkeys. In A. Chiarelli and R. Corruccini (eds), *Primate Behavior and Sociobiology Selected Papers*, (Part B), New York: Springer-Verlag.

—— 1982c: The development of social competence by rhesus monkeys. In P. Turillazzi, L. Rosenblum and S. Suomi (eds), *Normal and Abnormal Social Development in Primates*, Rome: Programme di Instituto di Sanita.

—— 1983: Social development in rhesus monkeys: Consideration of individual differences. In A. Oliverio and M. Zappella (eds), *The Behavior of Human Infants*, New York: Plenum Press.

Suomi, S. J., Eisele, C. D., Grady, S. A. and Tripp, R. L. 1973: Social preferences of monkeys reared in an enriched laboratory environment. *Child Development*, 44, 451-60.

Suomi, S. J. and Harlow, H. F. 1975: The role and reason of peer relationships in rhesus monkeys. In M. Lewis and L. A. Rosenblum (eds), *Friendship and Peer Relationships*, New York: John Wiley.

Sussman, R. W. 1977: Socialization, social structure, and ecology of two sympatric species of lemur. In S. Chevalier-Skolnikoff and F. E. Poirier (eds), *Primate Bio-social Development: Biological, social, and ecological determinants*, New York: Garland Publishing.

Swartz, K. and Rosenblum, L. 1981: The social context of parental behavior: A perspective on primate socialization. In D. Gubernick and P. H. Klopfer (eds), *Parental Care in Mammals*, New York: Plenum Press.

Swayamprabha, M. S. and Kadam, K. M. 1980: Short communications: Mother-infant relationship in the slender loris (*Loris tardigradus lydekkerianns*). *Primates*, 21, 561-6.

Symmes, D. and Biben, M. 1985: Maternal recognition of individual infant squirrel monkeys from isolation call playback. *American Journal of Primatology*, 9, 39-46.

Tattersall, I. 1982: *The Primates of Madagascar*. New York: Columbia University Press.

Taub, D. M. 1980a: Testing the 'agonistic buffering' hypothesis. I. The dynamics of participation in the triadic interaction. *Behavior Ecological Sociobiology*, 6, 187-97.

—— 1980b: Female choice and mating strategies among wild Barbary macaques (*Macaca sylvana* L.). In D. G. Lindburg (ed.), *The macaques: Studies in ecology, behavior and evolution*, New York: Van Nostrand Reinhold Company.

Taylor, H., Teas, J., Richie, T., Southwick, C. and Shrestha, R. 1978: Social interactions between adult male and infant rhesus monkeys in Nepal. *Primates*, 19, 343-51.

Thau, R. B. and Lanman, J. T. 1975: Endocrinological aspects of placental function. In P. Gruenwald (ed.), *The Placenta*, Great Britain: Medical and Technical Publishing Co., Ltd.

Thorington, R. W. 1968: Observation of squirrel monkeys in a Columbian forest. In L. A. Rosenblum and R. W. Cooper (eds), *The Squirrel Monkey*, New York: Academic Press.

Tilford, B. L. and Nadler, R. D. 1978: Male parental behavior in a captive group of lowland gorillas (*Gorilla gorilla gorilla*). *Folia Primatologica*, 29, 218-28.

Vaitl, E. A. 1977: Social context as a structuring mechanism in captive groups of squirrel monkeys (*Saimiri sciureus*). *Primates*, 18, 861-74.

Van Lawick-Goodall, J. 1967: Mother-offspring relationships in free-ranging chimpanzees. In D. Morris (ed.), *Primate Ethology*, Chicago: Aldine Publishing.

Vogt, J., Carlson, H. and Menzel, E. 1978: Social behavior of a marmoset (*Saquinus fuscicollis*) group I: Parental care and infant development. *Primates*, 19, 715-26.

Walters, J. 1980: Interventions and the development of dominance relationships in female baboons. *Folia Primatologica*, 34, 61-89.

White, L. E. and Hinde, R. A. 1975: Some factors affecting mother–infant relations in rhesus monkeys. *Animal Behavior*, 23, 527–42.

Wilson, E. O. 1975: *Sociobiology: The new synthesis*. Cambridge: Belknap Press.

Wolters, H. J. 1977: Some aspects of role taking behavior in captive family groups of the cotton-top tamarin *Saguinus oedipus oedipus*. In H. Rothe, H. J. Wolters and J. P. Hearn (eds), *Biology and Behavior of Marmosets*, Gottingen, West Germany: Eigenverlag Hartmut Rothe.

Woodcock, A. 1983: The social relationships of twin common marmosets (*Callithrix jacchus*) and their breeding success. *Primates*, 24, 501–14.

Woolridge, F. L. 1971: *Colobus guereza*: Birth and infant development in captivity. *Animal Behaviour*, 19, 481–5.

Young, G. H. and Hankins, R. J. 1979: Infant behaviors in mother-reared and harem-reared baboons. *Primates*, 20, 87–93.

Zucker, E. L. and Kaplan, J. R. 1981: Allomaternal behavior in a group of free-ranging patas monkeys. *American Journal of Primatology*, 1, 57–64.

7 Human Mother-to-Infant Bonds

W. Sluckin

In chapter 1 a distinction was drawn between generalized care-giving behaviour and specific parental behaviour. It should be said that adult human beings tend to exhibit some degree of care-giving towards all young children in general and women, in particular, often act as temporary substitute mothers, and can do this extremely effectively. In contrast, parental behaviour, in our usage of the term, refers to caring specifically for one's own, or one's adopted, children. It implies parental feelings of love and an affectional attachment between parents and their children. In this chapter we are concerned with the formation of such parent-to-infant attachments or bonds, and particularly mother-to-infant or maternal bonds. Very many mothers behave as if they were already strongly attached to their infants at the time of birth, but some, especially primiparous women, report a feeling of indifference towards their new-born children (Robson, 1981). As time passes, very few mothers show no signs of attachment to their babies and their young children, and indeed, to their adult offspring. It is of both theoretical and practical interest to know *when* and *how* these maternal (and paternal) bonds are formed.

PERINATAL AND POSTNATAL EVENTS

How a new-born baby is cared for varies greatly with culture, custom and current fashions. Even within western culture there have been striking variations in perinatal practices. For instance, there was at one time a widespread belief that labour pains were conducive to the growth of motherly love for the baby; hence the opposition to the administration of pain-killing drugs during labour. More recently, the view that pain in childbirth is not inevitable has been advocated, and it has been argued that it can be avoided by 'natural' means. Leaving that issue aside, some studies have suggested that pain-killing drugs given during labour could have adverse long-term effects on the mother, her infant and their mutual relationship. Indeed, this tended to be the prevalent view until some fairly recent findings strongly indicated that pain-killers do no harm (Rosenblatt et al., 1979). In fact, the evidence suggests that, unless very high doses of drugs are given, pain-killers result in a less anxious birth and so may well be psychologically beneficial both in the short and in the longer term.

In Victorian times mothers who could afford to do so were advised to rest after giving birth by staying in bed for several weeks. Opinion has now swung over in the opposite direction; women are told that, after parturition, exercise of all the muscles is highly desirable. Although strenuous exercise would, of course, not be encouraged, women are expected to be up and about soon after birth as possible.

The methods of feeding new-born babies also vary – although breast feeding by the natural mother is still the traditional way. However, among the well-to-do, new-born babies have often been given to wet-nurses to suckle. The fashion for bottle-feeding, which sprang up several decades ago, is still prevalent in many parts of the world, but it is now on the wane among the middle classes in the West and breast feeding is once again thought to be best, both for the mother and her baby. Infant dress has also been subject to changing views; swaddling clothes were at one time the rule throughout Europe, and although no one swaddles new-born infants in Western countries anymore, in some parts of the world the custom persists. What is of especial interest is that no clear evidence has been found to suggest that swaddling clothes are harmful to the infants or have long-term adverse effects, although it is rather unlikely that putting babies into swaddling clothes is in any way beneficial.

Views as to how close a new-born infant should be kept to its mother are of particular interest in this chapter. Dr Dick-Read's (1942) book, *Childbirth without Fear*, was very influential in the forties and fifties. It advocated, among other things, the practice of 'rooming in', that is, of keeping the new-born baby in the same room as its mother. The aim was to enable the mother to see her baby at all times and to encourage her to attend to the baby's needs; this is what most mothers would wish to do. It could be asked why this practice had to be encouraged in the first place, since it appears to be 'natural' practice. The reason was that stringent requirements of hygiene, and especially avoidance of diarrhoea and respiratory infections, had led to the practice of isolating new-born infants into special, separate wards. Dr Dick-Read and others argued that such arrangements impeded the healthy growth of mother–infant relationships. Dr Spock (1945), too, was an enthusiastic advocate of what he called the 'rooming-in plan'. There was, of course, nothing revolutionary about this 'plan' – it was no more than a return to age-old practices.

How close to the mother should the new-born baby be? Close bodily contact occurs at the time of breast feeding. In the seventies arguments were advanced for insisting on, or even enforcing, close bodily contact, the so-called skin-to-skin contact, soon after birth. For instance, Leboyer (1975), an influential writer, also favoured close bodily contact between mother and neonate; but – and this is significant – early skin-to-skin contact between mother and baby came to be regarded as not merely desirable but as necessary for the proper development of the mother's love for her infant and her attachment to it. This was largely due to the very effective writings of Klaus and Kennell (for example, Klaus and Kennell, 1976a,b).

THE IDEA OF MATERNAL BONDING

The question that Klaus and Kennell and their followers thought they had conclusively answered concerned the nature of development of the mother-to-infant attachment. They used the term 'bonding', rather than 'attachment', because it conjured up the onset of a strong and lasting tie that the mother formed to her infant. In a nutshell, it was initially asserted that the mother was bonded to her infant in a rapid manner through skin-to-skin contact during a sensitive (or critical) period of short duration soon after giving birth to her baby.

One of the problems that this 'bonding' doctrine presented was how to assess, at a later stage, that a mother is bonded or otherwise to her child? Unless we have some reliable means of ascertaining the extent of the mother's attachment to her child, we cannot conduct a conclusive follow-up study comparing mothers who had been 'bonded' to their infants by close contact after confinement with mothers who had been separated from their babies after birth (that is, those that had not undergone the treatment of 'bonding'). The question of assessment of maternal attachment is central to any attempt at evaluating the soundness of the doctrine that 'bonding' occurs rapidly, through skin-to-skin contact, during a critical period following parturition.

To judge whether a mother is attached to her child, whether she loves it, is a more complex problem than meets the eye since attachment is multi-faceted and is not a unidimensional quantity. Some mothers are undemonstrative but nevertheless devoted to their children; some are outwardly loving but neglect their children – both types could be said to be attached and loving, but their attachments are qualitatively different. Touching and fondling an infant, smiling at it, making friendly sounds, and so on, are actions indicating affection, but these signs are not enough in themselves to be taken as evidence of a specific attachment or a special relationship. Instead of observing the mother's behaviour (or in addition to observing it) we could ask the mother how she feels about her child. However, deeds are more convincing than words. So, after all, we have to observe the mother to see how well she looks after her child and also judge whether she is 'bonded' to her child by the way she shows affection by – as Klaus and Kennell (1976b) put it – 'fondling, kissing, cuddling and prolonged gazing'. Clearly, however, such forms of behaviour may often be directed towards any children rather than specifically towards one's own. So additional signs of attachment are required as evidence of 'bonding', but agreement as to what these signs may be is a matter of debate.

We know that affectionate behaviour has many components, but how do the various indices of affection intercorrelate? Dunn and Richards (1977) set out to answer this question empirically. They carried out a longitudinal study of 77 mother–child pairs and assessed the mothers' affectionate behaviour in a number of ways. They then correlated these different categories of behaviour and found that the correlations were not at all high. It emerged, for instance, that touching the baby, often interpreted as a sign of maternal affection, did not correlate with other measures customarily so used. The investigators were unable to establish an entity which could be described as 'warm mothering'.

But this result does not mean that affection, or indeed attachment, can never be assessed. It does mean, however, that such assessments are difficult and to a considerable degree subjective. To arrive at a reasonably objective indication of a mother-to-infant bond in each case, a scale of attachment which is administered in a standard manner is needed. The administration of the test must be done by observers who are not only unbiased but who also have no knowledge of the extent of the mother's contact with her infant directly after the confinement (Herbert et al., 1982). So, the assessment of maternal attachment is not methodologically a simple matter; and hence the value of follow-up studies of 'contact' and 'no-contact' mothers will depend, to a great extent, on the research methods adopted.

The bonding doctrine bases itself partly on evidence from animal-behaviour studies and partly on evidence from follow-up studies of human mothers who either had little or no contact with their new-born infants or who had experienced extended contact with them. We shall examine the evidence from the two types of research in the next section of this chapter. At the level of plausibility, however, the bonding doctrine leaves much to be desired. In brief, it would be odd if natural selection produced a bonding mechanism in the human species which is strictly dependent on a short-duration sensitive period, for such a mechanism would make the well-being and survival of children very hazardous. Therefore, one would expect selection pressure to be in the direction of a protracted mother-to-infant bonding period. Still, considering the plausibility of the bonding doctrine, it may be worth noting that critical periods are generally atypical of behavioural development. As far as children are concerned, critical periods, strictly speaking, cannot be said to exist; although there is evidence for various sensitive periods in behavioural development (Rutter, 1980; Bateson, 1983). However, these more or less sensitive periods occur during infancy and childhood, and not later in life. Thus, a critical period in adulthood for mother-to-infant bonding would be a very remarkable exception. Nevertheless, its existence could not and should not be ruled out and all assertions concerning mother-to-infant bonding must ultimately depend on the available empirical evidence.

THE BONDING DOCTRINE: EMPIRICAL EVIDENCE

Those who propound the bonding doctrine believe that its roots are in biology and, in particular, in ethology (Klaus and Kennell, 1976a,b). It is argued that in at least some animal species the mother's early exposure to her infant brings about maternal bonding. Studies of the so-called maternal imprinting in goats are often cited, especially those by Klopfer and Gamble (1966) and Klopfer and Klopfer (1968). As a matter of fact, investigations of maternal attachment in female sheep and goats go back to about the mid-1950s (see review in Herbert, Sluckin and Sluckin, 1982). What initially emerged was that a few minutes' contact after birth ensured mother-to-infant attachment. On the other hand, some days of enforced proximity to an infant animal *other* than the mother's own also ensured a stable adoption. Subsequently, studies by Klopfer and co-workers cited above, appeared to show that maternal attachment in goats occurred soon after parturition through a rapid familiarisation of the she-goat with her kid's smell. Incidentally, the most recent research has shown that vision is at least as important as smell in maternal bonding in goats.

Despite these findings, Klopfer (1971) pointed out that mother-to-infant ties in mammals 'need be neither very specific, nor rapidly formed, nor yet stable'. Later studies showed that even for goats, mother-to-kid bonding 'may not occur as rapidly as previously reported' (Gubernick, Jones and Klopfer, 1979). It transpires that a she-goat, having given birth, will accept *any* kid which is not too strongly tainted with the smell of another female goat – in other words, attachment will develop to any kid which is free from the smell of other mother-goats. It appears that such 'maternal labelling' is mediated by the mother's milk. Thus, 'bonding' in goats is essentially the absence of rejection (Gubernick, 1980, 1981). It is important to note that in other mammals nothing has been found which could genuinely lend support to the bonding doctrine.

Monkeys, apes and other non-human primates are of particular interest, but especially anthropoid apes, because they are the nearest relations of the human species. However, there is no evidence in those animals of mother-to-infant bonding, as envisaged by the bonding doctrine. More important still, even if the so-called maternal olfactory imprinting were an established fact in animals such as goats and sheep, this would be of little relevance to maternal bonding in the human species. Behaviour patterns of ungulates are quite different from those of primates, and extrapolations from the former applied to the latter are of very dubious value. If the bonding doctrine had been supported by observations of ape behaviour, then that would have given some credence to the doctrine. As it is, what really matters are research findings from studies of human mothers: what happens to them after parturition and how attached they are (in so far as this can be reasonably accurately assessed) to their infants some days, weeks, months and years thereafter.

The first thing to note is that, despite the fairly widespread acceptance of the bonding doctrine and the equally widespread adoption of the bonding practices in maternity hospitals, the sum-total of research in this field is very modest. This is probably so partly because rigorous studies are difficult to carry out and partly because the idea of bonding has taken such a strong root in the late seventies and early eighties that further research seemed unnecessary. What, then, does a close examination of existing research findings reveal?

An early investigation by Leifer et al. (1972) involved three groups of mothers of normal, full-term babies, mothers of premature babies who handled them to some extent from birth onwards, and mothers of premature babies who were placed in an incubator and so not handled by the mothers for up to twelve weeks after birth. Maternal behaviour was in all cases observed and assessed and it was found that all mothers displayed good mothering. There was no clear evidence to indicate that mothers separated from their premature infants were psychologically damaged.

However, Klaus et al. (1972) found certain differences in the quality of mothering between those mothers who had extended contact with their infants soon after birth and those who had no such extended contact. More specifically, at one month after birth the early-contact mothers looked at, and fondled, their babies for longer periods than did mothers in the other group. Comments on two problems may be pertinent here: first, eye-contact and fondling are not sufficient as criteria for the existence of bonding; second, no comparisons were drawn on later occasions between the two groups of mothers.

Aware of the doubts posed by the second problem, Kennell et al. (1974) did look again at the mothers in the two groups, in each case one year postpartum. Early-contact mothers were then found to be more helpful during the physical examination of their babies and their answers to interview questions were somewhat more forthcoming. As far as mother–baby interactions are concerned, there were some, but not many, differences. In particular, the two groups did not differ with regard to mother–infant play. Thus, the results of this study cannot be regarded as conclusively establishing that bonding had occurred in one group and not in the other. Nor does the third and final study settle anything; here five mother–infant pairs selected at random from each of the two groups were followed-up at two years of age (Ringler et al., 1975). The relevant finding was that the extended-contact mothers had more to say to their children than the mothers in the control group; they asked more questions, used more adjectives, gave fewer commands, and so on. Whether this indicated stronger bonding must remain uncertain. Generally, the Klaus-Kennell team reported quite weak effects of early contact, and such effects as were found could have been due to other factors, for example, the knowledge the women involved in the experiment had of being treated differently, and the expectations they might have had about the efficacy of the special treatment they had been given.

Although the early experimental results were inconclusive, the bonding doctrine was being presented in the late seventies with confidence and conviction; so much so, that the belief in the efficacy of skin-to-skin contact, called 'bonding', became widespread and well-established. Some research workers were at that stage interested in the duration of the sensitive period for bond-formation. Hales et al. (1977) reported that a group of Guatemalan mothers who immediately after delivery had skin-to-skin contact with their babies, were significantly more affectionate at 36 hours postpartum than control-group mothers. De Chateau and Wiberg (1977), however, found that a group of Swedish mothers rather similarly treated did *not* differ from control-group mothers in the display of affectionate behaviour towards their infants. Subsequently the mothers involved in the experiments were assessed again at three months and at one year postpartum. At one year these mothers differed from controls in a number of ways; they held their infants in closer body contact, they touched and caressed them more often and they generally appeared to show greater warmth towards their children. These findings seemed to provide good support for the bonding hypothesis; yet the research workers suggested certain other possible explanations for their results. By then, although it was recognised that further studies were needed, the expectations of at least some researchers were that further confirmation of the bonding hypothesis would repeatedly be forthcoming.

This was certainly the belief of the Swedish research group of Carlsson and co-workers. In an early study, Carlsson et al. (1978) reported that close contact between the mother and her new-born baby made it easier for her to feed the baby at four days after birth. The explanation for this not very surprising finding did not necessarily implicate bonding. Later, Carlsson et al. (1979) made a study of fifty mother-infant pairs to see how various amounts of initial mutual contact affected the mother's nursing behaviour at six weeks after birth. The authors reported being surprised that at six weeks the nursing behaviours of the extended-contact and limited-contact groups were alike; they had expected 'bonding' to bring about a marked difference between the two groups. A further study by this research team (Schaller et al., 1979) found that extended-contact mothers touched their babies more than control-group mothers during the first week postpartum, but there was no difference in this type of behaviour five weeks later. As a result, the research group lost confidence in the bonding doctrine.

Svejda et al. (1980) considered previous findings concerning bonding to be of doubtful validity on a number of counts. They therefore paid strict attention to the methodology of their own investigation of thirty mother–infant pairs. They adopted a double-blind experimental design and used a random assignment of pairs to the experimental group (mother–infant contact for one hour immediately after birth and ninety minutes at each feeding) and the control group (brief contact after birth and thirty minutes at each feeding). A wide range of responses indicating attachment were examined. The conclusion, in brief, was that no differences in maternal behaviour between the two groups were found. Thus, this carefully conducted study gave no support to the bonding hypothesis. Other investigations, primarily concerned with other influences on mothering (for example, age of mother) also reported that early extra mother–infant contact made no difference to the subsequent mothering behaviour (for example, Jones et al., 1980). It is noteworthy that adoptive mothers, who obviously have had no early contact with their adopted children, are in no way inferior in mothering and degree of attachment to natural mothers (Tizard, 1977).

One of the implications of the bonding doctrine is that child-abuse by mothers could be the consequence of the lack of initial close bodily contact between the mother and her infant, for the lack of early bonding would result in inadequate affectional attachment on the part of the mother and so she could develop a tendency to treat her child in a harsh manner. This proposition was considered by Gaines et al. (1978) who studied 240 women drawn from populations of abusing, neglectful and normal mothers. The researchers reported that 'the hypothesized relationship between mother–neonate bonding and maltreatment was not supported'. Cater and Easton (1980) studied 80 cases of child abuse to see if they could be ascribed to the separation of the abusing parents from their new-born infants. The findings were inconclusive; the researchers thought that although the lack of early parent–infant contact might have played some role, other factors (such as the immaturity of parents, instability of domestic arrangements, or psychiatric illness in the family) must have been decisive in the cases of child abuse investigated.

There is little doubt that early mother–infant contact, if all is well, is good for the mother and for the baby. Among other things, it encourages lactation and it can give the mother a sense of psychological well-being. However, as Egeland and Gaughn (1981) put it, 'to imply that lack of contact between mothers and their new-borns is indicative of a current failure to bond with the infant, or is predictive of later breakdown in the mother–infant bond, is a disservice to the millions of mothers and infants who have developed perfectly healthy bonds and attachments under the current hospital regimen'. They go on to say that 'when parent–child bonds do break down, as in cases of abuse and neglect, researchers should look for multiple causes rather than pinpointing some very early event as the predisposing single trauma'.

Leiderman (1981) reported on his own earlier studies; he found that mothers of premature infants who had been separated from them after birth for as long as two to three months, displayed maternal-attachment behaviour that was no different from that of full-term mothers who had not been initially separated from their babies. He concluded that if there is a sensitive period for bonding then it is not a short-lasting one. He further argued that 'particularly dangerous is the belief that if proper mother–infant contact is not made in the neonatal period, and if the early mother–infant bond is deficient, subsequent social bonding to other individuals is more difficult'.

Leiderman (1981) was also generally cautious in his criticisms of the bonding doctrine. Chess and Thomas (1982) were more outspoken. They reviewed the literature on mother-to-infant bonding and concluded that there was no adequate evidence for bonding during a short sensitive or critical period; they, therefore, considered 'bonding' to be a 'mystique' rather than a 'reality'. Bonding as presented in academic and popular literature, implying 'instantaneous glueing', has also been vigorously criticised by Redshaw and Rosenblatt (1982). They, too, found the evidence for the critical period for bonding 'inconclusive, to say the least'. Likewise, Lamb (1983) argued that the mother's attachment to her infant is brought about by a host of conditions other than short-duration postpartum contact.

When the second edition of the book by Klaus and Kennell (1976a) appeared under a new title (Klaus and Kennell, 1982) some of the criticisms of the bonding doctrine had made an impact on the views of the authors. Thus, they declare in the new preface: 'We were distressed when the word *bonding* became too popular too rapidly and was confused with a simple, speedy, adhesive property rather than the beginning of a complex human psychobiological process'. The term, sensitive period, is used in the new edition in preference to the more extreme notion of critical period. Generally, the authors 'back-pedal' to some extent with regard to the importance of post-birth 'bonding' procedures. Nevertheless, they still insist that 'at least 30 to 60 minutes' of early close physical contact in privacy between mother and neonate has beneficial and lasting effects on the mother-to-infant attachment. It is less clear whether the authors believe that the well-being of the child in the years to come is influenced by this early mother–infant contact.

THE DEVELOPMENT OF MOTHERLY LOVE

The affective reactions of mothers to their new-born infants vary a great deal: some feel overwhelming love for their babies while, at the other extreme, some mothers react with emotional indifference (Robson and Kumar, 1980). In most cases the feelings of indifference vanish within a few days, but in some cases the onset of love is slow. Since there are no grounds for believing that emotional mother-to-infant bonds are formed through direct contact during a short-lived critical stage after parturition, then the questions arise as to *how* and *when* these bonds do develop. Furthermore, it may be asked whether maternal attachment and behaviour *are* essentially different in the human species from paternal attachment and behaviour. We are, of course, concerned here not with general protective behaviour associated with warm feelings towards all young and helpless creatures, but with specific attachments and love towards one's own infants. Before more is said about it, we may note that over a decade ago Harlow (1971) viewed the growth of love as a gradual learning-like process, and earlier, Dr Spock asserted that 'love for the baby comes gradually' (Spock, 1957). This had tended to be the common assumption; but, with the advent of the bonding doctrine, the 'gradualist' view became unfashionable.

It goes without saying that the tendency in adults to form specific attachments to their own (or adopted) infants has genetic roots. For the tendency to be realised, the mother or father must be, in the first place, *exposed* to the new infant. In addition to exposure, conditioning mechanisms may well contribute to the development of bonds. There is no evidence for rapid 'glueing' of the mother to the infant, but there are strong indications of a steady development of parental love and attachment (Sluckin et al., 1983). So-called 'exposure learning', is an important factor in this process, and may well be the dominant one.

The term 'exposure learning' refers to preferences acquired by a subject for objects to which the subject is exposed. The mode of operation of this acquired preferrence is particularly clear in the case of imprinting, that is, in the development of attachments by very young nidifugous birds (chicks, ducklings, goslings, and so on) to their mothers or other figures (for example, moving objects) in their environment (Sluckin, 1964, 1972). More generally, there is strong empirical evidence that human preferences or likes for anybody or anything are *initially* a direct function of familiarity – that is, that the process of familiarisation with a figure brings about a liking for the figure. But a well-known saying asserts that 'familiarity breeds contempt', and there is a good evidence that this, too, is in a sense true. For, while up to a point exposure results in liking, continuing exposure beyond a certain point results, in some situations, in a decline in liking (Sluckin et al., 1982). But in many situations the peak of liking is never reached.

Just as the young of the so-called precocial birds and mammals develop attachments through exposure learning, so human infants also become attached primarily in the same way to their parents. This is the view of Bowlby (1969) whose attachment theory is now highly influential in developmental psychology. Bowlby (1977) considers that 'learning to distinguish the familiar from the strange is the key process in the development of attachment'. He incidentally also draws attention to the fact that 'attachment can develop despite repeated punishment from the attachment figure.

We know that exposure learning occurs in some circumstances in adults as well as in infants (Sluckin et al., 1982). It is likely that the sheer act of exposure of the mother to her new-born infant plays a role in the growth of her motherly love (Sluckin et al., 1983). Her love and attachment can have some punishing consequences; in caring for her infant she has to do much hard work, to endure many frustrations and to encounter ingratitude. Nevertheless maternal attachment is in most cases strong and unwavering.

There is little doubt that, apart from exposure learning, that two well-known forms of conditioning play an important role in developing and strengthening the attachment of the mother to her infant. Consider, in the first place, Pavlovian or classical conditioning. The unconditional stimuli which evoke general caregiving responses in adults have been described by ethologists (Eibl-Eibesfeldt, 1970, for instance). They include disproportionately large eyes characteristic of babies, relatively short and thick limbs, a rounded body shape, a large head relative to the body, a large forehead relative to the face, and so on. Certain features of behaviour, such as clumsiness, also help in releasing care-giving responses. Now, one's own baby has also certain specific (non-releasing) characteristics, such as its voice, perhaps its clothes, and so on. These could become the conditioned stimuli through association with the unconditioned stimuli, such as large eyes and short limbs. These conditioned stimuli can then by themselves evoke care-giving affectionate behaviour. In this way, it is suggested, one's child and everything associated with it, stimulate parental love. Thus, one's own children are the source of both unconditioned and conditioned affection-evoking stimuli. This Pavlovian conditioning process may account, to some extent, for the gradual development of mother-to-infant bonds. In so far as this is so, the build-up goes hand in hand with the development of parental attachment through exposure learning.

The other type of conditioning, instrumental or operant, may also contribute to the growth of mother-to-infant attachment. In this type of conditioning, activities which are reinforced or rewarded become firmly established while the unreinforced actions wane. As far as the mother–infant relationship is concerned, whenever the mother does anything which is visibly beneficial to the child, her actions are thereby reinforced. As the mother gradually becomes acquainted with her new infant, she gets to know how best to satisfy her infant's needs. The most effective ways of feeding the infant, of keeping it warm and clean, become more frequent and more strongly rooted. Her mothering, her love and attachment steadily grow stronger. Some infants, those who fail to thrive, are less rewarding and the mothers of such children may be less loving, and even rejecting, because their efforts are not properly reinforced. When reinforcement is manifest, operant conditioning can make a significant contribution to the gradual development of effective mothering behaviour-patterns and mother-to-infant bonds.

Undoubtedly the specific maternal attachment develops as a result of many forces pulling together. One of the influences on young mothers is seeing how other, more experienced mothers behave, and so empathising with those mothers' feelings. New mothers will have seen their own mothers in action and they imitate mothering behaviour prevalent in their cultural environment. If strong mother-to-infant attachment is typical of the prevalent culture, young mothers will tend to follow the norm. All in all, well-known forms of learning, including exposure learning, observational learning and imitation, together with classical and operant conditioning, can adequately account for the development of mother-to-infant bonds (Sluckin et al., 1983), whereas the bonding doctrine (in its more uncompromising form) is both theoretically implausible and remains unsupported by empirical evidence.

MATERNAL AND PATERNAL ATTACHMENTS

The implication of what has been said so far in this chapter is that a learning process exemplified by maternal bonding need not be restricted to the human female. Indeed, the development of maternal love and attachment can be, and as a general rule is, paralleled by the growth of father-to-infant bonding. Very often the father's responses to his newborn infant resemble very strongly maternal behaviour (Sullivan and McDonald, 1979). Likewise, attachment resulting from exposure to the baby and from other forms of learning develops in the father, if he is closely involved in child rearing, just as it develops in the mother.

However, in practice, paternal attachment is often less pronounced than maternal attachment, although certainly not invariably or inevitably so. In the first place, in primate species (such as the anthropoid apes) males are less responsive to infants than are females. Quite possibly the human male is genetically programmed to be less nurturant than the female. Nevertheless, the males of our nearest ape relations act in a protective manner towards the young of their species (and are also protective towards their females). We may surmise that even if the human male is genetically programmed to display less care-giving behaviour than the female, he is still quite strongly inclined by nature to be care-giving and protective towards babies and children.

Perhaps a more important cause of the apparently less strong paternal bonding, as compared with maternal bonding, is our cultural environment. Until relatively recent times European males were not expected to perform such domestic tasks as feeding young infants, cleaning them up, and so on. This, in turn, limited the exposure of fathers to their offspring, thereby also restricting other modes of learning. Nevertheless, fathers have always shown very marked signs of attachment to their infants. At the present time, the situation with regard to the division of domestic duties between males and females appears to be changing quite rapidly; men are now expected to attend to the needs of their children to a much greater extent. The consequence of this could only be that maternal and paternal attachments will be much alike, when perhaps it would be more appropriate to use the generic term, parental attachment. The whole complex topic of the role of the father in child development is extensively discussed by Lamb (1981) and his co-authors.

It is probably not out of place at the end of this section to mention some abnormalities in the development of parental attachments. One type of pathology is the rejecting attitude which some mothers display towards their infants. Herbert and Iwaniec (1977) have drawn attention to 'children who are hard to love'. In other cases, parental rejection is a consequence of particularly difficult situations (financial problems, emotional difficulties between husband and wife, and so on) at the time of birth. Rejection can take various forms; it does not always entail neglect or physical cruelty. For instance it can be 'emotional abuse', to use a phrase adopted by the professionals (social workers, psychiatrists, psychologists) concerned with child abuse. Another type of pathology of parental attachment is excessive and over-protective attachment which interferes with the emotional and social development of the child. Such abnormally overprotective attachments may occur after a long period of infertility, or when the mother has had several miscarriages, or when one or more of her children have died; why such exaggerated attachments occur is, of course, a matter for speculation. What is not in doubt is that there exist both rejecting and over-protective parents and that both types need skilled help.

SOME PRACTICAL IMPLICATIONS

The consideration of the nature of parental (and especially maternal) attachment in the human species leads to a wide range of practical problems. These concern procedures followed by obstetricians and midwives in maternity hospitals, advice given to mothers by paediatricians and health visitors, policies adopted with regard to child welfare (for example, adoption, fostering, child abuse) and so on. Starting with birth, there is little doubt that in the case of healthy full-term babies, rooming-in and physical contact between mother and infant are highly desirable. However 'bonding' the mother to her new-born, as the procedure is often called, is not a condition of successful mother-to-infant attachment in the days, months and years to come.

There are different situations when immediate skin-to-skin contact is undesirable, or just inconvenient, and there is no evidence that this hinders the formation of mother–infant bonds.

One such situation is when, for medical reasons, a caesarian section has to be performed. Instead of giving birth to a baby in the normal way, a fairly major abdominal operation is undergone by the mother. She will have been under an anaesthetic, or at least under the influence of an epidural injection which numbs all or almost all sensation from the waist down. In either case the mother is hardly fit for skin-to-skin contact. Fortunately, missing this stage and so not being forced to 'bond', does no harm to the future mother–infant relationship. It is of practical importance that a mother who has given birth by caesarian section should not be made anxious by believing that her maternal attachment might be diminished.

Another situation where the procedure of 'bonding' is inadvisable is in the event of premature birth. Pre-term, low-weight babies are best cared for in special-care nurseries; without such special attention many babies would not survive. This type of care entails a separation of the infant from its mother and again the bonding doctrine suggests that this would endanger maternal bonding. In fact, there is no evidence that, after a period of enforced separation, when parents and babies are reunited the development of parental attachment is impaired. There is no good reason why mothers of premature babies, after being reunited with them, should not care for these babies in a competent manner, and should not love them just as much as full-term mothers love their babies.

'Bonding failure' has been blamed for a variety of problems, for instance, when infants fail to thrive or when children turn out to be autistic, or when babies are 'battered' and abused. The so-called non-organic failure-to-thrive syndrome (at one time known as marasmus) has greatly puzzled paediatricians; this is because children affected by this syndrome are abnormally slow in gaining weight and in growing, while being free from infections and sometimes despite seemingly adequate food intake. Therefore it has been suggested that the mother–infant relationship might be at fault, and this has the implication of inadequate maternal bonding. However, studies such as those of Vietze et al. (1980) and of Iwaniec and Herbert (1982) have shown that failure to thrive is a complex condition with varied antecedents and that it cannot reasonably be ascribed solely to a bonding failure.

Autistic children, first so-labelled over 40 years ago, are typically uninterested in some ways in what is going on around them, are unable to form normal social relationships and exhibit some peculiar mannerisms and obsessions. An autistic child tends to be retarded in some respects but not in others. Quite often, unlike a normal child, it shows no special attachment to its mother. At a later date Bettleheim (1967) claimed that childhood autism was caused by the mother's unconscious rejection of her child. Then gradually the view that bonding failure was at the root of the problem started to gain ground. The suggestion was that the autistic child retreated into a world of its own as a reaction to the mother's lack of affection for it. Obviously, such theorising contributed to the distress of parents of autistic children. When research began, no evidence was found to support the view that parents of autistic children differed from parents of normal children in the demonstration of parental warmth, emotional responsiveness and so on (Cox et al., 1975). Bonding failure is not now generally claimed to be the cause of autism, and at present a great deal can be done to alleviate this particular condition (for example, Wing, 1971).

The most worrying practical consequences of the bonding doctrine is its impact on the problem of child abuse. For it has been said that physical and emotional child abuse occur when the mother is not 'bonded' to her child (Lynch and Roberts, 1977; Argles, 1980). Small-scale, retrospective studies of abused children cannot be regarded as conclusive; their interpretation is always open to doubt. A larger-scale study of both abused and control children (Gaines et al., 1978), mentioned in an earlier section of this chapter, has concluded that there was no evidence of any connection between mother–neonate 'bonding' and subsequent child battering. Cater and Easton (1980), also mentioned earlier, investigated the effect of separation of the abusing parents from their new-born infants. As we have seen, they found that baby battering had a great deal to do with 'unstable domestic arrangements, and psychiatric disturbance and immaturity in the parents'. There are cases when an abusing mother seems to the social worker or health visitor or doctor to be loving towards her child and, hence, properly 'bonded'. In such cases the 'at risk' child will not be taken into care, when in fact the only safe thing to do is to take it away from the mother. To place emphasis on the apparent 'bonding' may lead astray the helping professionals. In any case, the idea of bonding 'carries with it the implications of irreversibility and lays a heavy burden on mothers' (Eisenberg, 1981). Predictions of child abuse based on bonding can be very misleading (Montgomery, 1982).

If true bonding really did occur primarily through skin-to-skin contact soon after birth, then the prospects for adoptive parents and their adopted children would be very poor. However, it is well known that adoptions can be most successful. This is not just hearsay; the view is backed up by clear evidence. Tizard (1977), for instance, carried out a study of children who had been initially in care for some years. Some of them were then adopted while others were returned to their families; and the former were found to have settled down better than the latter. Of course, it is best for adoptions to take place as early as possible – apart from being less unsettling to the child, adoptive parents then start caring for their charges in every way much as do natural parents. Adoptive parents are generally eager to be good parents, and they do come to love their adopted children and show all the signs of a strong attachment to them. Foster parents, too, can become very attached to their charges. When, however, the foster child is not emotionally accepted in its foster home, then there is a risk of disturbance to its development (Tizard, 1977).

It is important to remember that child-care practices have varied throughout history and vary a great deal at the present time. Furthermore, despite extensive research, there is much doubt as to how modes of child-rearing influence the development of the child's personality. Thus, we do not know whether the psychological development of children is really greatly affected by feeding methods (breast feeding versus bottle feeding, or fixed-interval feeding versus on-demand feeding) or by early or late weaning. It has long been known that the best parents tend to be those who treat their children according to what, at the time, the community generally believes is best for the child (Behrens, 1954); for such mothers tend to be relaxed and confident in their mothering. There is a risk that mothers will become anxious when they are rigidly instructed about the *details* of their mothering activities – a case in point is mandatory skin-to-skin contact immediately after birth. It is this anxiety that is prone to have adverse effects on the mother–child relationship rather than any particular 'faulty' mothering procedure.

The bonding doctrine and the practice of mother-to-infant 'bonding' were well-established in countries such as Britain and America by about 1980. Soon after, however, sceptical voices began to be heard. As mentioned in an earlier section, Leiderman (1981) indicated that practical action based on the belief in a short-lived critical period for maternal bonding could be physically and psychologically harmful to women who have just emerged from a confinement. Chess and Thomas (1982), Herbert et al. (1982), Redshaw and Rosenblatt (1982) and Lamb (1983) all came independently to similar conclusions. Sluckin et al. (1983) were particularly concerned in their book with the various practical aspects of the bonding doctrine; and the authors' message to mothers was that there was no need to worry about their feelings towards their new-born babies or the closeness of their physical contact with their babies. It is to be hoped that those concerned with perinatal and early postnatal procedures – obstetricians, midwives, paediatricians, health visitors and others – will act with caution when giving directions to mothers and when advising them on their relationship with their new offspring.

REFERENCES

Argles, P. 1980: Attachment and child abuse. *British Journal of Social Work*, 10, 33–42.
Bateson, P. 1983: The interpretation of sensitive periods. In A. Oliverio and M. Zappella (eds), *The Behavior of Human Infants*, New York: Plenum Press.
Behrens, M. L. 1954: Child rearing and the character structure of the mother. *Child Development*, 25, 225–38.
Bettelheim, B. 1967: *The Empty Fortress*. New York: Free Press.
Bowlby, J. 1969: *Attachment and Loss, Vol. 1: Attachment*. London: Hogarth Press.
—— 1977: The making and breaking of affectional bonds: 1. Aetiology and psychopathology in the light of attachment theory. *British Journal of Psychiatry*, 130, 201–10.
Carlsson, S. G., Fagenberg, H., Horneman, G., Hwang, C.-P., Larsson, K., Rodholm, M., Schaller, J., Danielsson, B. and Gundewall, C. 1978: Effects of amount of contact between mother and child on the mother's nursing behavior. *Developmental Psychobiology*, 11, 143–50.
Carlsson, S. G., Fagenberg, H., Horneman, G., Hwang, C.-P., Larsson, K., Rodholm, M., Schaller, J., Danielsson, B. & Gundewall, C. 1979: Effects of various amounts of contact between mother and child on the mother's behavior: a follow-up study. *Infant Behavior and Development*, 2, 209–14.
Carter, J. I. and Easton, P. M. 1980: Separation and other stress in child abuse. *Lancet*, 1, 972–4.
Chess, S. and Thomas, A. 1982: Infant bonding: mystique and reality. *American Journal of Orthopsychiatry*, 52, 213–22.
Cox, A., Rutter, M., Newman, S. and Bartak, L. 1975: A comparative study of autism and specific developmental language disorder. II. Parental characteristics. *British Journal of Psychiatry*, 126, 146–59.
De Chateau, P. and Wiberg, B. 1977: Long-term effects on mother–infant behavior of extra contact during the first hours postpartum. *Acta Paediatrica Scandinavia*, 66, 145–51.
Dick-Read, G. 1942: *Childbirth without Fear*. London Heinemann.
Dunn, J. B. and Richards, M. P. M. 1977: Observations on the developing rela-

tionship between mother and baby in the neonatal period. In H. R. Schaffer (ed.), *Studies in Mother–Infant Interaction,* London: Academic Press.
Egeland, B. and Gaughn, B. 1981: Failure of 'bond formation' as a cause of abuse, neglect and maltreatment. *American Journal of Orthopsychiatry,* 51, 78–84.
Eibl-Eibesfeldt, I. 1970: *Ethology: The Biology of Behavior.* New York: Holt, Rinehart and Winston.
Eisenberg, L. 1981: Social context of child development. *Paediatrics,* 68, 705–11.
Gaines, R., Sandgrund, A., Greer, A. H. and Power, E. 1978: Etiological factors in child maltreatment: a multivariate study of abusing, neglecting and normal mothers. *Journal of Abnormal Psychology* 87, 531–40.
Gubernick, D. J. 1980: Maternal 'imprinting' or maternal 'labelling' in goats? *Animal Behaviour,* 28, 124–9.
—— 1981: Mechanism of maternal 'labelling' in goats. *Animal Behaviour,* 29, 305–6.
Gubernick, D. J., Jones, K. C. and Klopfer, P. H. 1979: Maternal 'imprinting' in goats? *Animal Behaviour,* 27, 314–15.
Hales, D. J., Lozoff, B., Sosa, R. and Kennell, J. H. 1977: Defining the limits of the maternal sensitive period. *Developmental Medicine and Child Neurology,* 19, 454–61.
Harlow, H. F. 1971: *Learning to Love.* San Francisco: Albion Press.
Herbert, M. and Iwaniec, D. 1977: Children who are hard to love. *New Society,* 4, 109–11.
Herbert, M., Sluckin, W. and Sluckin, A. 1982: Mother-to-infant 'bonding'. *Journal of Child Psychology and Psychiatry,* 23, 205–21.
Iwaniec, D. and Herbert, M. 1982: The assessment and treatment of children who fail to thrive. *Social Work Today,* 13 (22), 8–12.
Jones, F. A., Green, V. and Krauss, D. R. 1980: Maternal responsiveness of primiparous mothers during the postpartum period: age differences. *Paediatrica,* 65, 579–84.
Kennell, J. H., Jerauld, R., Wolfe, H., Chester, D., Kreger, N., McAlpine, W., Steffa, M. and Klaus, M. H. 1974: Maternal behavior one year after early and extended postpartum contact. *Developmental Medicine and Child Neurology,* 16, 172–9.
Klaus, M. H., Jerauld, R., Kreger, N., McAlpine, W., Steffa, M. and Kennell, J. H. 1972: Maternal attachment – importance of the first postpartum days. *New England Journal of Medicine,* 286, 460–3.
Klaus, M. H. and Kennell, J. H. 1976a: *Maternal–Infant Bonding.* St. Louis: Mosby.
—— 1976b: Parent-to-infant attachment. In D. Hull (ed.) *Recent Advances in Paediatries,* vol. 5, London: Churchill Livingstone.
—— 1982: *Parent–Infant Bonding.* St. Louis: Mosby.
Klopfer, P. H. 1971: Mother love: what turns it on. *American Scientist,* 59, 404–7.
Klopfer, P. H. and Gamble, J. 1966: Maternal imprinting in goats: the role of chemical senses. *Zeitschrift für Tierpsychologie,* 23, 588–92.
Klopfer, P. H. and Gamble, J. 1966: Maternal imprinting in goats: the role of chemical senses. *Zeitschrift für Tierpsychologie,* 23, 588–92.

Klopfer, P. H. & Klopfer, M. S. 1968: Maternal imprinting in goats: fostering of alien young. *Zeitschrift für Tierpsychologie,* 25, 862–6.

Lamb, M. E. (ed.) 1981: *The Role of the Father in Child Development.* New York: John Wiley and sons.

—— 1983: Early mother–neonate contact and the mother–child relationship. *Journal of Child Psychology and Psychiatry,* 24, 487–94.

Leboyer, F. 1975: *Birth without Violence.* London: Wildwood House.

Leiderman, P. H. 1981: Human mother–infant social bonding: is there a sensitive phase? In K. Immelmann, G. W. Barlow, L. Petrinovich and M. Main (eds), *Behavioral Development,* Cambridge: Cambridge University Press.

Leifer, A. D., Leiderman, P. H., Barnett, C. R. and Williams, J. A. 1972: Effect of mother–infant separation on maternal attachment behavior. *Child Development* 43, 1203–18.

Lynch, M. A. and Roberts, J. (1977). Predicting child abuse: signs of bonding failure in the maternity hospital. *British Medical Journal,* 1, 624–36.

Montgomery, S. 1982: Problems in the perinatal prediction of child abuse. *British Journal of Social Work,* 12, 189–96.

Redshaw, M. and Rosenblatt, D. B. 1982: The influence of analgesia in labour on the baby. *Midwife, Health Visitor and Community Nurse.* 18, 126–32.

Ringler, N. M., Kennell, J. H., Jarvella, R., Novojosky, B. and Klaus, M. H. 1975: Mother-to-child speech at two years – effects of early postnatal contacts. *Journal of Paediatrics,* 86, 141–4.

Robson, K. M. 1981: A study of mothers' emotional reactions to their newborn babies. Unpublished Ph.D. thesis, University of London.

Robson, K. S. and Kumar, H. A. 1980: Delayed onset of maternal affection after childbirth. *British Journal of Psychiatry,* 136, 347–53.

Rosenblatt, D., Redshaw, M., Packer, M. and Lieberman, B. 1978: Drugs, birth and infant behaviour. *New Scientist,* 81, 487–9.

Rutter, M. 1980: The long-term effects of early experience. *Developmental Medicine and Child Neurology,* 22, 800–15.

Schaller, J., Carlsson, S. G. and Larsson, K. 1979: Effect of extended postpartum mother–child contact on the mother's behavior during nursing. *Infant Behavior and Development,* 2, 319–24.

Sluckin, W. (1964, 2nd edn 1972). *Imprinting & Early Learning.* London: Methuen.

Sluckin, W., Hargreaves, D. J. and Colman, A. M. 1982: Some experimental studies of familiarity and liking. *Bulletin of the British Psychological Society,* 35, 189–94.

Sluckin, W., Herbert, M. and Sluckin, A. 1983: *Maternal Bonding.* Oxford: Basil Blackwell.

Spock, B. 1945: *The Common Sense Book of Baby and Child Care.* New York: Duell, Sloan and Pearce.

Spock, B. 1957: *Baby and Child Care.* New York: Pocket Books.

Sullivan, J. and McDonald, D. 1979: Newborn oriented paternal behavior. In J. G. Howells (ed.) *Modern Perspectives in the Psychiatry of Infancy,* New York: Brunner/Mazel.

Svejda, M. J., Campos, J. J. and Emde, R. N. 1980: Mother–infant 'bonding': failure to generalize. *Child Development,* 51, 775–9.

Tizard, B. 1977: *Adoption: A Second Chance.* London: Open Books.
Vietze, P. M. O'Connor, S. M. Falsey, S. and Altemeier, W. A. 1980: Newborn behavioural and interactional characteristics of nonorganic failure-to-thrive infants. In *High-Risk Infants and Children: Adult and Peer Interactions.* London: Academic Press.
Wing, L. 1971: *Autistic Children.* London: Constable.

8 The Role of the Father in the Human Family

C. Lewis

During the 1970s fathers became a vogue subject of study among developmental psychologists. So much research was carried out that lengthy reviews (for example Lamb, 1976a, 1981) and bibliographies (Price-Bonham, 1976; Price-Bonham et al., 1981) had to be updated frequently. As a result this chapter will focus specifically upon a brief period in the life-cycle – the first year of the child's life. The father's role during this time is of special interest, since obviously parental 'input' is most important to the child's survival and development.

I will attempt to show that the part played by fathers is dependent upon a complexity of factors and that we cannot simply talk of fathering 'behaviour' as the variations in paternal styles are many. The chapter is divided into four sections. The first reviews the most popular area of recent research – the spate of observational investigations of father–infant interaction. I shall argue that a major reason why interest in this area has waned since the turn of the decade lies in its failure to provide a broad enough conceptual picture of the father's role. The rest of the chapter will consider fathers from another angle – men's own accounts of their relationships with their young children. The second section examines just how much men participate in family life, focusing upon the part they play in child care. This will hopefully give some indication of fathers' general commitment to their families. The third section looks more closely at the ways in which paternal roles develop over the first year of fatherhood. Such an examination suggests that the influences upon men are complex and that a full understanding of the paternal role must take into account key 'social' factors, such as the man's marital relationship, his work and more general cultural influences upon him and his wife. In the final section I shall return to some issues raised in the first, which con-

With thanks to Wynford Bellin, David Clark Carter, Heather Hughes, Graham Luke and Rosemary Smith.

The Role of the Father in the Human Family

cern the man's psychological role. I shall suggest that this is very much influenced by the same factors which determine the level of his practical involvement in family life.

THE FATHER'S PSYCHOLOGICAL ROLE: OBSERVATIONAL EVIDENCE

As I mentioned above, in the past ten years developmental psychologists have shown great interest in the father's part in the child's early emotional development. Numerous observational studies have been carried out in the laboratory and more 'ecologically' valid settings, like the home, to discover whether fathers' styles of interaction influence the child's development. This research has given rise to some interesting data, but as I shall argue here, has not yet produced a coherent theory of the father's psychological role. Firstly, it suffers from a legacy of matricentric research. Second, since the research is based primarily upon observational data, theoretical interest has been upon interactional styles and their supposed 'effects', rather than upon the relationships and roles of fathers in the family. I shall deal with each of these issues in turn.

The dominant model of early emotional development in the last fifteen years has been Attachment Theory (Bowlby, 1969). This suggests that in very early life 'bonds' form between a mother and her offspring during sensitive or critical periods and that the emotional relationship (attachment) which develops between them during the first year is vital to the child's psychological stability. Central to Bowlby's original thesis (Bowlby, 1954) and perhaps even to his formal theory is the idea of 'monotropy', which posits that the child becomes attached to one primary figure (his/her mother) and that other attachments are only of secondary importance (Bowlby, 1969: pages 366–8). As a result of this theory (the ideas of which of course are by no means peculiar to Bowlby) a great deal of research and theory about fathers have been concerned with testing the notion of monotropy, by comparing maternal and paternal attachments to their offspring.

The results of much of this research are perhaps not surprising. For example, in 1974 Greenberg and Morris, impressed by the research of Klaus and Kennell on early maternal bonding (see chapter 6), interviewed thirty fathers very soon after the delivery of their children. Twenty-nine were reported to feel 'high' and twenty-seven wanted to take part in the care of the child. Greenberg and Morris also found that men tended to perceive their own newborns as prettier than other babies and they often expressed a desire to touch them. As a result these authors devised the term 'engrossment' to describe the father's 'bond' to his offspring. They suggested that such feelings reflect an innate potential within men to form attachments. However, it seems hardly surprising that men feel overjoyed at the arrival of parenthood and the biological speculations of Greenberg and Morris do not necessarily result from their data.

More recent work on early paternal sensitivity to children has employed more scientific procedures. Two areas have been examined in detail. In the first place researchers have been keen to see whether babies activate the same responsiveness in males and females (for a review see Frodi, 1980). They usually show films of babies crying to men and women and measure their psychophysiological responsiveness, in terms of heart rate, blood pressure and skin conductance. In general, men and women appear to react in similar ways (Boukydis and Burgess, 1982; Frodi et al., 1976). Secondly, researchers have examined the ways in which parents greet their newborns, since in many species parental behaviour is programmed to protect neonates and enhance responsive behaviour toward the young. Again studies show few differences between parents in the ways they greet their new children (Parke and O'Leary, 1976; Rodholm and Larson, 1979, 1982). Mothers and fathers tend to exhibit sterotypical patterns of handling their young, starting with their trunk and proceeding to their limbs and faces.

However, such behaviour is not simply released chemically as a result of the onset of parenthood. Even volunteer medical students tend to display the same ways of touching newborn babies (Rodholm and Larson, 1982). Despite the scientific controls used in these studies of parental responsiveness, they tell us only that men appear to react in a 'parental' way to stimuli which induce such a reaction in women. They reveal little about the origins of 'parental responsiveness' — if indeed such a term has any validity.

A second area of research which has compared parents' attachments to their offspring has examined the closeness of the father–child relationship later on in infancy. The first experiments to do this used what has become 'the' measure of attachment; Ainsworth's 'Strange Situation' (Ainsworth and Wittig, 1969). In this procedure, the child is placed on the laboratory floor with his mother and/or father. Observers measure her reactions to the arrival of a stranger, the departure of each parent (leaving the child alone with the stranger) and their return, in a fixed procedure of three minute episodes. Under the stress caused by the stranger's presence and the comings and goings of her parents the child demonstrates something about her confidence with a stranger and her 'attachment' to her parents (Ainsworth et al., 1978).

While the design of the 'Strange Situation' is neat, the many experiments which have compared mother–infant and father–infant 'attachments' have told us little that was not already available in the literature. For example, Schaffer and Emerson (1964) interviewed 60 mothers about the nature of their child's relationships between the ages of six and eighteen months. They found that, contrary to Bowlby's predictions about the centrality of motherhood, other family members were reported to have close relationships with their infants during this period. At six months many children (69 per cent) appeared to have a closer attachment to their mothers. By a year, however, a majority of children (59 per cent) showed no preference for either parent and only a few (17 per cent) retained their 'matricentricity'. The 'Strange Situation' experiments have replicated these findings. Attachment behaviour, as measured in the laboratory, develops in the second half of the first year, is found in most children at one year and peaks at eighteen months. Most studies find that children do not distinguish between their parents – either can serve as an emotional haven (Feldman and Ingham, 1975; Kotelchuck, 1972; Willemsen et al., 1974) – although there is some evidence to suggest that under severe stress children may show a preference for their mothers (Lamb, 1976b; Cohen and Campos, 1974).

Recent research has attempted to examine the dynamics of parent–infant interaction and the 'effects' of fathers' behavioural styles (Pedersen, 1980a). These more naturalistic studies usually take place in the infant's home and at regular intervals over a period of time. Researchers measure the parents' actions towards the child and the child's responses and initiations. Their studies suggest that mothers tend to do the vast majority of the care of their infants while fathers play more social roles (Clarke-Stewart, 1980; Pedersen et al., 1980). However, in the main the interaction styles of mothers and fathers are largely similar (Pedersen, 1980c). This is perhaps not surprising as parents appear to copy one another's styles (Parke and Sawin, 1980) and both use strategies of interaction which are appropriate to the child's developmental level.

While these studies show similarities between parents they also suggest a few differences between mothers and fathers. Early on in the child's life the fathers in these studies appear to differ. For example, Yogman (1977, 1982) filmed parents at play with their ninety-six-day old infants and noted that paternal styles were more intense. Fathers tended to have longer bouts with their babies – for example, their vocalizations lasted an average of eight seconds – and a proportion of these were highly stimulating. Fourteen per cent of these paternal 'turns' consisted of 'tapping games', like walking their fingers up their infants' arms. Mothers spent a quarter of the time playing 'verbal games' and their interactions tended to be shorter – for example, their vocalizations were 2.8 seconds long – and less phrenetic than those of their husbands.

The types of difference which Yogman found in parents of young children appear to have parallels with some parental styles later in the child's development. For example, mothers tend to engage in more verbal interaction (Pedersen et al., 1979; Lamb et al., 1982). Lamb (1980) finds that they tend to play 'conventional games', particularly those with toys.

In contrast he noticed that, when observed, fathers engage in stimulating games, particularly idiosyncratic, rough and tumble play.

That these differences have been found between parents has led researchers to suggest that the father may have a distinct role to play in the child's development. From about the age of one the father comes to be seen as the person to turn to for play (Clarke-Stewart, 1978; Lamb, 1977a, b) and this has been thought to have two possible effects. Firstly, some studies suggest that the father has a special role in the infant's sex-role development. Within days of the child's birth parents ascribe sex-appropriate traits to their offspring and fathers appear to do this more than mothers (Rubin et al., 1974). When still in hospital after delivery observed fathers handle their first born sons more than daughters, although for children born after these patterns do not persist (Parke and O'Leary, 1976). Lamb's (1977b) American data suggested that at the end of the second year clear father-son links are established and are important.

Secondly, other evidence suggests that fathers have a part to play in their childrens' early intellectual development, particularly during the second year. For example, at fifteen months of age the amount children explore the toys that surround them has been found to correlate with paternal involvement (Belsky, 1980), particularly those men who both play vigorous games with their children and also show affection. However, correlations like these do not necessarily tell us about the *causes* of paternal interest and child exploration (Parke, 1978). It may be that stimulating fathers produce brighter children, but alternatively men with bright children may themselves participate more because their children stimulate them — a possibility suggested by Clarke-Stewart's (1980) data.

Such contradictory findings as these reveal two kinds of weakness in the recent spate of observational studies – methodological and theoretical. First, the literature lacks detailed consideration of the methodologies of the studies (Lewis, 1982a). They examine predominantly middle-class, volunteer families for brief periods of time. Their observations normally occur in the early evening, when the father has just returned home from work and when the parents are usually preparing to put the child to bed. While this is a typical time when fathers interact with their children, there are many others when they might not behave in such a stereotypical way – they may well do less rough and tumble play in the early morning or at weekends. When attempting to measure parent-child interaction these researchers tend to investigate large numbers of behaviours. This gives rise to multiple correlations which unfortunately are susceptible to chance findings (Belsky, 1981; Pedersen, 1980c). Perhaps the most worrying aspect of this recent research is that it takes little account of the impact of the observer on the family interaction styles. Parents may well be influenced by the observer to act in what they consider appropriate ways, but little attempt has been made to control for this 'effect' on family interaction.

Secondly, theoretical considerations of the nature of fathers' psychological roles are few in number. So dominant has Bowlby's concept of monotropy been that much discussion has been concerned once more to show that fathers are as competent as mothers (Parke, 1979). What little consideration there is of the origins of paternal involvement tends to be speculative. For example, the psychoanalyst Burlingham (1973) suggested that fathers play rough and tumble games to express their omnipotence – the child in effect symbolises the man's erect penis. More recent researchers have addressed the issue within the traditional nature-nurture debate. For example, Clarke-Stewart (1980) suggests on the basis of the sorts of finding discussed above that fathering styles are both 'biologically induced' and 'culturally supported'. As Pedersen et al. (1979) comment, explanations based on these criteria are not easily testable.

As a result of these interpretative problems, recent efforts to understand the nature of fathering 'behaviour' have examined paternal involvement in 'non-traditional' families, which break from the 'traditional' family pattern – a nuclear unit, with the father the bread-winner and the mother as 'primary care-giver' (Lamb, 1982; Lamb et al., 1983; Russell, 1983). These studies usually focus upon couples where the mother has a job and the father either is the primary care-giver or shares the care with the mother.

Again the results of this research are interesting, but are contradictory and difficult to interpret. For example, two studies have examined parent–infant interaction in traditional and role-reversed families with a baby only a few months old. In the first, Field (1978) found that the interaction styles of American primary care-giver fathers resembled those of mothers more than traditional fathers. For example, full-time fathers tended to talk in high pitched voices – a style characteristic of mothers. In contrast, Lamb et al. (1982) in a study of parent–infant interaction in Sweden, found that parents tended to adopt a style characteristic of their sex, regardless of their care-giving roles. Indeed, this study went on to find, in contrast to Lamb's American 'Strange Situation' data, that Swedish infants appeared to show a strong attachment preference for their mothers (Lamb et al., 1983).

There are many possible reasons why Field and Lamb obtained such different results. Their studies took place in different countries and they used different methods to study parental behaviour – for example, Field used video-recordings to examine the minutiae of mothers' and fathers' interaction styles, while Lamb observed more global patterns of parental behaviour. Whatever the reasons for the differences in their results, the contradictions between them suggest that the selection of special samples, like primary care-giver fathers, does not enable us to infer any more from data which is purely observational. Indeed it creates the additional problem of understanding how such special groups perceive both the research process and indeed their roles as parents. Full-time fathers in the United States may well be under more pressure to appear to be involved with their children than similar men in Sweden.

The trend toward studying different family structures has given rise to more questions than answers. It has underlined the point, made many times (Pedersen, 1980c), that we must attempt to see the world through the eyes of the participants of studies in order to understand more about family relationships and roles. Indeed the interesting features of many of the studies of non-traditional families are the problems faced by parents attempting to establish alternative family types. Many who plan to reverse roles in the end do not (Lamb and Levine, 1983). Follow-up studies of non-traditional families show that a large proportion revert to the traditional pattern, since social pressure to do so is great (Russell, 1982; 1983).

In the early 1980s there was a dramatic drop in the number of published works on father–infant interaction. At the same time theoretical discussion tended to suggest the need to set observational data into a wider social context (for example, Entwisle and Doering, 1980; Belsky, 1981). Richards (1982), for example, suggested that we need to broaden our focus to examine the 'social institution of fatherhood'. In the rest of this chapter I will examine how parents themselves describe the role of the father in families with a young baby.

PATERNAL INVOLVEMENT IN INFANT CARE

Any attempt to measure paternal involvement in the home is bound to encounter problems, since there are no absolute guidelines about which measure can or should be used to construct an overall picture of father participation. Many studies have examined father participation, but, as recent critiques point out (McKee, 1982; Oakley, 1974), these tend to rely upon maternal accounts of paternal involvement, and they ask a limited number of questions about only a few aspects of the domestic division of labour – usually those in which men are more participant. Even when fathers or both parents are used as respondents, we can never be absolutely sure that the picture which a father paints is a valid one. For example, one recent review of studies of paternal involvement found that when compared with interviews, questionnaires are more likely to elicit egalitarian responses about fathers' domestic roles and are less likely to reveal any problems which new parents are facing (Cronenwett, 1982). However, as a means of examining the amount that fathers participate, it seems worthwhile to start this analysis with a consideration of their overall involvement. The data discussed below are derived from a study of 100 married fathers and thirty of their wives. Most fathers were interviewed alone, although in some households this was not possible. They all had a one-year-old child when they were interviewed. They were selected at random from the birth records in Nottingham, although care was taken to distribute the sample equally across the class spectrum[1] and to choose both first- and second-time parents. A description of the sample is shown in table 1. The interview which formed the basis of this study discussed fathers' specific involvements in child care, housework and 'psychologically' with their one-year-olds. Each activity was discussed in detail and a precise account of the fathers' involvement taken. Here we shall examine their participation in child care. Table 2 gives a full breakdown of the items used to construct this measure of paternal involvement. This consisted of fathers' responses to specific questions about feeding, putting to bed, getting up in the morning, nappy changing and taking single-handed responsibility for the child.

Figure 1 plots the distribution of fathers' scores on this scale. It shows that, in keeping with both previous research which used mothers as respondents (Pedersen and Robson, 1969) and other recent work with

Table 1 Description of the sample by occupational/social class background (UK Registrar-Generals' classification, 1970)

	Middle class		Working class	
	I and II (professional/ managerial)	IIIn (white collar, e.g. clerical)	IIIm (skilled manual, e.g. trades)	V (unskilled labourers)
Numbers of first-born children	15	15	15	15
Numbers of second-born children	10	10	10	10
Total (n = 100)	25	25	25	25

[1] The social class divisions are based upon the Registrar General's Classification of Occupations (1970; HMSO) with three modifications. First, I have combined social classes I and II (upper and lower professional–managerial). Secondly, I have divided class III into white-collar (shops and clerical) and manual (skilled trades). Thirdly, I have excluded men from his class IV (semi-skilled occupations).

Table 2 Breakdown of items used to construct the fathers' 'child care' scale

Questions and response types	Score
Questions 81 and 82: Regularity of father feeding N solids throughout the week.	
Less than once a week	0
Once a week	1
More than once a week, but less than wife	2
Shared with wife in time available	3
Question 83 'Do you ever prepare food for him? . . . How often?'	
Never	0
Less than weekly	1
1-3 meals per week	2
At least four meals per week	3
Question 97 'Who usually puts N to bed?'	
Father once a week or less	0
Father 1-3 times per week	1
Shared father and mother (50-50)	2
Father more than mother	3
Question 105 Father's morning role with N	
Occasional (not regular) caretaking	0
Occasional caretaking/regular play	1
Regular caretaking (shared with wife)	2
Question 110 'How often does father change N's nappies?'	
Less than weekly	0
Less than daily	1
At least once daily	2
Question 112 'Do you ever take care of N on your own?'	
Less than monthly	0
Less than weekly	1
1-3 times per week	2
4 or more times per week	3
Maximum possible score =	16

fathers (Russell, 1983), reported father involvement varies greatly from highly participant men to those who appear to do virtually nothing.

Figure 1 only gives a summary of fathers' descriptions of one area of domestic activity and it should be noted that their involvement in child care does not correlate highly with their participation in housework

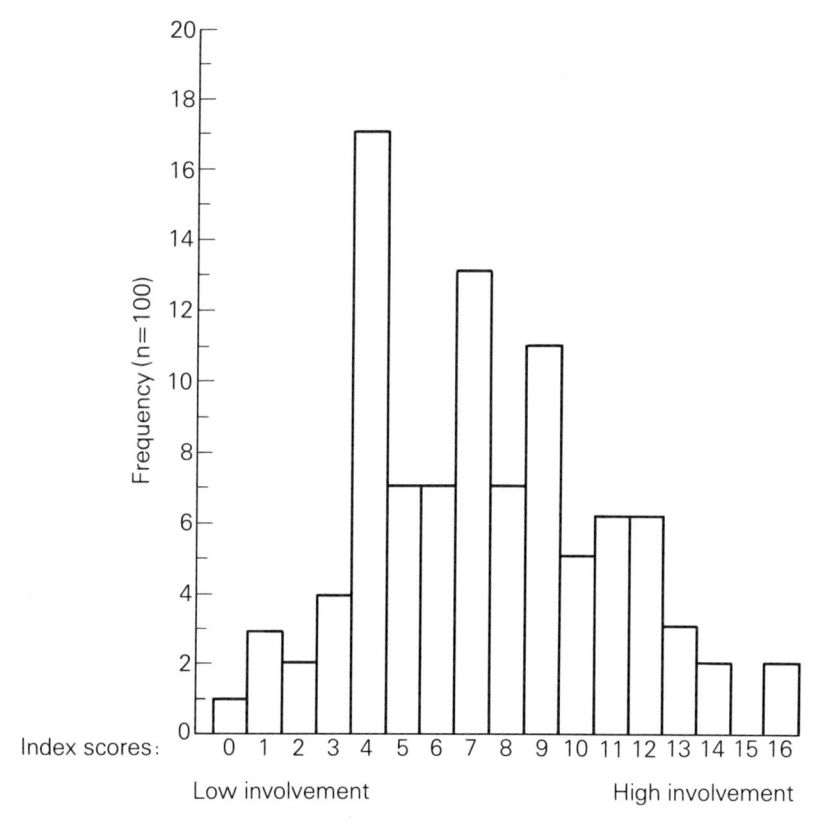

Figure 1 The frequency distribution of scores in the scale measuring father's participation in child care

– some men specialize in caring for the child while others devote more time to housework. This finding replicates Oakley's (1972) data.

Given that the variations in paternal involvement are great, it seems pertinent to examine just what appears to cause such differences between men. Perhaps as a result of the diversity of means by which information about paternal involvement has been gathered, previous studies have suggested a complexity of factors which may be correlated with the level of their participation. The following can be categorized as psychological factors: the man's sex-role identity, his relationship with his own father, his attendance at delivery and his age. Other factors are more social: the father's social class position and the family's work patterns outside the home.

Attempts to distinguish participant fathers from uninvolved men on psychological grounds have tended to produce contradictory evidence. For example, some studies have suggested that men who are more involved in child care tend to identify with an 'androgynous' sex role; that is they tend to hold both 'masculine' and 'feminine' values (Russell, 1978). However, as Russell points out, correlations such as these do not tell us about the direction of effects – whether androgynous men become involved fathers or the reverse – and other research has failed to find such associations (Alter, 1978; Russell, 1983).

Similarly, some authors have argued that fathers who participate highly in the care of their children do so because they themselves had warm, nurturant and involved fathers (Herzog, 1982; Kelly and Worrell, 1976; Keylor, 1978; Manion, 1977). The measures used in this study revealed a statistically significant association between a man's recall of his father's practical involvement in child care and his own participation[2]. However, we cannot conclude that identification is the main cause of involved fatherhood. Other research finds no such significant relationship (Blendis, 1982) and at the same time the opposite case has been put – some highly involved fathers in non-traditional families claim to participate more than typical men because their own fathers were aloof or rejecting (Eiduson et al., 1982).

Some studies have found correlations between two other 'psychologocal' factors and paternal involvement which failed to be replicated in these data, on either the child care scale outlined above, or any of the individual measures within the scale. In the mid-1970s it was found that highly involved men were more likely to have attended the delivery of their children (Manion, 1977; Richards et al., 1977). Despite a groundswell of opinion at the time which predicted that father-attended delivery would increase a man's bond and hence his participation with his offspring (Klaus and Kennell, 1976; Lind, 1974), Richards was at pains to stress that there may not be a causal link between the two correlates. The data from this study do not show a significant correlation between birth attendance and later involvement. This supports the belief that attenders in the early 1970s may well have been highly committed to parenting before the arrival of their offspring. Today a much higher proportion of fathers attend delivery and this seems to have little or no effect on their practical involvement in the home.

Pressman (1980) found, in a sample of forty men, that father-participation is greater in younger fathers. However, she did not distinguish

---

[2] For this analysis, fathers were divided into two groups – those who recalled their fathers as being involved in child care as a matter of routine, and those who did not. A comparison of these two groups revealed that men in the former reported doing more child care [$F(1,90) = 5.007$, $p < 0.05$].

The Role of the Father in the Human Family 239

between older fathers and those with older childern – the men in her study had children at any age below ten. This analysis, where the age of the child was controlled for, failed to find a correlation between the father's age and his contribution to child care. Pressman also found that men with children under two participated more than other men. So perhaps this latter correlation is the more important as older children need less necessary care and attention.

The area of most discussion in the literature concerns the social factors which may correlate with increased paternal involvement. Researchers have been particularly concerned with the participation of men in different social class-groups. Again studies have produced contrasting findings. Large-scale surveys have often found that men in low socio-economic groups do more in the home than their counterparts of high socio-economic standing (Robinson, 1977). In contrast specific studies of families with young children suggest that father-involvement may decrease in the blue-collar professions (Cleary and Shepperdson, 1981), particularly when it comes to housework (Oakley, 1972).

The contradiction between these findings again appears to result from the diversity of samples and methods used in studies of paternal involvement and they may reflect differences in the amounts men participate over the life-cycle of the family. They certainly suggest the need for research which encompasses both the involvement of people across the class spectrum and at various stages of family development.

With its focus upon an early period of the family life-cycle this study found no overall class difference in their child care or housework. However, certain class differences emerged when considering particular child care tasks. For example, non-professional white-collar workers (social class III, white-collar) do more child care in the mornings before going to work than members of the other social groups. Men in professions (social classes I and II) put their one-year-olds to bed significantly less often than their counterparts in other social class groups. At the same time they bottle feed them more often. These sorts of difference suggest that we should examine just what members of different social groups do with their offspring, as well as their overall contribution.

While the type of a man's occupation does not closely correlate with his involvement in child care, two interrelated factors appear to do so. First, paternal involvement has been associated with a couple's division of labour outside the home. Many previous studies have shown that fathers participate more if their wives have jobs (Berk and Berk, 1979; Blood and Wolfe, 1960; Eriksen et al., 1979; Pressman, 1980), although Russell (1983) points out that most of these studies show an increase in fathers' *relative* contribution rather than the hours they devote to child care each week. In contrast to Russell's observation, table 3 shows that in this study there was a positive, linear relationship between the amounts the wives worked and their husbands' participation in child care. At the same time table 4 shows that a father's involvement in child care is related to the number of hours he is available at home – a finding which replicates that of Walker and Woods (1976). Four men spent much of the day at home especially to look after their children. They all did so to enable their wives to work. A finance company representative, for example, worked in the evenings so that his wife could resume her job as a primary school teacher. Like the other fathers in this group, he was able to do this because his job allowed him to work outside the house when his wife was at school. These fathers are few in number and must be treated as a special group for two reasons. First, current employment patterns prevent the vast majority of men from even contemplating such an arrangement. Second, as the studies of non-traditional families also found, the role patterns of these families appeared to be temporary, in that three of them were making plans to revert to a more traditional family pattern.

Even when this group of highly participant men was removed from this analysis, there was a direct relationship between the hours that a man was available during his child's waking day and his actual involvement in his/her care. The fathers who do least child care are, perhaps not surprisingly, those who work more than 45 hours per week. These do slightly less than the men who work a 'normal' working week (indeed an individual comparison between these two groups fails to reach statistical signifance). Men who work shifts, but do not do overtime (and who are therefore at home during the child's waking day) tend to do more caregiving than the other two groups. It is perhaps not surprising that the

Table 3 Mean scores on the 'child care' scale for fathers according to the amount of time their wives were employed per week[a]

Wife's employment status	N	Mean score	Standard deviation
Not working	67	6.55	3.25
Less than 15 hours per week	18	8.06	3.28
More than 15 hours per week	6	8.80	2.56
Full time	5	12.00	4.30
Overall scores	96	7.26	3.28

[a] Comparison between groups: $F(3,92) = 5.33$, $p < 0.05$.

Table 4 Mean scores on the 'child care' scale for fathers according to their own employment patterns[a]

Father's employment status	N	Mean score	Standard deviation
Long hours (more than 45 per week)	25	5.80	3.08
Normal hours (35-45 hours per wk)	46	6.82	2.77
Shifts	12	8.08	3.25
Unemployed/ disabled	8	9.62	2.82
At home in day	4	14.50	1.73
Overall scores	95	7.27	3.05

[a] Comparison between groups: $F(4,90) = 7.75$, $p < 0.0001$.

eight unemployed or disabled men in the sample obtained higher scores on the scales than the three groups of employed fathers.

Like other analyses of father participation, this exercise suggests that psychological correlates with high father-involvement are difficult to pinpoint. On the other hand 'social' factors, like a couple's division of labour outside the home, seem to be better predictors of a man's practical investment in parenthood. This finding is supported by more general evidence from cross-cultural studies. These suggest that father involvement is highly correlated with a culture's dominant means of production − men in hunting communities, for example, do very little child care (Katz and Konner, 1981; Sanday, 1981).

While general patterns of the domestic division of labour are of interest, it seems necessary to consider just how men and women come to adopt such role patterns. The next section will examine how fathers' involvements in care-giving develop over the child's early life.

THE DEVELOPMENTAL CONTEXT OF PATERNAL INVOLVEMENT

Close examination of paternal involvement in child care suggests two further things about the nature of contemporary fathers' participation in family life. First, fathers appear to carry out certain tasks more than others. Second, even if we examine the activities in which men are often more participant and consider them only during the times when they are available, in comparison to their wives contemporary fathers do far less child care.

Table 5 gives a breakdown of paternal involvement in individual child care tasks. In absolute terms these are hard to compare with one another, since activities like nappy changing have to be performed regularly throughout the day while other chores, like getting up to the child at night, vary greatly from child to child over the first year of life. The table therefore gives a breakdown of each task according to three criteria. 'Little or no involvement' meant that the father would involve himself less than weekly in chores like nappy changing, bathing or putting to bed and he would get up to the child at night much less than his wife. 'Some involvement' meant that the father performed an activity more than sporadically, but less than his wife when he was not at work. 'A lot' of involvement meant that a father participated as much as his wife in each activity in his available time at home. Table 5 shows that according to these criteria fathers tend to feed their one-year-olds and put them to bed on a regular basis. They are much less likely to involve themselves in nappy changing, bathing and taking care of the child alone. Indeed a large number of men hardly ever or never do these jobs. The selectivity of their involvement highlights the point that men are much less committed to the practical aspects of parenting than their wives. There are many popular stereotypes regarding just why men do so little, but each contains only a grain of truth. A more complex model of paternal involvement is needed to depict the historical, cultural, marital and developmental aspects of his role. Father-involvement seems upon close inspection to be dependent upon a number of interacting factors which critically change over time. This analysis will therefore examine the way paternal involvement unfolds over the first year of the child's life.

The cultural subordination of men's involvement in family life is clearly visible, even during the period around the child's arrival. As I mentioned in the previous section, this is a time which has attracted much attention from psychologists interested in the early relationship and attachments between parents and their offspring. On the surface it seems that fathers have become much more central characters during this period, since in most western cultures father-attended delivery has become institutionalized (Lewis, 1982b).

I also suggested above that the attendance by men at childbirth has been taken as a sign of increased participation in men's practical involvement in family life. Certainly the increased trend in father-attended delivery has been accompanied by greater participation by husbands in the early days after birth. In 1959 thirty per cent of an equally matched

Table 5 Levels of fathers' involvement in caretaking tasks in families with a one-year-old child

	Bottle feeding %	Solid feeding %	Meal preparation %	ACTIVITY Putting to bed %	Care during the night %	Bathing %	Nappy changing %	Looking after alone %
Little or no involvement	36	32	51	26	13	62	40	53
Some involvement (less than wife)	11	13	29	24	52	9	32	33
A lot of involvement (shared with wife when home)	53	55	20	48	35	29	28	14

sample of Nottingham fathers helped during this period, while in 1979 the figure was seventy-seven per cent. Indeed in the recent study in fifty per cent of homes the father was the only attendant to the mother on her return home from hospital. While changes in personnel have taken place it would be wrong to assume that they necessarily reflect any increased psychological commitment by new fathers to the care of their offspring. A number of factors have contributed to this shift. First, since 1959 women have entered the labour force in large numbers (Ratner, 1980). This means that the traditional source of support, a woman's mother or sister, is far less likely to be on hand to help the new mother. Second, men have longer holidays and are more likely to have time off to support their wives (English Tourist Board, 1979). Third, the hospitalization of childbirth has in effect diminished the importance of the new mother's attendant. By the time she is discharged home she is considered to be in good enough health to be looked after by anyone (even her husband!). No longer do women have to rely upon the experienced female kin in order to learn the skills of parentcraft.

Even during this brief period the role of the father is usually limited. Some countries, like Sweden, have implemented complex 'paternity leave' schemes so that men can be at home with their wives after delivery, and also so that couples may reverse the traditional role pattern if they choose (Lamb and Levine, 1983). In Great Britain no such governmental provision exists and of the few firms who give their employees time off, many do so at reduced rates of pay and usually only for a few days (Bell colleagues, 1983; Daniel, 1980; Equal Opportunities Commission, 1980).

So, seventy-two per cent of the fathers in this sample took some or all of their annual leave after the arrival of the child. Within this group significantly more non-professional white-collar workers, (Class III, white-collar) took more than one working week, than members of the other social groups represented in the study. Professional men claim to have too many commitments to take a lot of time off and manual workers said they needed the money or did not have paid leave available to them. A large number, which was difficult to estimate, take sick leave.

It is difficult to measure the exact participation of fathers in the period when they and their wives are at home with the young baby. Just as there are great variations in the amount of child care which men do with their one-year-olds, paternal involvement varies greatly. Few fathers share the care of the baby at this time. The two men in this study who took complete control, did so as their wives were ill. In general fathers suggest that they have a clear role to play when they attend to their wives. As this father suggests, most take on responsibility for the home and in the case of families with more than one child, the older children:

Very much a sort of . . . er . . . becoming, I suppose, not a nurse . . . what would you call it – a skivvy I suppose. I was very much . . . cleaning up after Carrie (wife), looking after that thing there (looks at the dog), looking after myself . . . cooking meals etcetera . . . so generally being house 'man' I suppose, or house 'husband'. (Sales representative)

Even during the early days of parenthood fathers often appear to be cut off from their children. They suggest three reasons why their wives take care of the child, while they work around the mother–baby pair. First, many mention the cultural prescription of this division of labour. Their wives have been central characters during the pregnancy and delivery, so they assume that 'mothering' comes naturally to females.

Second, the cultural assumption that women 'mother' during this period, may well be exacerbated by the way in which we manage the care of mothers and babies just after delivery. Lomas (1964) suggests that the segregation of the new mother in hospital has the symbolic effect of excluding the father or denying his involvement, and this view is often expressed by fathers themselves. On a more practical level mothers in hospital go through a rapid learning process, however brief their stay. First-timers learn *in situ* the skills of feeding, changing and bathing under the watchful eye of the nursing staff. Women who stay longer in hospital have an obvious advantage over their spouses. The following mother, for example, compares her experiences with her husband's after her Caesarian delivery:

I suppose in a way it's a bit different, because I was in hospital a fortnight and by the time I came out I'd been used to bathing the baby and feeding . . . the feeding was all right. I think we managed quite well . . . it's just that I was exhausted all the time. I'd see to the baby and fell asleep. David (husband) did most of the housework. He had a week off and me mother helped the next week. (Housewife, whose husband is a floor layer)

This mother also points out that it seems easier and most logical for the domestic work load to be divided in this way – the third reason suggested by parents for the father's diminished early role. In order to understand the domestic division of labour at any time we must consider not only the availability of each family member but also the ways in which patterns of child care develop. The very early period of the child's life, when her father is likely to be at home, is only short and transitory. After this brief interval the father usually has to return to work leaving the mother to cope with the baby. It, therefore, seems advantageous for the mother to gain as much experience with the baby as possible and also to rest in preparation for her role as primary care-giver.

In addition, when a married mother breast feeds, her husband can only participate in activities like feeding when the baby's routine allows. In these families fathers are less likely to involve themselves in other activities such as night-waking because their involvement is necessarily minimal:

Well the first three months . . . as I say she was awake every couple of hours . . . um, at first I used to get up and wake up, sort of properly, but as we got used to it, I tended to sort of wake up and cast a bleary eye and then turn over and go back to sleep (Prompt: So you lost your enthusiasm for getting up?) Well there just didn't seem any need. I had to go to work . . . I mean I never went and slept in the other room or anything . . . In the first month I was up and watching and really with it, but as it went on I . . . (Wife: There wasn't a lot you could do was there?). (Sales director and his wife who is part-time physiotherapist)

Like this father many men become less available to their children, both practically and psychologically, after this brief period when they are at home, particulary when the demands of home and work are at odds with one another. As the first year of parenthood unfolds the father's involvement becomes dependent upon how his own and his child's routines fit in with one another. While one study found that paternal involvement increased during the first three months (Katsh, 1981) most suggest that after this initial high profile, father participation decreases over this period (Bernstein and Cyr, 1957; Oakley, 1979; Rebelsky and Hanks, 1971). Later in the first year fathers come more into the picture (Richards et al., 1977; Wandersman, 1980), although as the above discussion and other studies (Beail, 1982; Cowan and Cowan, 1981; Kotelchuck, 1972; Moss 1981; Shereshevsky and Yarrow, 1973) suggest, there is usually a large difference between parents' roles.

As I mentioned at the start of this section many contrasting reasons have been put forward to explain this clear division of labour. Traditionally it has been accepted that such an arrangement occurs naturally – according to a simple biological imperative. In recent years in the wake of the recent re-emergence of feminism, a more critical eye has been cast upon the part played by fathers in domestic 'politics':

It is clear that birth produces a peak of masculine domesticity; many fathers may be quite heavily involved in the early days, but this level of participation falls off as babies become older, life becomes more routine-like and the novelty of fatherhood is eroded by time and sleepless nights (p. 211) . . . All the old dodges are dragged in: you do it better than me; we have different standards; I don't know how to do it; you go out and earn the money then (p. 219). (Oakley, 1979)

The feminist view, that men simply 'dodge' any involvement in child care is indeed supported by a great deal of the evidence, in that fathering is regarded in many families as an 'optional' activity. For example, table 5 shows that forty per cent of fathers change their one-year-olds never or

The Role of the Father in the Human Family   247

hardly ever. Men often excuse their non-participation in ways similar to those suggested by Oakley, above:

Never . . . I've never changed his nappy. I'm hopeless at it, for one thing. (Prompt: Is there any particular reason why you haven't?) Oh yes; I could vomit on the spot, I could, at the sight of pooh, I could. (Prompt: You do?) I do, yes. On certain occasions where I've had to lend a hand . . . you know if Rachel . . . when on occasions she's got herself in a bit of a state and we have to lend a hand . . . it's terrible! I couldn't even . . . we used to have a guinea pig and I couldn't even clean the guinea pig out . . . It's just one of those things I cannot do. (General studies lecturer)

It is undeniable that men in many respects shirk away from child care, but the feminist perspective by itself does not explain why most fathers become far less involved than their wives. At least three other interacting factors must be taken into account, and the latter two derive from the first – the father's usual absence from the home while at work.

The analysis of overall father-involvement reported above showed that the strongest predictor of a married father's participation is his working arrangements outside the home. The time and energy taken in working and travelling to and from employment necessarily impinge upon a man's availability and his inclination to take part in the care of the children. Table 3 shows that in a sample of 100, twenty-five worked long hours as a matter of course. It does not show that many other men worked overtime on a more sporadic basis. Recent research shows ironically that at the stage of the life-cycle when the amount of domestic labour increases – when the children are young – men are far more likely to seek additional hours of employment in order to keep their families financially solvent (Moss, 1980). Moss reported, for example, that fathers with preschool children are four times more likely to work overtime than their childless contemporaries. As a less financially rewarding exercise, the employment of wives is much more likely to be perceived as a bonus to the father's income, or as a means of giving her a 'break' away from the children (Moss, 1980; Ratner, 1980).

So the popular stereotype that men are the 'providers' for the family holds for large numbers of families. This has led many sociologists (Parsons and Bales, 1955, for example) and psychoanalysts (Bowlby 1954) to ignore the role of men as fathers and strongly contributed to a neglect of the relationship between men's working and domestic lives. Only recently have psychologists (Pleck, 1977; O'Brien 1982) begun to realize the conflicts which men have between these two areas of their lives.

Second, child care is not only a job that is a 'chore' and which women cannot shirk away from. As Davidoff (1976) suggests, domestic chores can be perceived in a number of contrasting ways. Like any other task they can be regarded on the one hand as a 'drudge' but may also be labelled a 'science' (a constructive way of fulfilling one's ambitions), or the hallmark of good parenting. On one level, mothers gain psychological advantages from their major role in child care, and the amount husbands participate depends greatly on the extent their wives are prepared to let them do so. The negotiation process by which couples come to organize the family division of labour surely lacks psychological research (Pedersen, 1980b) and only a few family sociologists (Backett, 1982; La Rossa, 1977; La Rossa and La Rossa, 1981) have treated this aspect of paternal involvement seriously. It was clear in this study that fathers could attempt to participate 'too much', at least as far as their wives were concerned, and that mothers felt ambivalent about paternal involvement:

At the beginning I used to interfere . . . I think that's perhaps the right word . . . *too much*. You know, if she wanted to do something such a way I'd say 'I think this is a better way'. And I learned, through getting my fingers burnt, a little bit, that perhaps that wasn't the best way to go about it . . . with regard to feeding. After the first couple of weeks I was back at work, you know . . . I hadn't got a say in it anyway 'cos I wasn't here. (Hairdresser)

(Interviewer: Were you keen for Ron to give James his bottle?) No, I really wanted to do it myself this time, though I really felt he *ought* to. Just to show interest. I really contradicted myself a lot in my feelings over that. I didn't want anybody else to feed him but me, but I used to think he ought. (Part-time nursing auxiliary whose husband is a goods representative)

Third, the child's needs and abilities change continually, especially over the first year of life. He/she demands varying amounts of attention from his/her care-givers. Paternal involvement appears to slot into neatly prescribed routines generally agreed upon by both parents and in keeping with the child's demands. Table 5 shows that the most frequent activities which fathers participate in tend to be feeding and putting the child to bed. About half the sample did these on a regular basis and many commented that the child's evening routine was set by the father's normal arrival home. At the age of one the child's bedtime is relatively flexible. At later ages it is not and one recent survey found that more than half of fathers of five-year-olds arrive home after the child's bed time (Jackson, 1984). So, men do not simply shirk their responsibilities. Indeed the evidence suggests that they attempt to keep up with changes in the child's daily pattern. For example, table 6 depicts two 'landmarks' in the early life of the child. The left-hand column shows when the 100 children were first fed with a bottle. The right-hand column shows when fathers recall having first fed their children. Taken together the figures suggest that fathers appear keen to participate in child care when the opportunity arises.

While they may want to try their skill at certain tasks, men do not become involved on a regular basis. One reason for this lies in their unavailability. For example, table 5 shows that 62 men hardly ever bathed their one-year-olds. Forty of these were home for less than two baths per

Table 6 Ages at which the child was first bottle-fed and the father gave his first bottle[a]

Age of child	N started bottle feeding at	F started bottle feeding N at
Less than 2 weeks	55	42
2-3 weeks	4	10
1-2 months	15	17
3-5 months	10	12
After 6 months	2	3
Father has never fed N	–	1

[a] 14 of sample never had a bottle, so were never bottle-fed by father.

week. Such infrequent availability has its effect upon both parents. Mothers get into a routine of bathing the child, while fathers are constantly aware that they are less practiced and less skilled.

This analysis of paternal involvement in child care suggests that the practical aspects of fathering are part-time and often of secondary importance. Men's participation is only necessary if their wives work. That their direct involvement seems slight does not automatically imply that they are psychologically detached from their families' daily routines. Indeed there is evidence to suggest that fathers may play an important supportive role to their wives. For example, one recent study found that a new mother's adaptation to successful breast feeding was best predicted by her views on her husband's specific encouragement about feeding and his general support (Switzky et al., 1979).

THE FATHER'S PSYCHOLOGICAL ROLE IN CONTEXT

In this chapter I have attempted to contrast parents' accounts of the father's role with psychologists' observations of men's interactions with their infants. The above analysis of paternal involvement in child care suggests that the vast majority of men play only a small part in their infants' daily routine. They continue to act as occasional 'helpers' to their wives long after the first days of parenthood (Cleary and Shepperdson, 1981; Oakley, 1974). However, observational studies suggest that the father becomes 'engrossed' with his new offspring and that their relationship develops into a close attachment, which at one year postpartum is easily seen, for example, in the infant's behaviour in the 'Strange Situation'. In this concluding section I will attempt to reconcile these two contrasting views, by describing parents' accounts of two aspects of the father's psychological commitment to his child. First, I will examine briefly the man's relationship with his infant immediately after delivery. Second, I will look more closely at the nature of parent–infant attachments as they develop. Parents' accounts suggest that, in keeping with his diminished domestic role, the father's psychological influence in the family is usually qualitatively different from that of the mother.

If we start by examining briefly the period following delivery, fathers appear to react to the child's arrival in more complex ways than by simply becoming engrossed. The early days of parenthood are often not easy and sociological accounts have often described the period as a transitional 'crisis' (Le Masters, 1957). While this description may exaggerate the problems faced during this time (Hobbs, 1965, 1968) it is clear that the crisis perspective is at least partly valid.

Just as his working life appears to impinge upon his commitment to child care, a father often finds that the conflict between the demands of employment and home influences his psychological state. The available evidence suggests that men are likely to suffer from symptoms of depression, similar to those experienced by their wives. Two recent studies have found such symptoms in one third of new fathers (Atkinson, 1979; Zaslow et al., 1981). Indeed research suggests that we should consider the family as a system of interconnected influences during this period. For example, psychiatric evidence suggests a link between a mother's postpartum depression and her husband's support – or lack of it! (Kaplan and Blackman, 1969). Similarly, easier adjustment to early motherhood has been correlated with paternal assistance (Feiring, 1975).

The stresses and strains of early parenthood seem to have an influence upon mothers' and fathers' attitudes towards the baby. Oakley (1979) found that seventy-five per cent of mothers felt 'not interested' in their babies in the first days after delivery. Likewise, fathers do not simply fall in love with their new offspring. Fifty per cent of these men (a group which included significantly more first-timers) reported that their early contact with their babies was an 'odd' experience. Indeed many men have very little to do with the child during the first few weeks. The one study which has measured early verbal interactions (using continuous tape recordings over twenty-four-hour periods), suggests that in the United States fathers in the mid-1960s spent, on average, 37.7 seconds per day interacting with their infants in the first three months. There is no evidence to indicate that great change has occurred since this date[3].

---

[3] A colleague (Myron Korman, at the University of Nottingham) and I are currently replicating the Rebelsky and Hanks study with a new sample to see whether their findings have any relevance today. While contemporary fathers interact more than those in the former study, the difference between them is only slight (Korman and Lewis, in preparation).

So, while most mothers learn quickly how to cope with handling the new arrival, men often find it hard to establish a relationship, particularly if they work long hours. Repeated studies have reported that fathers remain wary of handling their babies when small (Cleary and Shepperdson, 1981; McKee, 1979; Oakley, 1979):

> I spend a bit more time with her now than what I did when she was born, 'cos I enjoy the fact that she can walk around and she's a bit stronger. As I've said before, I was a bit frightened when she was a bit weak ... you know, when her head was flopping about the place. I was a bit worried ... and I would, er ... I wouldn't stay away purposefully, but I'd seem to edge away ... sit in the single chair sooner than sit on the settee with her, you know. (Transport manager's assistant)

In the first section of this chapter I reported that in many observational studies fathers seem to specialize in social and boisterous play with their children. While cross-cultural evidence in Sweden (Lamb et al., 1983) and Fiji (Katz and Konner, 1981) shows that this role is not ubiquitous, parents' accounts of fathers' styles suggest that a majority of British men take on the role as playmate (Lewis et al., 1982; Newson and Newson, 1963). Likewise in this study both parents dwelt at length upon the ways in which fathers involved their children – particularly in social games, like peep bo, and rough and tumble play.

While many men appear to perceive their initial feelings towards the child as somewhat distant, a year after delivery their relationship is usually upon a much stronger footing. Like the transport manager's assistant quoted above, sixty-three per cent of the men reported that they had become emotionally more involved as the year unfolded. Fathers may spend very little time care-giving, but many suggested that much of their available time is spent in play. Eighty of these men claimed to spend at least half-an-hour per day engaging their one-year-olds in such activities. Indeed the contemporary father uses play as a means of forging a relationship with his children in their brief daily encounters and, as this quotation suggests, gradually both parties come to regard one another as more approachable:

> I think when they're first born and they're lying there and they are having a go at you in their own little way, and then they go through a stage when they're just sitting there and looking around ... to me, I don't know ... it's like having a little dog, that you're sat there looking at ... there's nothing you can do with it, you know ... he just sits and laughs at you or ignores you completely and just sits and looks at something in his own little way. But now he's crawling round, he's getting to the stage where any time now he'll be wandering off of his own, his two little feet running off somewhere. But he's at a better stage now. He's at a nice stage now. (Prompt: So it's a stage you find much nicer than the ...) Oh, I enjoy it much, much more, yeah ... he's giving you things now, he'll come and put his head in your knee which I presume is his own way of showing affection. (Goods representative)

Parents' accounts suggest further that the nature of their play often differs in important ways. In 'traditional' families, mothers usually fit 'playful' activities into their daily domestic routines. Fathers' play is more concentrated, it may be the main point of contact between a man and his child and furthermore it usually is more 'public' in that mothers are normally around when it takes place. When discussing each of their play-styles mothers and fathers differed on one point. They were asked if any of their play routines assumed any special significance to the child. When the thirty wives were compared with their husbands, significantly more mothers labelled one or more of their husbands' activities as 'special' than fathers did about their own activities. This difference is in keeping with some observational data; for example, Clarke-Stewart (1978) found that when their husbands come home, mothers tended to withdraw to let their husbands have their 'special' playtime with the baby. While fathers become involved with their infants through play, these parents suggest that as a result of the nature of their daily parental contact with them mothers and fathers tend to develop different kinds of closeness to them. For example, table 7 summarizes fathers' descriptions of two aspects of their child's preference for a companion. It shows that if the child has a preferred playmate, he/she is more likely to choose the father. On the other hand, a majority of one-year-olds will turn to their mothers when frightened. That mothers and fathers agree about these matters suggests that their assessments are reliable.

Indeed, when describing the differences between each parent's relationship with their children, many fathers suggested that their psychological role, like their practical involvement, is of secondary importance. Their emphasis upon play makes many men become 'mates' to their children rather than essential care-givers. Repeatedly in interviews fathers suggested that they can be substituted:

You know anyone could come in and once he'd got to know them he'd be . . . I think he seems to be equally happy with them, or could easily be. (Prompt: Is this the same with the mother?) I don't know it's a bit different with the mums 'cos of

Table 7 Who does the child turn to (a) for fun (b) when frightened? (Fathers' reports)

Child turns to	For fun	If frightened
1 Mother	17	64
2 Either parent	52	31
3 Father	31	4
	100	99 (1 missing)

the amount of time spent with her. They spend a lot of time with them . . . I could be substituted . . . that's what I mean. He recognises me, things like that, but give him five minutes and he'd be happy with someone else. I suppose I could flit out of his life without him bothering a great deal. (Panel beater)'

(Interviewer: Can a father and his son be as close as a mother and her son at this age?) Yes, but in a different way . . . with a mother it's a bond which is obviously . . . all mothers . . . you have with all mothers. Whereas with a father its a, he's a mate, probably a friend (Prompt: that's the difference, then?) Yeah, I think one's a love thing and one's a friend thing. (Warehouse labourer)

Table 7 shows that, like their practical involvement, there are great variations in fathers' relationships with their one-year-olds. That such differences exist shows just why fatherhood is a cultural institution which remains opaque and so hard to define. In contemporary society some men do become highly involved with their children, while most make a token effort to keep up with the practical skills of child rearing and concentrate on becoming 'super pals'. However, many have an underlying feeling that their paternal relationships are not on as sound a footing as their wives' maternal 'bonds'. These patterns continue long after infancy (Lewis et al., 1982).

Research on fathers should now maintain its broad perspective. Instead of simply comparing fathering with mothering, the interaction of work, community provision and family life has to be a central feature of our analysis of men's family roles. Any reinstatement of primarily biological factors must allow for the constraints which culture and socioeconomic conditions have been shown to impose. We are only beginning to realize the implications of Margaret Mead's (1950, 1962) point that fatherhood is a biological necessity, but a social invention.

REFERENCES

Ainsworth, M. D. S. and Wittig. B. A. 1969: Attachment and exploratory behaviour in one-year olds. In B. M. Foss (ed.), *Determinants of Infant Behaviour*, London: Methuen.

Ainsworth, M. D. S., Blehar, M. C., Waters, E. and Wall, S. 1978: *Patterns of Attachment: A Psychological Study of the Strange Situation*. Hillsdale, New Jersey: Lawrence Erlbaum.

Alter, J. L. S. 1978: *The Relationship Between Self Esteem and the Male Sex Role*. Unpublished PhD. thesis, Catholic University of America. (U. M. 78-16858).

Atkinson, A. K. 1979: *Postpartum Depression in Premiparous Parents: Caretaking Demands and Prepartum Expectations*. Unpublished PhD. thesis, Wayne State University. Original reference not used. See *Dissertation Abstracts International* (1980), 40 (5326B).

Backett, K. C. L. 1982: *Mothers and Fathers: A Study of the Development and Negotiation of Parental Behaviour*, London: Macmillan.

Beail, N. 1982: The role of the father in child care. Paper presented to the *British Psychological Society*, Social Psychology Section, Edinburgh, September.
Bell, C., McKee, L. and Priestley, K. 1983: *Fathers, Childbirth and Work: A Report of a Study*. Manchester: Equal Opportunities Commission.
Belsky, J. 1980: A family analysis of parental influence on infant exploratory competence. In F. Pedersen (ed.) *The Father-Infant Relationship: Observational Studies in the Family Setting*, New York: Praeger.
—— 1981: Early human experience: a family perspective. *Developmental Psychology*, 17, 3-23.
Berk, R. A. and Berk, S. F. 1979: *Labour and Leisure at Home*. Beverley Hills: Sage Publications.
Bernstein, R. and Cyr, F. 1957: A study of interviews with husbands in a prenatal and child health programme. *Social Casework*, 38, 473-80.
Blendis, J. 1982: Mens' experiences of their own fathers. In N. Beail and J. McGuire (eds), *Fathers: Psychological Perspectives*, London: Junction Books.
Blood, R. and Wolfe, D. 1960: *Husbands and Wives*. New York: Free Press.
Boukydis, C. F. Z. and Burgess, R. L. 1982: Adult physiological responsiveness to infant cries: effects of temperament of infant, parental status and gender. *Child Development*, 53, 1291-8.
Bowlby, J. 1954: *Child Care and the Growth of Love*. Harmondsworth: Penguin.
—— 1969: *Attachment and Loss: Volume 1 Attachment*. Harmondsworth: Penguin.
Burlingham, D. 1973: The pre-Oedipal infant-father relationship. *Psychoanalytic Study of the Child*, 28, 23-47.
Clarke-Stewart, K. A. 1978: And daddy makes three: the fathers' impact on mother and young child. *Child Development*, 49, 466-79.
—— 1980: The father's contribution to childrens' cognitive and social development in early childhood. In F. A. Pedersen (ed.), *The Father-Infant Relationship: Observational Studies in the Family Setting*, New York: Praeger.
Cleary, J. and Shepperdson, B. 1981: *The Fynone Fathers*, Supplementary Paper 2, Motherhood in Swansea Project, University College, Swansea.
Cohen, L. and Campos, J. 1974: Father, mother and stranger as elicitors of attachment behaviours in infancy, *Developmental Psychology*, 10, 146-54.
Cowan, C. P. and Cowan, P. A. 1981: Couple role arrangements and satisfaction during family formation. Paper presented at the *Society for Research in to Child Development*, Boston, March-April 1981.
Cronenwett, L. R. 1982: Father participation in child care: a critical review. *Research in Nursing and Health*, 5, 63-72.
Daniel, W. W. 1980: *Maternity Rights: The Experience of Women*. London: Policy Studies Institute, (588).
Davidoff, L. 1976: The rationalisation of housework. In D. L. Barker and S. Allen (eds), *Dependence and Exploitation in Work and Marriage*, London: Longman.
Eiduson, B. T., Kornfein, M., Zimmerman, I. L. and Weisner, T. S. 1982: Comparative socialisation practices in traditional and alternative families. In M. Lamb (ed.) *Non-Traditional Families*, Hillsdale, New Jersey: Erlbaum.
English Tourist Board 1979: *Forecasts of Tourism by British Residents, 1985-1995*.

Entwisle, D. and Doering, S. 1980: *The First Birth: A Family Turning Point*, Baltimore: Johns Hopkins University Press.

Equal Opportunities Commission 1980: *Study 230*, 18, Manchester, November 1980.

Erikson, J. A., Yancy, W. L., and Eriksen, E. P. 1979: The division of family roles, *Journal of Marriage and the Family*, 41, 301–13.

Feiring, C. 1975: *The Influence of the Child and Secondary Parent on Maternal Behaviour: Toward a Social Systems View of Early Infant–Mother Attachment*, Unpublished PhD thesis, University of Pittsburgh. (U. M. 76–347).

Feldman, S. and Ingham, M. 1975: Attachment behavior: a validation study in two age-groups, *Child Development*, 46, 309–30.

Field, T. 1978: Interaction behaviors of primary versus secondary caretaker fathers. *Developmental Psychology*, 183–4.

Frodi, A. M., Lamb, M. E., Leavitt, L. and Donovan, W. L. 1976: Fathers' and mothers' responses to infant smiles and cries. *Infant Behavior and Development*, 1, 187–98.

Frodi, A. 1980: Paternal–baby responsiveness and involvement. *Infant Mental Health Journal*, 1, 150–60.

Greenberg, M. and Morris, N. 1974: Engrossment: the newborn's impact upon the father. *American Journal of Orthopsychiatry*, 44, 520–31.

Hobbs, D. F. 1965: Parenthood as crisis: a third study. *Journal of Marriage and the Family*, 27, 367–72.

—— 1968: Transition to parenthood: a replication and an extension. *Journal of marriage and Family*, 30, 413–17.

Jackson, B. 1984: *Fatherhood*, London: Allen and Unwin.

Kaplan, E. and Blackman, L. 1969: The husband's role in psychiatric illness associated with childbearing. *Psychiatric Quarterly*, 43, 396–409.

Katsh, B. S. 1981: Fathers and infants: reported caregiving and interaction. *Journal of Family Issues*, 2, 275–96.

Katz, M. M. and Konner, M. J. 1981: The role of the father: an anthropological perspective. In M. E. Lamb (ed.), *The Role of the Father in Child Development*, New York: Wiley.

Kelly, J. A., and Worrell, L. 1976: Parental behaviors related to masculine, feminine and androgynous sex-role orientations, *Journal of Consulting and Clinical Psychology*, 44, 843–51.

Keylor, R. 1978: *Paternal interaction with two-year-olds*. Unpublished PhD thesis, Boston University.

Klaus, M. H. and Kennell, J. H. 1976: *Maternal–Infant Bonding*. St Louis: C. V. Mosby.

Kotelchuck, M. 1972: *The Nature of the Child's Tie to His Father*. Unpublished PhD thesis, Harvard.

Lamb, M. E. 1976a, 1981: *The role of the father in child development*. New York: Wiley.

—— 1976b: Effects of stress and cohort on mother and father–infant interaction. *Developmental Psychology*, 12, 435–43.

—— 1977a: Father–infant and mother–infant interaction in the first year of life. *Child Development*, 48, 167–81.

—— 1977b: The development of mother–infant and father–infant attachments in the second year of life. *Developmental Psychology*, 13, 637–48.

—— 1980: The development of parent–infant attachments in the first year of life. In F. Pedersen (ed.), *The Father–Infant Relationship: Observational Studies in the Family Setting*, New York: Praeger.
—— 1982: *Non-Traditional Families*. New Jersey: Lawrence Erlbaum.
Lamb, M. E., Frodi, A. M. Hwang, C-P., Frodi, M., and Steinberg, J. 1982: Mother– and father–infant interaction involving play and holding in traditional and non-traditional Swedish families. *Developmental Psychology*, 18, 215–21.
Lamb, M. E., Frodi, M. Hwang, C-P., and Frodi, A. 1983: Effects of paternal involvement on infant preferences for mothers and fathers. *Developmental Psychology*, 54, 450–8.
Lamb, M. E. and Levine, J. A. 1983: The Swedish parental insurance scheme: an experiment in social engineering. In M. E. Lamb and A. Sagi (eds), *Fatherhood and Family Policy*, Hillsdale, New Jersey: Lawrence Erlbaum.
La Rossa, R. 1977: *Conflict and Power in Marriage: Expecting the First Child*. Beverley Hills: Sage.
La Rossa, R. and La Rossa, M. M. 1981: *Transition to parenthood: How Infants Change Families*. Beverley Hills: Sage.
Le Masters, E. E. 1957: Parenthood as crisis. *Marriage and Family Living*, 19, 352–5.
Lewis, C. 1982a: The observation of father–infant relationships: an 'attachment' to outmoded concepts? In L. McKee and M. O'Brien (eds), *The Father Figure*, London: Tavistock.
—— 1982b: 'A feeling you can't scratch?' The effect of pregnancy and birth on married men. In N. Beail and J. McGuire (eds), *Fathers: Psychological Perspectives*, London: Junction.
Lewis, C., Newson, J. and Newson, E. 1982: Father participation through childhood and it's relation to career aspirations and delinquency. In N. Beail, and J. M. McGuire (eds), *Fathers: Psychological Perspectives*, London: Junction.
Lind, J. 1974: Observations after delivery of communications between mother–infant–father. Paper presented at the *International Congress of Paediatrics*, Beunos Aires 1974.
Lomas, P. 1964: Childbirth ritual. *New Society*, 4, (118). 31 December, 13–14.
Manion, J. 1977: A Study of fathers and infant caretaking. *Birth and the Family Journal*, 4, 174–8.
McKee, L. 1979: Fathers' participation in infant care. Paper presented at the *British Sociological Association Fatherhood Conference*, Warwick, April 1979, 368–72.
—— 1982: Fathers' participation in infant care: a critique. In L. McKee and M. O'Brien (eds), *The Father Figure*, London: Tavistock.
Mead, M. 1950, 1962: *Male and Female*. Harmondsworth: Penguin.
Moss, P. 1980: Parents at work. In P. Moss and N. Fonda (eds), *Work and the Family*, London: Temple Smith.
—— 1981: *Transition to Parenthood Project*. Annual Report, Thomas Coram Research Unit, unpublished manuscript, January 1981.
Newson, J. and Newson, E. 1963: Infant care in an urban community. London: Allen and Unwin.
Oakley, A. 1972: Are husbands good housewives? *New Society*, 19, February 1972, 337–40.

—— 1974: *The Sociology of Housework*. London, Martin Robertson.
—— 1979: *Becoming a Mother*. Oxford: Martin Robertson.
O'Brien, M. 1982: The working father. In N. Beail and J. McGuire (eds), *Fathers: Psychological Perspectives*, London: Junction.
Parke, R. D. 1978: Parent–infant interaction: progress, paradigms and problems. In G. P. Sacket and H. Haywood (eds), *Application of Observational-Ethological Methods in the Study of Mental Retardation*, Baltimore, University Park.
—— 1979: Perspectives on father–infant interaction. In J. D. Osofsky (ed.), *Handbook of Infancy* 1979.
Parke, R. D. and O'Leary, S. 1976: Family interaction in the newborn period: some findings, some observations and some unsolved issues. In K. Riegal and J. Meacham (eds), *The Developing Individual in a Changing World*, The Hague: Mouton.
Parke, R. D. and Sawin, D. B. 1980: The family in early infancy: social interaction and attitudinal analysis. In F. A. Pedersen (ed.), *The Father–Infant Relationship: Observational Studies in a Family Context*, New York: Praeger.
Parsons, T. and Bales, R. F. 1955: *Family, Socialisation and Interaction Process*. New York: The Free Press.
Pedersen, F. A. 1980a: *The Father–Infant Relationship: Observational Studies in the Family Setting*. New York: Praeger.
—— 1980b: Issues related to fathers and infants. In F. A. Pedersen (ed.), *The Father–Infant Relationship: Observational Studies in the Family Setting*, New York: Praeger.
—— 1980c: Overview: answers and formulated questions. In F. A. Pedersen, (ed.), *The Father–Infant Relationship: Observational Studies in the Family Setting*, New York: Praeger.
Pedersen, F. A., Anderson, B. J., and Cain, R. L. 1980: Parent–infant and husband–wife interactions observed at age five months. In F. A. Pedersen (ed.), *The Father–Infant Relationship: Observational Studies in the Family Setting*, New York: Praeger.
Pedersen, F. A. and Robson, K. 1969: Father participation in infancy. *American Journal of Orthopsychiatry*, 39, 466–72.
Pedersen, F. A., Yarrow, L., Anderson, B. and Cain, R. L. 1979: Conceptualization of father influences in the infancy period. In M. Lewis, and L. Rosenblum (eds), *The Social Network of the Developing Infant*, New York: Plenum.
Pleck, J. 1977: The work–family role system. *Social Problems*, 24, 417–27.
Pressman, R. A. 1980: *Father Participation in Childcare: An Exploratory Study of Factors Associated with Father Participation in Child Care Among Fathers with Working Wives and Their Children*, Unpublished PhD thesis, Boston University School of Education, University Microfilms, UM 80-24146.
Price-Bonham, S. 1976: Bibliography of literature related to roles of fathers, *Family Coodinator*, 25, 489–512.
Price-Bonham, S., Pittman, J. F., and Welch, C. O. 1981: The father role: an update. *Infant Mental Health Journal*, 2, 264–89.
Ratner, R. S. 1980: The policy and the problem: overview of seven countries. In R. S. Ratner (ed.), *Equal Employment Policy for Women*, Philadelphia: Temple University Press.

Rebelsky, F. and Hanks, C. 1971: Fathers' verbal interaction with infants in the first three months. *Child Development*, 42, 63-8.
Richards, M. P. M. 1982: How should we approach the study of fathers? In L. McKee and M. O'Brien (eds), *The Father Figure*, London: Tavistock.
Richards, M., Dunn, J. and Antonis, B. 1977: Caretaking in the first year of life: the role of fathers and mothers social isolation. *Child Care, Health and Development*, 3, 23-36.
Robinson, J. P. 1977: *How Americans Use Time: A Sociological Analysis of Everyday Behaviour*. New York: Praeger.
Rodholm, M. and Larson, K. 1979. Father-infant interaction at the first contact after delivery. *Early Human Development*, 3, 21-7.
Rodholm. M. and Larson, K. 1982: The behaviour of human male adults at their first contact with a newborn. *Infant Behaviour and Development*, 5, 121-30.
Rubin, J. L., Provenzano, F. J. and Luna, Z. 1974: The eye of the beholder: parents' views on sex of newborns. *American Journal of Orthopsychiatry*, 4, 512-19.
Russell, G. 1978: The father role and its relation to masculinity, feminity, and androgyny. *Child Development*, 49, 1174-81.
—— 1982: Shared-caregiving families: an Australian study. In M. Lamb (ed.) *Non Traditional Families*, Hillsdale, New Jersey: Erlbaum.
—— 1983: *The Changing Role of Fathers*. Milton Keynes: Open University Press.
Sanday, P. R. 1981: *Female Power: Male Dominance*. Cambridge: Cambridge University Press.
Schaffer, H. R. and Emerson, P. 1964: The development of social attachments in infancy. *Monographs of the Society for Research in Child Development*, 29, (3 series 94).
Shereshefsky, P. and Yarrow, L. 1973: *Psychological Aspects of a First Pregnancy and Early Post Natal Adaptation*, New York: Raven.
Switzky, L. T., Vietze, P. and Switzky, H. 1979: Attitudinal and demographic predictions of breast-feeding in mothers of six-week-old infants. *Psychological Reports*, 45, 3-14.
Walker, K. and Woods, M. 1976: *Time Use: A Measure of Household production of Family Goods and Services*. Washington, D.C.: American Home Economics Association.
Wandersman, L. 1980: The adjustment of fathers to the first baby. *Birth and the Family Journal*, 7, 155-62.
Willemsen, E., Flaherty, D., Heaton, C. and Ritchey, G. 1974: Attachment behaviour of one-year-olds as a function of mother versus father, sex of child, session and toys. *Genetic Psychology Monographs*, 90, 305-24.
Yogman, M. 1977: The goals and structure of face to face interaction between infants and fathers. A paper presented to the *Society for Research in Child Development*, New Orleans, March 1977.
—— 1982: Games fathers and mothers play with their infants. *Infant Mental Health Journal*, 2, 241-8.
Zaslow, M., Pedersen, P., Kramer, E. Suwalsky, J. and Fivel, M. 1981: Depressed mood in new fathers: interview and behavioral correlates. Paper presented to *The Society for Research in Child Development*, Boston, April 1981.

9 Substitute Parenting

M. Shaw

Human behaviour takes place, not in a social vacuum, but within a framework of laws, customs, attitudes and expectations which help to shape the behaviour of people in families and other social groupings, and, at a micro level, the interactions of parents and children. For much of everyday life, the influence of these external pressures on parental behaviour may go unnoticed, and it is often only when the normal rules cannot be applied or are called into question that their existence becomes apparent. One such situation is where, for some reason, the normal mechanisms for child-rearing break down and alternative measures have to be set in motion. In countries such as the UK and USA where there is strong emphasis on the importance of the nuclear family and of the birth parents in child-rearing the preferred solution to the problem of the child who cannot continue to live in the family of origin is its placement in a foster or adoptive family. The study of parenting behaviour in such families is not only of interest in itself but also helps to illuminate features of normal parent–child relationships which tend to be taken for granted.

The use of the term 'normal' should serve to introduce a note of caution against the use of convenient stereotypes. The notion of a 'normal' family – a household consisting of a married couple and the dependent children of their union – is firmly entrenched in popular (and much professional) thinking and is curiously resistant to research evidence showing this to be but one of a range of family types (Rapoport, Fogarty and Rapoport, 1982). Similarly, as will emerge later in this discussion, the terms 'foster' and 'adoptive' family embrace a wide range of circumstances and parenting tasks. Most research in foster home care relates to long-term, often quasi-adoptive care, while adoption research has concentrated on the 'traditional' adoption (family-building by otherwise childless couples), but it is unsafe to assume that either type is wholly representative of its own field.

LEGAL AND ADMINISTRATIVE FRAMEWORK

Before turning to the research, it is important to consider the legal and administrative framework of foster care and adoption, part of the social context within which these modes of child-rearing are conducted. Under the old English Common Law, children were regarded as the property of their parents, and parental (or, more accurately, *paternal*) rights over children were virtually absolute (Hoggett, 1981). The commonsense observation that not all parents were suitable to have the care of children was largely subordinated to the belief that the intervention of law into parent–child relationships would create more problems and hardship than it would solve. The substitute care of children was conducted informally with virtually no legal safeguards for foster parents, birth parents having the right to reclaim their children at any time, however long they had been cared for by foster parents to whom they might become firmly attached. Only in the late nineteenth and early twentieth centuries – for a variety of economic, political and moral reasons – did the welfare of children or the interests of foster parents begin to receive any real statutory recognition.

Adoption, which differs essentially from fostering in that it involves the virtually complete transfer of rights and responsibilities from one set of parents to another, came comparatively late to English law. Early adoption legislation in the UK concentrated mainly on the rights of adults and it established strict legal criteria to be applied and procedures to be followed in the process of transferring parental rights. The consent of birth parents to the adoption application was normally expected and could be dispensed with only under limited and strictly defined conditions, not simply because the child's welfare might be better served by assigning parental rights to the prospective adopters. Even more than in the case of fostering, the response of the law to the growing concern for the child's welfare in adoption proceedings has been slow and circumspect, largely because of the absolute nature of the outcome in relation to parental rights.

Children in the UK who are (or are likely to be) placed in foster or adoptive homes are presently largely the responsibility of local authority social services departments (in Scotland, social work departments), who play the major part in implementing child welfare legislation. Some children are placed directly in foster homes by private arrangement between their parents and the foster parents, but such placements have been largely neglected by researchers (Holman, 1973). There are increasing legal restrictions on the placement of children for adoption by other than a local authority or registered voluntary adoption agency.

Children come 'into care' (that is, normally, the care of a local authority) either by court order or by voluntary agreement with the parents. The court order may follow proceedings arising from the child's delinquency or other behaviour which is considered to put the child 'at risk', parental neglect or maltreatment, or matrimonial proceedings where it is felt that neither parent is likely to provide a satisfactory home for the child. A local authority into whose care a child is committed by court order becomes, for many practical purposes, the child's parent for the duration of the order. Where children come into care by parental agreement (often because of the parent's illness or other incapacity), the local authority's powers are more limited, although there are controversial powers to assume parental rights in certain circumstances.

Children in care who are placed in foster homes become subject to regulations intended to safeguard their welfare and setting out minimum requirements for social work supervision, medical examination, and so on. Children placed for adoption are also subject to social work supervision but differ from foster children in that contact with the agency normally ends with the making of the adoption order by the court.

Fuller accounts may be found elsewhere of the historical (Packman, 1981) and legal (Hoggett, 1981; Hoggett and Pearl, 1983) aspects of child welfare.

VARIETIES OF PARENT

The term 'parent' as commonly used contains at least three components: birth parent, legal parent and parenting parent (Fahlberg, 1984). The *birth parents* give the child life, physical appearance, an intellectual potential and certain personality characteristics and special talents. The *legal parents* carry responsibility for the child's maintenance, safety and security, and make decisions about the child's residence, education, medical treatment and so on. *Parenting parents* provide the day-to-day love, care, attention and discipline. For many children the three components of parenting are embodied in one set of parents. Adopted children, however, are likely to have two sets of parents: the birth parents and the adoptive parents, the latter combining the legal and parenting roles. For foster children there will at least be a split between birth parents who are also legal parents and the foster parents undertaking part of the parenting function. In foster care there is quite commonly a three-way split, with birth parents, parenting (foster) parents, and the local authority as legal parent.

The important differences between foster care and adoption, combined with the fact that researchers tend to focus on either one group or the other, make it convenient here to deal with the two categories of care separately, although similarities will also be noted.

PARENTING IN FOSTER HOME CARE

'Foster home care' is an umbrella term covering a wide range of situations which may be classified according to time-span, aims and objectives, and the distribution of parenting tasks and responsibilities. In terms of time, a foster home placement may last for days or weeks, during a brief family crisis such as parental illness; for several months, the result of a more prolonged crisis such as homelessness or a parent's imprisonment; for several years, while attempts are made to resolve a complex array of personal, marital or social problems affecting the family; or virtually indefinitely, when the chances of the child's return to the birth family may never be fully extinguished but are at best remote.

The aims and objectives of foster home care may similarly range from temporary caretaking to the provision of a 'permanent' home. A relatively recent development is the treatment-oriented foster home where, as well as providing care, the foster parents are actively involved in a programme intended to prepare the young person (often an adolescent) either to return to the birth family or to move on to independent living (Shaw and Hipgrave, 1983). This mode of fostering is geared more to facilitating a significant transition in the person's life than to providing a substitute family.

The distribution of parenting responsibilities is also related to the length of the placement. In the short-term crisis placement, the foster parents offer temporary caretaking, with the birth parents still seen very much as the primary parents. A placement of several months or years with return to the birth family as the eventual aim requires more clearly 'shared parenting' among the adults responsible. In a placement of indefinite duration, often coming close to adoption, there is likely to be a more significant shift of parental responsibility away from birth parents and towards the foster parents and the placing agency. It is on placements of the latter type that research attention has largely focused (Taylor and Starr, 1967; Carbino, 1980).

Foster care as a social system

Foster home care is often most usefully viewed in social systems terms, with the behaviour of individuals in each subsystem crucially affected by behaviour in other subsystems (Weinstein, 1960; Eastman, 1979; Anderson, 1982). The typical foster home placement will involve a minimum of four individuals – birth parent, foster child, foster parent and social worker – with six dyadic relationships as subsystems. The foster care system is superimposed on a pre-existing family system and may be further complicated by the presence of a foster parent couple, their own children, and siblings of the fostered child; not to mention the possibility of triadic alliances which serve to isolate others in the system. For the purposes of the present discussion, the major focus must be on the foster parent–foster child subsystem but brief mention will be made of the other subsystems to set this particular relationship in context.

1   *Birth parent–child*   Whatever the previous relationship between parent and child, future relations once the child is in the foster home will be coloured by the fact of separation and its meaning for parent, child and other significant people involved. The effects on children of separation from parents have been much researched (Rutter, 1982). Less attention has been given to the effects on parents of separation from their children. Filial deprivation is in some respects the mirror image of maternal deprivation, with parents going through the classic stages of protest, despair and detachment (Jenkins and Norman, 1972), a phenomenon with serious implications for the parent–child relationship.

2   *Birth parent–social worker*   When a child is placed in a foster home, birth parents and social worker may share responsibility for helping the child 'settle in', maintain an awareness of the family of origin and prepare for the return home, should such a course of action become possible. This ostensibly rational, co-operative enterprise offers scope for conflict arising from differences of perception between the parties concerned: the parent perhaps seeking only service for the child, while the social worker may also consider the parent to be in need of 'treatment' (however defined) as a prerequisite for the child's return home. For a bibliography of US writing on the area of work, see Maluccio and Sinanoglu (1981).

3   *Social worker–child*   The extent to which a corporate body such as a local authority can adequately act in a parental capacity is a matter for debate (Adcock, White and Rowlands, 1983). Nonetheless, there is broad agreement that the social worker has a continuing responsibility during the foster home placement to assess the child's needs and the extent to which these are being met in the placement; and to act as a link for the child, both literally with the family of origin and the outside world, and more symbolically with the past and the child's future (Weinstein, 1960).

4   *Social worker–foster parent*   Much of the foster care literature is concerned with this particular relationship, which is often characterized by confusion and uncertainty on both sides (George, 1970; Kline and Overstreet, 1972; Adamson, 1973; Shaw and Hipgrave, 1983). At one level, the relationship may be seen as a working partnership in which the social worker supplements and supports the service being offered by the foster parent to the child. Social workers have, however, traditionally viewed an application to foster less as an offer of service than as an indirect attempt on the applicants' part to resolve some underlying personal or family problem, a perception which has tended to place the foster parent in a client rather than colleague role (Shaw and Hipgrave, 1983). The result has been an over-emphasis on an inquisition-style study of applicants' motives and general suitability for fostering at the expense of an educative and supportive approach to the foster parents once they have embarked upon their task. Foster parents, for their part, have often

been sceptical of the contribution of social workers to the fostering enterprise, seeing them as having insufficient life experience to justify their authority over people who have, in many cases, already successfully brought up children of their own.

An aspect of the conflict between the 'professional' orientation of social workers and the 'commonsense' approach of foster parents relates to the notions of 'inclusive' and 'exclusive' modes of foster care (Holman, 1975). An inclusive approach permits and indeed encourages the participation of birth parents and social worker in the foster care system, acknowledging significant differences between fostering and bringing up one's own children. An exclusive approach, by contrast, plays down or attempts to exclude the contribution of birth parents and social worker, and seeks to assimilate the foster child into the foster family. Research studies show foster parents as tending towards an exclusive approach, whereas social workers officially, if not always in practice, stress the importance for the child's psychological development of being aware of – and, if at all possible, in regular contact with – the birth family (George, 1970; Shaw and Lebens, 1976).

5 *Birth parent–foster parent* There is ample evidence that this relationship also, where it exists at all, is often highly problematic (George, 1970; Thorpe, 1974; Aldgate, 1976; Shaw and Lebens, 1977). The birth parents bring to the relationship an externally imposed or felt sense of stigma, with associated feelings of guilt, resentment and powerlessness (Jenkins and Norman, 1975). The foster parents, although at one level perhaps more understanding and sympathetic towards the birth parents, may also share the popular hostility felt towards parents who, for whatever reason, have publicly 'failed' to provide adequate care for their children. A major ingredient in the mixed feelings experienced by foster parents is fear that the birth parents will one day remove the child, to whom the whole foster family may have become closely attached. Many foster parents never lose this fear, even when by any objective reckoning the birth parents have neither the legal right nor the desire to resume the care of their children. With these and other similar concerns in the minds of the participants, it is not surprising that the notion of 'shared parenting', with a common focus on the wellbeing of the child, rarely achieves the status of a feasible or even justifiable objective.

6 *Foster parent–foster child* As well as being the major focus for this chapter, this is in many respects the key relationship in the foster care system, the relationship which, if successful, justifies the existence of the system as a whole. Of the numerous issues involved in 'the art of being a foster parent' (Littner, 1978), there is space here to give only brief consideration to a few: foster parent characteristics and motivation; attitudes and behaviour; and foster parents and foster children as consumers of service.

Characteristics and motivation of foster parents

Much of the research interest in foster care has revolved round the question of foster parent motivation. This interest derives in part from the fact that ordinary child-rearing, for all its satisfactions, demands a high level of commitment of both time and energy from parents even when the process is relatively trouble-free. Why, therefore, should some people deliberately choose to risk the additional hazards of raising other people's children?

Most studies of foster parents, particularly in the USA, portray them as predominantly lower-middle or working-class couples with limited education, adhering to conservative, even 'Victorian' values, very home-centred, and correspondingly somewhat isolated socially (Fanshel, 1966; Paulson, Grossman and Shapiro, 1974). UK research suggests a similar picture, the husbands in skilled or semi-skilled occupations, and the foster mothers tending to be women in their forties or over whose own children have either left home or are about to do so (Adamson, 1973). Another UK study showed the foster mothers to be similarly home- and family-centred but dissatisfied with the size of their own family and engaging in fostering in order to increase the family to that which they felt to be right size (Wakeford, 1963). In the absence of any substantial recent study it is impossible to say how far this picture of foster parents holds true today. It may be difficult to combine fostering with a cosmopolitan, socially and geographically mobile, or generally more 'trendy' life-style, but it is probably unwise to extrapolate from (say) Fanshel's Pittsburgh 1966 to Pittsburgh 1984, still less to New York, London or Leicestershire 1984. It is also worth noting that studies of foster parents focus almost entirely on those *selected* by social workers, probably a very biased sample of those people motivated to apply to foster in the first place.

Most research in this area of foster care has been concerned to establish links between patterns of motivation and successful outcome. Much of the research of the 1950s and 60s was within a narrowly psychodynamic framework, with strong emphasis on 'underlying' or 'unconscious' motivation, a somewhat hazardous line of enquiry (Josselyn and Towle, 1952). Foster parents who expressed a love of children or enjoyment of their company would be scrutinized with suspicion lest their urge for gratification should damage a foster child. There was perhaps too little effort to compare the motivations of foster parents with those of 'natural' parents, or even, at a more commonsense level, to wonder whether anyone would seek to foster if they did *not* like children or enjoy their company. Views were divided as to the relevance of a happy or unhappy childhood in the foster parents' own past life, the dominant view being that a good experience of parenting provided a helpful though not indispensable model, though some writers (for example, Kay, 1966) argued that some degree of early deprivation was a significant ingredient in successful fostering.

A major difficulty in this line of enquiry is that motivation, a philosophically complex notion (Peters, 1959) is not a reliable predictor of future performance: as is readily apparent in many fields of activity, it is perfectly possible to be well-motivated and ineffectual (Josselyn and Towle, 1952). Conversely, one study found that placements which resulted from an agency's campaign actively to encourage specific families to take foster children were at least as successful as those recruited in the normal way, where the initiative to make the first approach is left with families themselves (Stanton, 1956).

A more useful line of enquiry, looking forward rather than backwards, was indicated in a study by Trasler (1960), in which likely success was related to foster parents' expectations of the foster child. Trasler argued that, for a placement to be successful, the foster parents' expectations should be open, flexible and reasonable for the child. Thus, a foster parent may well hope that a foster child will be a companion to the family's own child, but there is a serious risk of failure if the success of the placement hinges entirely on that one expectation. Similarly, an overemphasis on academic or career expectations which are geared more closely to the aspirations of the foster parent than to the abilities and interests of the foster child may endanger the placement as it becomes apparent that such expectations are not being fulfilled. Trasler's thesis is largely an elaboration of the familiar idea that love which is heavily conditional has a limited life expectancy, but it provides useful encouragement to foster parents (and to those who select them) to look ahead and consider realistically the roles which foster children will be asked to perform in the family's life, rather than to scrutinize their own past lives for features which usually turn out to have little predictive value.

Attitudes and behaviour

Fanshel's study of foster parents indicated two patterns of motivation related to the age of the foster children (Fanshel, 1966). Foster mothers who specialized in babies and young children tended to seek private gratification from their role as mothers; whereas those who took mainly older children reported more social satisfactions, a sense of doing something socially useful, as major rewards. Fanshel's study also identified two patterns of motivation likely to prove problematic for the child's welfare, the first being the 'Benefactress of Children', in which the foster mother sees herself as saving the child from neglectful parents and 'doing good' in a rather patronizing fashion. Difficulties are likely to arise when the child fails to show sufficient gratitude for the sacrifices which the foster

Substitute Parenting 267

mother claims to have made on the child's behalf. The second pattern, Anomie, involves the use of the child to assuage the foster parent's social isolation and alienation, factors which in themselves may create problems for the child in placement.

Studies of factors found in successful foster parents sometimes lapse into tautology, with conclusions such as that the best foster parents show acceptance of the child as a person. More usefully, Colvin (1962) devised a foster parent attitude test in which three factors emerged as significant: *achievement*, showing a desire to overcome obstacles, exercise power, and to strive to do something difficult as well and as quickly as possible; *nurturance*, an urge to nourish, help and protect the child; and *play*, an ability to relax and 'have fun'. More recently, Levant and Geer (1981) have reported on the use of a rating instrument in the assessment of prospective foster families which they believe shows promise in discriminating between high and low risk candidates.

In a study of new foster parents in their first experience of fostering, Cautley (1980) offered as predictors of success: familiarity with children; having had parents who provided good parenting models; willingness to work with the social worker and agency; and verbal evidence of parenting skills as applied to specific behaviour incidents. A feature of Cautley's study is the importance ascribed to the foster father in the likely success of the placement, factors including the foster father's sensitivity and the extent to which he indicates a child-centred rather than self-centred view in talking about what would be difficult in being a foster parent.

Just as many writers on child care in general all too easily allow their focus to narrow from 'parents' to 'mothers', indeed much of the research on foster parents already discussed relates primarily to the attitudes and experiences of foster mothers. In practice, social workers themselves have tended to pay less attention to foster fathers, who are less likely to be seen on daytime visits, and who in traditional households may – or are assumed to – regard discussions on child care as the province of the foster mother and the (generally female) social worker. As a consequence, foster fathers have often been regarded as rather shadowy, ineffectual members of their households, showing little interest in the foster children or, at best, content to offer support to the foster mother (Fanshel, 1966; Cautley, 1980). It has also been suggested that foster mothers prefer to exclude their husbands from the fostering task, which they see as their own domain (Miller, 1968). Fanshel's study of foster parenthood perhaps throws some light on social workers' and foster fathers' experiences of one another: foster fathers indicated that they quite enjoyed the social workers' visits but did not find them particularly useful and sometimes found it difficult to understand precisely what social workers were trying to achieve in their visits (Fanshel, 1966). A limited study by

Wiehe (1982) suggests that there may be differences in personality between foster mothers and foster fathers which make the latter less responsive to some aspects of the fostering task.

Fanshel's finding that foster fathers play a more active part in the children's lives than they are often given credit for is echoed in the study by Cautley referred to above. Cautley also argues that foster fathers should be seen as important sources of information concerning the progress and likely outcome of placements: more of the information provided by the husbands than by the wives at selection was of value in predicting later success; more of the information given by the foster father during the placement provided clues that the placement was in jeopardy; and the more the foster father was actively involved with the child, the more likely was the placement to work out well (Cautley, 1980). Cautley suggests that the particular value of the foster father's information may in part be understandable in terms of role differentiation: the foster father's identity being normally less closely tied to relationships in the family, he is more likely to provide clues that problems exist. The foster mother, on the other hand, has a greater investment in the success of the placement and may express dissatisfaction only at the point when she is ready to give up, or has already made the decision to do so. Cautley's study is unusually interesting in that it traces foster parents' experiences and reactions over a period of eighteen months, rather than at a single point in time. Her findings also support practitioners' experience that there is often a crisis point at six months, after the initial 'honeymoon' period, and that placements which are going to 'fail' tend to do so within about a year.

Foster parents and children as consumers

Several recent studies have sought to explore the views of foster parents as consumers and providers of service (Adamson, 1973; George, 1970; Shaw and Lebens, 1977; Hampson and Tavormina, 1980). For most foster parents the primary satisfactions in fostering are similar to those gained from rearing their 'own' children – enjoying foster children's companionship, watching their progress, and so on – often with the added satisfaction of exercising parenting skills to help a child recover from unhappy earlier experiences. The frustrations of fostering seem to derive less from the children themselves than from ambiguities in the role and from what are often seen as unhelpful intrusions by birth parents and social workers into decision-making. At the same time there are frequent complaints about lack of support from agencies and social workers, rapid turnover of personnel often being a serious impediment to building effective working relationships with the social workers.

A foster parent is quoted in one study as saying, 'you don't think of them as foster children – it's other people who make the distinction'

(Shaw and Lebens, 1977, 15). Consumer studies reveal considerable ambivalence about those aspects of fostering which draw particular attention to that distinction: social work supervision, the question of financial reward (as distinct from meeting basic costs), and the recent emphasis on training for foster parents. The studies cited above indicate that traditional foster parents resent such reminders that there is something different about bringing up other people's children. However, recent developments in the specialization of fostering and the professionalization of foster parents themselves suggest that a major shift may currently be taking place in foster parent attitudes (Shaw and Hipgrave, 1983). One small study has shown that 'professional' foster parents taking difficult adolescents in return for fee payment gained more satisfaction from a sense of carrying out a professional task and looked less for responsiveness and emotional gratification from the children than did a comparable group of more traditional foster parents (Fullerton, 1982).

Although this chapter is concerned with parenting activity, it is perhaps appropriate to refer briefly to the intermittent attention given in research to the foster child as consumer (Weinstein, 1960; Thorpe, 1974; Bush, Gordon and LeBailly, 1977). Bush et al. indicate that the expectations of foster children are similar to those which other children have of their 'own' parents: older children do not want another set of parents but many younger children seek the love and affection of parent figures and value being treated just like one of the family. The authors suggest that, instead of asking (as foster care experts frequently do) whether foster parents should act as 'real' parents, it would be better to investigate the circumstances under which a closer approximation to a birth parent–child relationship is appropriate. They also warn that, given the ordinariness of many foster children's wishes – love, care and understanding – we should beware of searching for 'a complex, subtle, professional sounding and quasi-scientific description of an adequate caretaker' (Bush, Gordon and LeBailly, 1977, 494).

PARENTING BEHAVIOUR IN ADOPTION

Turning from foster care to adoption, some preliminary general statement of their similarities and differences may be helpful. Important areas of similarity include the fact that normally foster and adoptive parents are engaged in parenting someone else's children; that they are selected and in varying degrees prepared for their parenting roles by social workers; and that in carrying out their tasks they are under official public scrutiny through the medium of social work agencies.

The differences between the two modes of parenting are equally important. People embarking on adoption are normally more explicitly engaged in family-building on a permanent basis than are foster parents; and, if all goes well, the children placed with adopters will eventually become their 'own', which cannot legally be said of foster children, however long the placement. A further key difference is that, as a mode of care, fostering has long been regarded primarily as a service to children, many of whom would otherwise grow up in some form of institutional care; whereas adoption has until very recently been seen primarily as a service to adults, particularly to childless couples unable to produce children of their own.

However, the distinction between these categories is not entirely clear cut; some foster parents do adopt their foster children or come to regard them as their own without proper legal authority. The unattractively named notion of custodianship, introduced in the UK by the Children Act 1975 to give foster parents greater security but not full parental rights, serves to blur the distinction still further. In recent years, too, the emphasis in adoption has shifted from providing babies for childless couples to finding families for children, with the result that an increasing proportion of children placed for adoption nowadays are not healthy white babies but handicapped, mixed race and older children (Triseliotis, 1980). As most adoption research has focused on infant adoptions, care must be taken in extrapolating from studies of traditional adoption to current policy and practice.

Adoption may be further distinguished from fostering or other forms of substitute care by reason of its higher 'success' rate. The measurement of success is not a central concern of this chapter (see Shaw, 1984a) but studies of foster home placements suggest depressingly high rates of breakdown – one quarter within a year (Gray and Parr, 1957), one-third to two fifths of long-term placements (Trasler, 1960) 48 per cent within five years (Parker, 1966) and 59.8 per cent within the same period (George, 1970) – as compared with adoption success rates averaging around 75 per cent (Mech, 1973). On the differing aspirations of foster care and adoption, Triseliotis comments that 'setting low sights and expectations appears to be endemic to foster child placement in contrast to adoption' (Triseliotis, 1978, 22).

Viewed in systems terms, the most obvious difference between foster care and adoption is that in the latter case there has traditionally been no (or very little) contact between birth parents and adoptive parents, and that, once the judicial process is complete, social worker and agency withdraw from the scene, leaving adopters and child to engage in a relationship which, to the outside observer, approximates closely to a birth parent–child relationship. Rather than embark on what would of necessity be a somewhat repetitive survey of subsystems within the adoptive family, discussion here will be limited to some major issues: childlessness and role handicap; the question of origins; parental satisfactions; and parenting the 'hard to place' child. As in the case of fostering, adoptive families are found in a variety of forms and circumstances: childless couples building their own families; remarried couples adopting children of a previous marriage; adoption by relatives of a child born illegitimately elsewhere in the extended family; and adoption by people who already have, or are capable of having, their 'own' children, often seeking 'hard to place' children, partly for reasons of social service or altruism. Once again, conclusions drawn from simple comparisons between 'adoptive' and 'ordinary' family life must be viewed with caution.

Childlessness and role handicap

As part of normal socialization, children learn that they will one day marry and, once married, produce children. Married couples who find themselves unable for any reason to produce children have to face not only their own sense of sadness and loss but also the mixed reactions of a disappointed society in the shape of relatives, friends and neighbours. From the few who know and share their own sadness, the couple may receive sympathy, but the commoner wider reactions documented in the literature range from unthinking cruelty to envious resentment and malicious criticism (Kirk, 1964; Humphrey, 1969).

A major function of adoption in the UK and USA has been to act as a remedy for the personal and social consequences of childlessness. However, as Kirk's analysis of adoptive parenthood in role theory terms makes clear, the remedy itself may be problematic (Kirk, 1964). While there are recognized and often well-rehearsed roles for expectant mothers and fathers and for new parents (with full supporting cast of grandparents, other relatives and friends), prospective adopters have no such ready-made script and must to a great extent 'play it by ear', taking in often ambiguous and misleading clues from the world around them. Important among the various role-handicaps experienced by prospective adopters are a lack of autonomy (their ability to become parents heavily dependent on the judgment of judicial and social work agencies) and a positive discouragement from making a full commitment to the child from the beginning, in case for any reason their eventual adoption application should prove unsuccessful. The latter handicap places new adoptive parents in the position of 'loving at arm's length', at least in the initial stages of the placement.

Kirk identifies further dilemmas for adoptive parents once they have embarked on their task, most significantly the dilemma of 'integration versus differentiation', balancing the need to integrate the adopted child fully into the family against the recognition that part of the child's heritage is distinct and separate from the adoptive family. Adoptive parents, in Kirk's view, cope with these dilemmas by means of strategies which fall broadly into one of two types, characterized as 'rejection-of-difference' and 'acknowledgment-of-difference'. As their labels suggest, rejection-of-difference strategies essentially involve denying the fact that adoptive parenthood is at all different from birth parenthood and may lead adopters to seek to play down or even deny the child's true status or the place of the birth parents in the child's background; whereas acknowledgment-of-difference strategies recognize that there are crucial differences as well as similarities between the two forms of parenthood. Kirk hypothesized that acknowledgment-of-difference strategies are conducive to open communication within the family and thus to more healthy adoptive family life; whereas rejection-of-difference strategies can be expected to make for poor or blocked communication, with consequent ill-effects on the family.

There is some support for Kirk's thesis in one small study of twenty couples where a significant relationship was found between capacity for object relationships and acknowledgement-of-difference as a coping strategy (Senzel and Yeakel, 1970) but it is surprising that Kirk's neat hypothesis has yet to be comprehensively tested by empirical research. There are interesting parallels here with the notion of inclusive and exclusive models of fostering. It seems likely that, rather than opting for the simple application of one type in pure form, some mixture of both types of coping strategy may be more effective, with, perhaps, the balance of ingredients varying at different stages in adoptive family life. Openness is certainly desirable but over-emphasis on an adopted child's separate identity can be interpreted by the child as rejection (Raynor, 1980).

The question of origins

There is virtually unanimous agreement within the adoption world that adopted children should have adequate information about their origins and be helped to a full understanding of their status. There is also substantial agreement that many adoptive parents find it very difficult to discuss these issues with their children (Goodacre, 1966; McWhinnie, 1967; Triseliotis, 1973). Talking about origins may reactivate painful memories of former childlessness, and for some adopters there is discomfort in being reminded about the birth parents, producing feelings compounded of disapproval, sympathy and 'survivor guilt' (Lifton, 1968). Perhaps most widespread of all is the fear that their children will cease to love them once they discover that the adopters are not their 'real' parents.

The result of such discomfort on the adopters' part is that some are unable to discuss the matter at all with their children, or else provide only minimal information, leaving the initiative for further questions and discussion with the children themselves. As with other 'delicate' subjects, children quickly sense their parents' discomfort and hold back from asking awkward questions, with the result that each party waits for the other to make the next move (McWhinnie, 1967). Follow-up studies show significant discrepancies in the perceptions of adopters and adopted children as to the quantity and quality of information imparted by parents to children on this subject (Jaffee, 1974; Raynor, 1980).

For all the emphasis in the literature on the importance of 'telling', there is, with rare exceptions (BAAF, 1972), a distinct lack of guidance for adopters on how to handle the issue at different stages in the child's life. The once popular story that the child was chosen by the adopters seems to have fallen into disuse, partly because it is untrue (the choice is the agency's) and partly because the idea of being specially chosen places a heavy emotional burden on the child – why was I chosen? what kind of person do they expect me to be? The person who has to live with the burden of being chosen is the adoptive parent, with all the high expectations which being chosen seems to imply (Mackie, 1982). Perhaps the major difficulty which adopters face in discussing adoption with their children is that it raises directly the question of their entitlement to become adoptive parents in the first place. Entitlement, a rather elusive term, probably most fully developed by Jaffee and Fanshel (1970), relates to the task facing adopters in overcoming the psychological insult associated with being infertile, and coming to accept themselves as the child's true parents.

Parental satisfaction in adoption

Reference was made earlier to the difficulty of defining and measuring success in adoption. Some researchers have made their own assessment (for example Kornitzer, 1968); more frequently, emphasis has been given to the levels of satisfaction reported by adoptive parents (Jaffee and Fanshel, 1970; Kadushin, 1970); while a minority have incorporated a measure of satisfaction indicated by adopted people themselves (Triseliotis, 1973; Raynor, 1980). The politics of adoption seem to dictate the degree of importance ascribed to the adoptive parents' views (Shaw, 1984a) and it has to be recognized that some level of parental satisfaction is a prerequisite for the continuation of the placement. The problems of prediction already referred to in the discussion on foster home care make for difficulties in identifying in advance the factors conducive to parental satisfaction – and thus success – in adoption.

In their attempts to reduce the risk of failure, adoption agencies for many years sought to recruit only married couples who could demonstrate an unblemished record in conventional living (Schapiro, 1956; Shaw, 1984). In a small comparative study of adoptive and 'ordinary' parents, Humphrey and Kirkwood (1982) reported that the adoptive parents were more satisfied with their lot, more content, for example, with their drastically reduced social lives. The adopters gave emphasis to sharing responsibilities and tasks but were otherwise conventional in their approach to married life. Considerable importance also used to be given to the careful matching of adopters and children in terms of appearance and temperament – an exercise, it is worth noting, which belongs to the rejection-of-difference school of thought. With the emergence of recent trends in adoption, less value has been attached to matching for appearance but the idea received a new lease of life with Raynor's finding that a feeling of likeness (physical, temperamental or simply of interests) is associated with the level of satisfaction expressed by adoptive parents (Raynor, 1980). As the author herself notes, however, resemblances were reported by the adopters rather than in any way independently evaluated, and it is just as likely that people come to see resemblances in those to whom they have become warmly attached as it is that resemblances lead to people developing warm attachments. Certainly, the levels of satisfaction expressed by adopters of children who may differ quite markedly from them by reason of ethnic origin (Gill and Jackson, 1983) or medical condition or physical handicap (Franklin and Massarik, 1969) are hardly inferior to the satisfactions expressed by more traditional adopters. Similarly, the recent increased interest in single people as adopters suggests greater willingness on the part of adoption agencies to deviate from their former idealized image of 'normal' family life (Kadushin, 1970a; Shireman and Johnson, 1976; Feigelman and Silverman, 1977).

Parenting 'hard to place' children

The studies discussed so far have in general reflected traditional adoption practice, that is, a service geared to providing infertile but otherwise healthy white couples with healthy white babies. Since the mid-1960s, improved contraception and abortion services, and the reduced stigma attached to single parenthood have reduced the supply of such babies, resulting not only in disappointment for prospective adopters but in a crisis for adoption agencies, who were faced with a choice more familiar to the world of industry and commerce – diversify or go out of business. Adoption agencies which chose to diversify often broadened the scope of their placement activities to include children previously defined as hard-to-place: namely children of mixed race, children older than the norm for placement, and those suffering from some medical condition or physical or mental handicap.

Pursuing the commercial analogy, changes in the type of product (child) on offer have forced adoption agencies to seek out different kinds of consumer (applicant). The infertile young couple seeking a healthy baby to help them merge inconspicuously into the local community of young families is unlikely to be satisfied by the offer of an older child with some obvious physical handicap and possibly also of a different ethnic origin. The question arises, therefore, as to who are the adopters who are willing to accept hard-to-place children.

Even during the long period when adoption agencies were intent on providing 'the perfect baby for the perfect couple' (Schapiro, 1956), some 'less than perfect' babies were placed by a more or less conscious process of matching with 'less than perfect' couples. Kadushin used the term 'marginal eligibility' in relation to applicants who fell short of agency demands in one or more respects: being over forty years of age; spouse being of differing religion or race; one partner having had a previous divorce; one or other having some health problem; or their having two or more children of their own (Kadushin, 1962). In their negotiations with the adoption agency, these less-than-perfect couples were under pressure to compromise on the kind of child they would accept if they were to compete in the adoption market.

A more recent development in this area of adoption has been the emergence of the 'preferential' adopter (Feigelman and Silverman, 1979), that is the fertile couple with children of their own already, but motivated partly by religious, social or humanitarian concerns to seek a further child by adoption. Those most inclined to adopt hard-to-place children tend to have autonomous life-styles independent of traditional family patterns and more flexible and interchangeable conceptions of husband and wife roles (Silverman and Feigelman, 1979). In a comparative study of couple adopting transracially (white couples adopting black children), the infertile couples tended to emphasize such reasons as the unavailability of suitable American children and the excessive delays and high costs of domestic adoptions; whereas significantly more preferential adopters emphasized social and humanitarian reasons (Feigelman and Silverman, 1979). An exception to the altruistic tendency in their motivation was that preferential adopters were more likely than the infertile adopters to mention the opportunity to choose the sex of their child as a very important motive for adopting transracially. The same study showed that although the infertile couples seemed to experience more distress arising from role handicap, it was the preferential adopters who more frequently mentioned adjustment problems in their adopted children. The authors suggest that it may be the relative lack of social support for preferential adopters which contributes to the children's poorer adjustment. Preferential adopters are subject to double discrimination: for adopting, when they could produce their own children; and for adopting transracially. One might also speculate as to whether there might be some overlap between preferential adopters and Fanshel's Benefactress of Children foster parents (Fanshel, 1966), with, possibly, a shared sense of disappointment in children who fail to show appropriate gratitude in response to altruistic endeavour.

In the 1960s and the early 1970s, there was a growing debate on the rights and wrongs of transracial adoption as a proper means of providing families for black children (Madison and Schapiro, 1973). The white liberal view that transracial adoption could assist black and white integration overlooked the politically significant fact that the traffic was invariably one-way; black children moving into white families, never the reverse. Black community leaders condemned transracial adoption as a form of racial and cultural genocide and the success of this opposition could be seen in the decline in the numbers of transracial adoptions in the USA throughout the 1970s.

Empirical studies of transracial adoptions showed them as comparing favourably with inracial adoptions (Grow and Shapiro, 1974; Simon and Altstein, 1977, 1981), but the children studied were still very young and unlikely as yet to be grappling with the important issues of ethnic identity which troubled the critics of transracial adoption (Chestang, 1972; Chimezie, 1975). One exploratory study of adopted children in their early teens found that there was no difference in overall self-esteem between transracially and inracially adopted children but that there may be a difference in their sense of racial identity, in which factors such as the family's nurturance of the child's black identity, the child's access to black peers and role models and the parents' attention to the child's black heritage are important factors (McRoy et al., 1982). The most recent follow-up study of black and mixed race children placed under the British Adoption Project in the 1960s showed these teenagers to be doing well by conventional standards of success. However, on the question of racial identity, the authors found that the children had few if any black friends, little or no contact with a black community and that the adoptive parents' awareness of the significance of the children's heritage tended if anything to diminish as the children grew older (Gill and Jackson, 1983). McRoy and colleagues report that no less than 60 per cent of the adopters in their study take a 'colour-blind' attitude towards their transracially adopted children (McRoy et al., 1984).

In earlier adoption research and practice, it was virtually axiomatic that children being adopted should be placed before the age of six months, placement after that age being an unduly hazardous venture (Pringle, 1967). More recent research indicates a considerable degree of success in placing older children from quite problematic backgrounds in adoptive families (Kadushin, 1970). The placement of older children raises new issues for adopters, not least the nature of parenting for children who have had contact with and may well remember previous parents (Gill, 1978; Goldhaber and Colman, 1978; Jewett, 1978; Ward, 1981). Traditional practice viewed the complete and final ending of contact between birth family and adoptive family as a prerequisite for adoption, the intention being to give the child a 'fresh start' in the new family. Recently, there have been signs of judicial and social work approval for arrangements facilitating continued contact with the birth family after the making of an adoption order, specifically for older children who are conscious of, and feel continuing loyalty towards, their dual heritage (Borgman, 1982; Sorich and Siebert, 1982).

Of the various categories of hard-to-place children, the group which has probably received least research attention consists of those who are in some way physically handicapped. Franklin and Massarik's 1969 study showed once again the rule of marginal eligibility in operation, children with moderate or severe medical conditions being more likely to be placed with older adopters, with fertile rather than infertile couples, and in families who already had two or more 'own' children. It was also noted that agencies attempting to place a child with a medical condition would search for parents with limited aspirations, 'rural, plain folk' (Franklin and Massarik, 1969, 598) rather than well-educated, achievement-oriented, urban, upper-class families. The authors considered this approach unduly simplistic, finding that securely placed, upper-class families were just as likely as lower-class families to achieve satisfying parenthood for these children. Failures were more often found in the middle-class group who, the authors speculate, were perhaps too preoccupied with their own upward social mobility to give full attention to the problems of a handicapped child. The achievement of many of the families in this study in caring for often quite seriously and multiply-handicapped children was all the greater when it is remembered that the study relates to a period before post-adoption support and counselling were considered appropriate, which adds force to the authors' view that social workers may previously have underestimated adoptive parents' fundamental adaptive capacities. Currently, the example of pioneering agencies in the US (Donley, 1975) and the UK (Sawbridge, 1983) in finding homes for older children with multiple health handicaps is increasingly challenging social work agencies to consider whether in principle there is any such creature as an 'unadoptable' or 'unfosterable' child, in contrast to the comparatively rigid criteria of suitability which tended to be taken for granted only fifteen or twenty years ago.

CONCLUSION

A review of the literature on parental behaviour in foster and adoptive families serves as a reminder that parenting is not a unitary, unvarying phenomenon conducted in a social vacuum. Equally, the limitation of the terms of reference for this chapter (and for others in this volume) to parental behaviour should not blind us to the importance of the child's contribution to what is essentially a process of interaction. Foster parenting, with its pervasive ambiguity and confusion in the aims and roles of its participants, offers the more obviously complex picture. Adoption seems to approximate more closely to 'normal' family life, but notions about children as property, the child's dual heritage, and (in traditional adoption, at least) the meaning for the adopters of their former childlessness, also serve to complicate the pattern.

Inevitably, a review of this kind draws attention to biases and gaps in existing research. Much of the literature in this field is anecdotal and prescriptive rather than empirically based and analytical, although the last fifteen years or so have seen a marked improvement in the rigour and sophistication of research activity (Maas, 1966, 1971, 1978). Future research could usefully take more account of the legal, professional and societal context within which foster care and adoption services operate. A particularly fruitful area of study would be the parenting implications of the current movement by foster parents and adopters into more professional, quasi-therapeutic modes of child care. More generally, increasing specialisation and sophistication in the activities of foster parents and adopters call for correspondingly greater clarity and specificity of focus in research. The quest for the ideal or typical foster parent or adopter has always been something of a vain pursuit but now more than ever it needs to be preceded by the formulation of precise questions – ideal for what kind of child, in what circumstances, and to what end? Answers to such questions should be as illuminating for general theories of parental behaviour as for specific areas of child welfare practice.

REFERENCES

Adamson, G. 1973: *The Caretakers.* London: Bookstall Publications.
Adcock, M., White, R. and Rowlands, O. 1983: *The Administrative Parent.* London: British Agencies for Adoption and Fostering.
Aldgate, J. 1976: The child in care and his parents. *Adoption and Fostering*, 84, 29–40.
Anderson, L. M. 1982: A systems model for foster home studies. *Child Welfare*, 61, 1, 37–47.
Borgman, R. 1982: The consequences of open and closed adoption for older children. *Child Welfare*, 61, 4, 217–26.
British Agencies for Adoption and Fostering (BAAF) 1972: *Explaining Adoption: A guide for adoptive parents.* London: BAAF.
Bush, M., Gordon, A. C. and LeBailly, R. 1977: Evaluating child welfare services: a contribution from the clients. *Social Service Review*, 51, 3, 491–501.
Carbino, R. 1980: *Foster Parenting: an updated review of the literature.* New York: Child Welfare League of America.
Cautley, P. W. 1980: *New Foster Parents: The first experience.* New York: Human Sciences Press.
Chestang, L. 1972: The dilemma of biracial adoption. *Social Work*, 17, 3, 100–5.

Chimezie, A. 1975: Transracial adoption of black children. *Social Work*, 20, 4, 296-301.
Colvin, R. 1962: Toward the development of a foster parent attitude test. In *Quantitative Approaches to Parent Selection*, New York: Child Welfare League of America, 41-53.
Donley, K. 1975: *Opening New Doors*. London: British Agencies for Adoption and Fostering.
Eastman, K. 1979: The foster family in a systems theory perspective. *Child Welfare*, 58, 9, 564-70.
Fahlberg, V. 1981: *Helping Children When They Must Move*. London: British Agencies for Adoption and Fostering.
Fanshel, D. 1966: *Foster Parenthood: A role analysis*. Minnesota: University of Minnesota Press.
Feigelman, W. and Silverman, A. R. 1977: Single parent adoptions. *Social Casework*, 58, 7, 418-25.
—— 1979: Preferential adoption: a new mode of family formation. *Social Casework*, 60, 5, 296-305.
Franklin, D. S. and Massarik, F. 1969: The adoption of children with medical conditions. *Child Welfare*, 48, 8, 459-67; 9, 533-9; 10, 595-601.
Fullerton, M. 1982: A study of the role of foster parents in family placement for adolescents. *Clearing House for Local Authority Social Services Research* 4, 45-138. Department of Social Administration, University of Birmingham, England.
George, V. 1970: *Foster Care: Theory and practice*. London: Routledge and Kegan Paul.
Gill, M. 1978: Adoption of older children: the problems faced. *Social Casework*, 59, 5, 272-8.
Gill, O. and Jackson, B. 1983: *Adoption and Race*. London: Batsford Academic.
Goldhaber, D. and Colman, M. 1978: Untitled review of research into adoption of older children. *Adoption and Fostering*, 2, 4, 41-8.
Goodacre, I. 1966: *Adoption Policy and Practice*. London: Allen and Unwin.
Gray, P. G. and Parr, E. 1957: *Children in Care and the Recruitment of Foster Parents*. London: Social Survey.
Grow, L. C. and Shapiro, D. 1974: *Black Children, White Parents: A study of transracial adoption*. New York: Child Welfare League of America.
Hampson, R. B. and Tavormina, J. B. 1980: Feedback from the experts: a study of foster mothers. *Social Work*, 25, 2, 108-13.
Hoggett, B. 1981: *Parents and Children*. 2nd edn London: Sweet & Maxwell.
Hoggett, B. and Pearl, D. S. 1983: *The Family, Law and Society: Cases and Materials*. London: Butterworths.
Holman, R. 1973: *Trading in Children: A study of private fostering*. London: Routledge and Kegan Paul.
—— 1975: The place of fostering in social work. *British Journal of Social Work*, 5, 1, 3-29.
Humphrey, M. 1969: *The Hostage Seekers: A study of childless and adopting couples*. London: Longman.
Jaffee, B. 1974: Adoption outcome: a two-generation view. *Child Welfare*, 53, 4, 211-24.

Jaffee, B. and Fanshel, D. 1970: *How They Fared in Adoption*. New York: Columbia University Press.
Jenkins, S. and Norman, E. 1972: *Filial Deprivation and Foster Care*. New York: Columbia University Press.
—— 1975: *Beyond Placement: Mothers view foster care*. New York: Columbia University Press.
Jewett, C. L. 1978: *Adopting the Older Child*. Massachusetts: Harvard Common Press.
Josselyn, I. and Towle, C. 1952: Evaluating motives of foster parents. *Child Welfare*, 31, 2. Reprinted in R. Tod (ed.), *Social Work in Foster Care*, London: Longman 18-38.
Kadushin, A. 1962: A study of adoptive parents of hard-to-place children. *Social Casework*, 43, 4, 227-33.
—— 1970: *Adopting Older Children*. New York: Columbia University Press.
—— 1970a: Single parent adoptions: an overview of some relevant research. *Social Service Review*, 44, 3, 263-74.
Kay, N. 1966: A systematic approach to selecting foster parents. *Case Conference*, 13, 2, 44-40. Reprinted in R. Tod (ed.), *Social Work in Foster Care*, London: Longman 39-50.
Kirk, H. D. 1964: *Shared Fate: A theory of adoption and mental health*. New York: Free Press of Glencoe.
Kline, D. and Overstreet, H. M. F. 1972: *Foster Care of Children: Nurture and treatment*. New York: Columbia University Press.
Kornitzer, M. 1968: *Adoption and Family Life*. London: Putnam.
Levant, R. F. and Geer, M. F. 1981: A systematic skills approach to the selection and training of foster parents as mental health paraprofessionals I: project overview and training. *Journal of Community Psychology*, 9, 224-30.
Lifton, R. J. 1968: *Death in Life*. New York: Random House.
Littner, N. 1978: The art of being a foster parent. *Child Welfare*, 57, 1, 3-12.
Maas, H. S. (ed.) 1966: *Five Fields of Social Service: Reviews of research*. New York: National Association of Social Workers.
—— 1971: *Research in the Social Services: A five-year review*. New York: National Association of Social Workers.
—— 1978: *Social Service Research: Reviews of studies*. New York: National Association of Social Workers.
Mackie, A. J. 1982: Families of adopted adolescents. *Journal of Adolescence*, 5, 171-9.
Madison, B. Q. and Schapiro, M. 1973: Black adoption - issues and policies - review of the literature. *Social Service Review*, 47, 4, 531-60.
Maluccio, A. N. and Sinanoglu, P. A. 1981: Social work with parents of children in foster care: a bibliography. *Child Welfare*, 60, 5, 275-303.
McRoy, R. G., Zurcher, L. A., Lauderdale, M. L. and Anderson, R. N. 1982: Self-esteem and racial identity in transracial and inracial adoptees. *Social Work*, 27, 6, 522-6.
—— 1984: The identity of transracial adoptees. *Social Casework*, 65, 1, 34-9.
McWhinnie, A. 1967: *Adopted Children: How they grow up*. London: Routledge and Kegan Paul.
Mech, E. V. 1973: Adoption: a policy perspective. In B. M. Caldwell, and H. N.

Ricciuti (eds), *Child Development and Social Policy: Review of child development research* 3, Chicago: University of Chicago Press, 467–508.
Miller, J. N. 1968: Some similarities between foster mothers and unmarried mothers and their significance for foster parenting. *Child Welfare*, 47, 4, 216–19, 235.
Packman, J. 1981: *The Child's Generation*. 2nd edn. London: Basil Blackwell and Martin Robertson.
Parker, R. A. 1966: *Decision in Child Care: A study of prediction in fostering*. London: Allen and Unwin.
Paulson, M. J., Grossman, S. and Shapiro, G. 1974: Child-rearing attitudes of foster home mothers. *Journal of Community Psychology*, 2, 1, 11–14.
Peters, R. S. 1959: *The Concept of Motivation*. London: Routledge & Kegan Paul.
Pringle, M. L. K. 1967: *Adoption: Facts and fallacies*. London: Longman.
Rapoport, R. N., Fogarty, M. P. and Rapoport, R. 1982: *Families in Britain*. London: Routledge and Kegan Paul.
Raynor, L. 1980: *The Adopted Child Comes of Age*. London: Allen & Unwin.
Rutter, M. 1982: *Maternal Deprivation Reassessed*. London: Penguin.
Sawbridge, P. (ed.) 1983: *Parents for Children: Twelve practice papers*. London: British Agencies for Adoption and Fostering.
Schapiro, M. 1956–7: *A Study of Adoption Practice*. New York: Child Welfare League of America.
Senzel, B. and Yeakel, M. 1970: Relationship capacity and 'acknowledgement-of-difference' in adoptive parenthood. *Smith College Studies in Social Work*, 40, 2, 155–63.
Shaw, M. 1984: Looking back: rejected applicants. *Adoption and Fostering*, 8, 2, xx.
—— 1984a: Growing up adopted. In P. Bean (ed.), *Adoption: Essays in social policy, law and sociology*, London: Associated Book Publishers Limited.
Shaw, M. and Hipgrave, T. 1983: *Specialist Fostering*. London: Batsford Academic.
Shaw, M. and Lebens, K. 1976: Children between families. *Adoption and Fostering*, 84, 17–27.
—— 1977: Foster parents talking. *Adoption and Fostering*, 1, 2, 11–16.
Shireman, J. F. and Johnson, P. R. 1976: Single persons as adoptive parents. *Social Service Review*, 50, 1, 103–16.
Silverman, A. R. and Feigelman, W. 1977: Some factors affecting the adoption of ethnic minority children. *Social Casework*, 58, 9, 554–61.
Simon, R. J. and Altstein, H. 1977: *Transracial Adoption*. New York: John Wiley and Sons.
—— 1981: *Transracial Adoption: A follow-up*. New York: Lexington Books.
Sorich, C. J. and Siebert, R. 1982: Towards humanizing adoption. *Child Welfare*, 61, 4, 217–26.
Stanton, H. R. 1956: Mother love in foster homes. *Marriage and Family Living*, 18, 4, 301–7.
Taylor, D. A. and Starr, P. 1967: Foster parenting: an integrative review of the literature. *Child Welfare*, 46, 7, 371–85.
Thorpe, R. 1974: Mum and Mrs So-and-So. *Social Work Today*, 4, 22, 691–5.

Trasler, G. 1960: *In Place of Parents.* London: Routledge and Kegan Paul.
Triseliotis, J. 1973: *In Search of Origins: The experiences of adopted people.* London: Routledge and Kegan Paul.
—— 1978: Growing up fostered. *Adoption and Fostering,* 2, 4, 11-23.
—— (ed.) 1980: *New Developments in Foster Care and Adoption.* London: Routledge and Kegan Paul.
Wakeford, J. 1963: Fostering – a sociological perspective. *British Journal of Sociology,* 14, 4, 335-46.
Ward, M. 1981: Parental bonding in older child adoptions. *Child Welfare,* 60, 1, 24-34.
Weinstein, E. A. 1960: *The Self-image of the Foster Child.* New York: Russell Sage Foundation.
Wiehe, V. R. 1982: Differential personality types of foster parents. *Social Work,* 27, 1, 16-20.

10 Parental Responsiveness and Child Behaviour

H. R. Schaffer and G. M. Collis

INTRODUCTION

The functions of parenting are many and varied, although among diverse species parents are commonly required to provide food, warmth and protection. In addition to these nurturant functions, it is believed that parent–offspring interaction serves a range of functions which, broadly speaking, can be considered to involve the transfer of information from parent to offspring – information that will have considerable value in the later life of the young. Some examples are: the learning of species identity (Immelmann, 1972); communication repertoires (Kroodsma, 1978); feeding techniques and preferences (McGrew, 1977); and social status (Cheney, 1977). Among human beings, most aspects of parent-to-offspring information transfer fall under the rubric of *socialization*, which is of major interest to developmental psychologists.

Both classes of function – nurturance and inter-generational transfer of information – require a degree of *responsiveness* on the part of the parent. Take, for instance, the feeding of the young. Typically, this involves more than merely providing food, for such provision must be modulated according to cycles of hunger and satiety expressed by the young and responded to by the parent. Even within feeding bouts parents respond to moment-to-moment changes in the behaviour of the young. (For example, while breast feeding, mothers respond to pauses between bursts of sucking by jiggling their babies (Kaye 1977)). Moreover, the composition of the diet will be changed to parallel the maturation and changing needs of the growing infant.

In the study of human development it is almost axiomatic that parents play a central role in the socialisation of children (although this does not rule out important roles for other adults and for peers). Yet to demonstrate the impact of parental behaviour on the long-term development of the child, by relating differences in parental behaviour to differences in developmental outcome, has proved to be astonishingly difficult. In the 1960s, reviews of the literature (Caldwell, 1964; Yarrow et al., 1968) repeatedly pointed to the conclusion that there was little robust evidence of a connection between parenting practices and later characteristics of the child.

The practical implications of this are enormous: on the positive side it is reassuring to know that choices between breast and bottle feeding, or between early or later toilet training, are unlikely to have major effects on later personality (Caldwell, 1964); on the negative side it means that there is extreme difficulty in exploring, and doing something about, 'cycles of disadvantage' (Rutter and Madge, 1976). While it is clear that disadvantage does carry over from one generation to another within families, it is not at all clear to what extent this is due to sub-optimal parenting or similar interpersonal processes within the family, rather than to extrinsic factors such as poverty, poor living conditions, discrimination of race or class, and so on, which also persist over successive generations, or even to hereditary factors.

Why is robust evidence of the effects of parenting so elusive? There is now broad agreement on several main points. First, with respect to various features of parent and child behaviour, a good case can be made for the view that congenital differences among children contribute to differences in parenting styles. Thus it is highly likely that characteristics of the child have some role in the aetiology of child abuse (Bell 1977). And it is very likely, too, that child characteristics help to mould the attitudes of parents towards child rearing, which may partly explain the failure of attempts to relate parental attitudes to children's personality development (Yarrow, Campbell and Burton, 1968). Thus child characteristics may be the origin and parental behaviour the outcome, rather than vice versa.

Second, it is erroneous to conceive of the infant as a passive recipient of stimulation, whether it be from parents or from any other source. To illustrate, the effect of parental behaviour on young infants is critically dependent on the state of the infant on a continuum from sleep and drowsiness through to active alertness and distress. The best-documented cases of state-dependent responsiveness relate to reflex-like responses of the kind used in neurological examinations of infants (Prechtl and O'Brien, 1982), but it is a matter of common observation that attempts to play simple games with babies, which will bring forth smiling and laughter when the babies are relaxed, can easily produce crying when they are restless. The same principle applies to higher level processes. For example, in verbal exchanges the impact of what is said to a child will depend crucially on a complex set of presuppositions and contextual features of the discourse. Small wonder, then, that attempts to facilitate development which, in effect, just bombard the child with stimulation, are destined for failure (Schaffer, 1977).

Third, it is clear that development is a continuing process, so that events that intervene between some early phase in development and an assessment of outcome later on will in all probability attenuate the measurable relationship between parenting in the early phase and the subsequent outcome. In animal studies intervening events tend to be held constant by experimental design — something that is not attainable in the real world of human development. Almost as a corollary to this point, less emphasis is now placed on early sensitive periods or phases. Experiments in which the intervening period between early experience and later outcome is protected from perturbation are weak tests of the robustness of the effects of experience at *particular* phases in development. Moreover, with respect to one of the paradigm cases illustrating the deleterious effects of early experience, it is now clear that the gross deficits produced by early social isolation in monkeys can be ameliorated by subsequent treatment (Suomi et al., 1974, Cummins and Suomi, 1976). In the human case too, rehabilitation from the effects of extreme conditions of deprivation is possible (Clarke and Clarke, 1976). While it is probably true that privileged phases for learning do occur in the course of development, their contribution to individual differences in later life is almost certainly not very great.

Finally, even though the *net* transfer of information may be from parent to offspring, this process is mediated by interactional processes in which the child influences the parent just as much as the parent influences the child, though perhaps in different ways. It is curious that reciprocal influence suffered so long from neglect as a topic of empirical investigation, as even extreme environmentalist theories of development involve two-way processes. For example, Skinnerian principles imply, on the one hand, that the behaviour of the child is conditioned by virtue of contingent responsiveness by the parent acting as reinforcement (Gewirtz, 1968), and on the other hand that parental behaviour can be conditioned by contingent responsiveness on the part of the child (Gewirtz and Boyd, 1977a). However, the principle of two-way responsiveness has implications far beyond the boundaries of conditioning theory. It is part of the very nature of social *interaction* that individual A does something to B who in return does something back to A and so on, and that what happens at one phase of the interaction influences subsequent interactions between A and B, and probably also the interactions of A and B with other persons (Hinde, 1979).

Thus a picture emerges of parent–offspring interaction as a continuing, dynamic, two-way process. It is not a process wherein, at certain phases in development, the parent just does something to the child which sets the course for development to follow irrespective of other circumstances. Nor is it a process in which other circumstances merely add noise in the form of minor but accumulating perturbations to the state of the developing child. Parent–offspring interaction is a flexibe system in which two (or more) component individuals, coupled together in mutual influence, can follow any of a variety of routes to broadly similar end points, compensating to some extent for minor perturbations and finding ways round developmental obstacles. Direct evidence that the interactive system acts to buffer the infant's development against disturbances from external sources is difficult to find. However, in rhesus monkeys it has been demonstrated that the effects on the infant of short term mother–infant separation depends a great deal on whether the mother is responsive or intolerant toward the infant as she tries to re-establish her own social relationships with other members of the group (Hinde and Davies, 1972; McGinnis, 1980).

It is, of course, easier to use terms like 'system', 'coupling' and 'mutual influence' than it is to spell out precisely how the system works. However, one implication of viewing development in terms of parent–offspring interaction is clear, namely that it becomes necessary to reconceptualise parental behaviour; instead of examining it as a phenomenon in its own right it needs to be seen as part of a dyadic unit, that is as a contributing rather than as a determining force. Thus the sort of questions that arise concern the way in which the parent interweaves his/her behaviour with that of the child's and the manner whereby the two jointly forge integrated dyadic sequences.

This does not imply, of course, that the two are in any sense to be seen as equal partners. Inevitably the child's initial state of immaturity means that the parent must complement that immaturity; he/she has to provide support for the child's efforts to act on his environment and make sense of the world, and accordingly he/she must be able to recognise what it is that the child is attempting to achieve and communicate. The parent's contribution, that is, needs to be thought about in terms of the degree and extent to which the child's particular requirements are responded to. Sensitive responsiveness is thus a key aspect of parental behaviour. As Lamb and Easterbrooks (1981) put it, there is now 'remarkable agreement that parental sensitivity is an extremely important influence on early child development', and in similar vein Wachs and Gruen (1982) conclude from their recent summary of research on early experience that 'the essential implications of the literature reviewed in this book is that both the physical and social environments must be highly responsive to the infant's and young child's needs and behaviour if positive development is to occur.' It is therefore essential to examine this aspect more closely.

THE NATURE OF SENSITIVE RESPONSIVENESS

A number of writers have attempted to convey by means of qualitative description the meaning of this characteristic. Ainsworth, in particular, has provided some vivid illustrative material. Take the following:

The sensitive mother is able to see things from her baby's point of view. She is tuned-in to receive her baby's signals; she interprets them correctly, and she responds to them promptly and appropriately. Although she nearly always gives the baby what he seems to want, when she does not she is tactful in acknowledging his communication and in offering an acceptable alternative. She make her responses temporally contingent upon the baby's signals and communications. The sensitive mother, by definition, cannot be rejecting, interfering or ignoring. The insensitive mother, on the other hand, gears her interventions and initiations of interactions almost exclusively in terms of her own wishes, moods and activities. She tends either to distort the implications of her baby's communications, interpreting them in the light of her own wishes or defences, or not to respond to them at all. (Ainsworth et al., 1971, 41)

This description concentrates on the extremes of the distribution; a continuum is, however, envisaged as representing the individual differences to be found in this respect. Details of a nine-point rating scale are accordingly given by Ainsworth et al. (1974), ranging from 'highly sensitive' through 'inconsistently sensitive' to 'highly insensitive'.

Let us note that there are two assumptions underlying such an effort to account for variation in parental behaviour. The first is that sensitive responsiveness is a personality trait and that variations in the relevant behaviour is to be explained in terms of stable dispositions characterising individual parents. There has been a great deal of discussion in recent years concerning the status of traits (Mischel 1977), with increasing doubts being expressed as to the value of such a static concept and more and more attention being paid instead to the person-by-situation interaction. The latter acknowledges that there may well be individual dispositions to act in particular ways, but challenges the notion that they alone allow one to predict behaviour. Efforts to predict, it is argued, must take into account the context in which that behaviour occurs; knowing where an individual stands along some continuum as assessed in one particular situation does not enable one to generalise to all other situations in which that individual may find himself. As we shall see, the influence of context on sensitive responsiveness is particularly marked; it stems from two sources, namely the characteristics of the child with whom the adult is interacting and the social and physical situation in which that interaction occurs. The former is, of course, particularly pertinent to any account that stresses the dyadic nature of what transpires between parent and child.

The second assumption refers to the unitary nature of sensitive responsiveness. Parents are ordered along one continuum; that continuum is conceived as representing a single entity. Yet a close reading of Ainsworth's description above shows what a highly complex phenomenon we are dealing with. Making up this phenomenon there are at least three different constituents: responding *promptly*; responding *consistently*; and responding *appropriately*.

Promptness is required in that infants have very limited abilities to appreciate the contingencies of events with their own behaviour; according to Millar (1972) an interval of only 3 seconds is required to disrupt the contingency learning of six-month-old infants. Where the adult takes appreciably longer to answer the infant's signals there will be no opportunity for the child to learn that his behaviour has communicative significance, that is, that he can thereby affect his environment and in particular the behaviour of other people. At its most extreme this is to be found among institutionalised children brought up under conditions of personal deprivation (Provence and Lipton, 1962); under such conditions the child has no opportunity to acquire that sense of effectance (Lewis and Goldberg, 1969) which is derived primarily from the realisation that the environment is responsive to his messages and that he can therefore control the environment.

The second characteristic, *consistency*, may also be regarded as a prerequisite to normal development. A child's environment must be predictable; he must be able to learn that his behaviour will produce particular consequences under particular conditions. Lack of predictability can stem from two sources, namely inter-individual and intra-individual inconsistency. The former is especially likely to characterize the experience of children exposed to a multiplicity of caretakers, as shown by Tizard's (1978) finding that the group of children in care investigated by her had had an average of 50 mother-figures by the time they reached the age of 4½. A certain diversity of caretaker styles may well be useful for any child to experience; however, in the early years in particular, that diversity needs to be kept within limits which are certainly exceeded in the case of Tizard's group. No two individuals are alike in their expectations and reactions *vis-à-vis* any one child; under normal family circumstances, however, such differences are not too great and can be tolerated by most children. It is only when a large number of caretakers, often with minimal previous acquaintance of the child's specific requirements and habits, are encountered that the environment assumes a bewildering and unpredictable character. Lack of intra-individual consistency produces a similar effect: when a child never knows whether one and the same adult will respond to him with anger or with affection or whether she will simply ignore him, a sense of insecurity seems a likely outcome. Clinical case studies dealing with 'problem parents' frequently refer to this feature; little systematic research, on the other hand, has been conducted on this topic.

The third characteristic refers to the *appropriateness* of the parent's behaviour. Responding promptly and consistently are of little use when the nature of the response fails to recognise the particular message that the child is attempting to convey. To meet an infant's crying with laughter is usually quite inappropriate; where such crying indicates

Parental Responsiveness and Child Behaviour   289

distress such other responses as picking up and soothing are intuitively selected by the parent as the 'right' ones. There is, of course, no automatic pairing of child signal and adult response. For one thing, there are individual differences among children which parents learn about through experience: one infant's distress may be stilled by rocking, cuddling and other forms of physical contact, while another may require toys, food and visual distraction (Schaffer and Emerson, 1964). What is appropriate in one case is not necessarily so in any other case. And for another, cultural requirements also dictate appropriateness. A Japanese mother, having very different values and goals from those which guide the child rearing habits of her Western counterpart (Caudill and Schooler, 1973; Caudill and Weinstein, 1969), is more likely to respond to her infant's signals by physical than by verbal means; rocking, carrying and soothing are thus resorted to in situations where the European or American mother would respond by talking, singing or crooning. There is enough flexibility in the parent–child system to permit such variations; it means that parents are able to express cultural norms or personal idiosyncrasy by guiding the child in one direction rather than another, thus ensuring their socialisation through exposure to relevant experience. It does mean, however, that the judgement of appropriateness has to be undertaken with care in order to make full allowance for individual and cultural norms. An infant's pain cry must inevitably be responded to by measures designed to alleviate the pain; the precise form of these measures, however, may vary from one society to another and from one parent–child couple to another.

Thus sensitive responsiveness contains a number of constituents that can be conceptually and empirically separated. What is lacking so far is any evidence that these constituents belong together, that is, that they necessarily co-vary in such a manner that they may be regarded as making up one unitary entity. Rating scales such as Ainsworth's merely ask the raters to make global judgements; they may thus average out the various components or be swayed unduly by one or another constituent to some unknown extent. Until the necessary research has been carried out we must therefore exercise caution in using the label 'sensitive responsiveness'; at present it is not justified to consider it more than an umbrella term, though one that usefully draws attention to a quality of parenting that needs to be singled out for attention.

THE MANIFESTATIONS OF SENSITIVE RESPONSIVENESS

While sensitive responsiveness is applicable to the interactions of parents with children at all ages it has mostly been investigated at the earliest stages of development. We shall therefore examine its manifestation at that period with respect to a number of interactive situations and so convey something of its nature and variety.

Infant crying

Given the newborn infant's state of immaturity it is essential he is equipped from the beginning with some means of signalling to others that he is in need of their attention; equally his caretakers for their part need to be prepared readily to respond and provide the requisite help and attention. The cry is a most potent signal for this purpose; it alerts the parent and, by bringing him/her to his side, ensures the infant's survival in the face of dangers that he cannot deal with on his own (Bowlby, 1969). Interference with such a basic communicative mechanism would have serious consequences for the child's immediate well-being, as well as for the formation of the bond that normally develops from interactions such as these.

Crying, as demonstrated by spectrographic analysis, is a highly complex form of behaviour. According to Wolff (1969) three different types of cry can be distinguished according to their temporal characteristics, namely a basic (or hunger) form, which starts arhythmically and at low intensity but gradually becomes louder and more rhythmical; a 'mad' or angry cry, characterised by the same temporal sequence as the basic pattern (namely cry-rest-inspiration-rest) but distinguished from it by differences in the length of the various phase components; and the pain cry, which is sudden in onset, loud from the outset, and is made up of a long cry, followed by breath holding and then by short gasping inhalations. Not all investigators agree with this three-fold categorisation (for example, Wasz-Hockert et al., 1968); on the other hand there is agreement that crying has signal value in that it both arouses the listener and also conveys some rather more specific information as to the source of distress.

The arousal value is well illustrated in a study by Lenneberg, Rebelsky and Nichols (1965) of infants born to deaf parents. These parents did not feel compelled to attend to their crying infants even when they could see the distressed state the child was in. It seems that the auditory signal is required for this purpose; the sound conveys an urgency that no other form of stimulation can produce. The alerting effect is illustrated by listeners' physiological reactions to cry sounds: for instance, Frodi et al. (1978) recorded diastolic blood pressure and skin conductance from parents during videotaped presentations of smiling and crying infants. Smiling produced little change in automatic arousal; hearing a crying infant, on the other hand, elicited considerable increases in the physiological measures. It is interesting to note that no difference was found in this respect in the reactions of mothers and fathers.

In so far as different types of cry convey different information, with the pain cry in particular requiring a more urgent response, the ability to distinguish the various cries becomes a matter of importance. According to Wolff (1969), mothers were generally able to differentiate the three kinds of cry described by him, responding immediately and with alarm to the pain cry, with considerably less sense of urgency to the mad cry, and (especially if they had previous experience of caring for babies) reacting in a much more relaxed and varied manner to the basic cry pattern. Wasz-Hockert et al. (1968), though using different cry types, found similar results: their subjects too were able correctly to identify the different cries well above chance level, with experienced women being more accurate than inexperienced women. These studies suffer from various methodological shortcomings (Murray, 1979); nevertheless, they illustrate how infant signal and adult responsiveness are innately organized into a complementary system designed to ensure the child's protection and survival.

There is, however, nothing stereotyped about this system, for variations occur in both the infant's and the parent's contribution. As far as the infant is concerned, individual differences of a fairly stable nature in such parameters as the duration of specific cries and the interval between them have been noted (Prechtl et al., 1969), enabling parents to recognise their own infant's cry and distinguish it from that of other babies. In addition, neurologically damaged infants are frequently characterised by deviant cries, differing in pitch, temporal patterning or sheer volume from normal crying and sometimes producing such an aversive effect on their parents that caretaking arrangements may be seriously interfered with (Lester and Zeskind, 1979).

The kinds of variation that one can find in parental responses to infant crying have been documented by Ainsworth and Bell (1969) and Bell and Ainsworth (1972). A sample of 26 white middle-class American infants and their mothers were investigated over the course of the infants' first year, and even in this small and culturally homogeneous group the range of individual differences among the mothers was striking. Some mothers were said to respond with great sensitivity to the baby's cries, answering promptly and appropriately and thus quickly terminating the cause of distress; other mothers, on the other hand, ignored the infant and allowed him to become tense and frantic while they remained unresponsive or, alternatively, reacted in some inappropriate manner. In between these extremes there were other patterns, such as mothers who behaved in a somewhat erratic and unpredictable manner. In quantitative terms, the most responsive mothers ignored 45 per cent of their infants' cries during the first quarter of the first year; the least responsive mothers ignored as many as 97 per cent of cries. In the last quarter the range narrowed somewhat, namely from 13 to 63 per cent. The individual differences among the mothers tended to be fairly stable throughout this period, as seen especially in the measures of delay but also in the extent to which crying was ignored altogether. Ainsworth and Bell also note that some infants gave clearer signals than others and that this no doubt accounted for some of the variability found in the mothers' behaviour. However, whether that variability in turn affected the infants' behaviour remains a matter of controversy. According to Bell and Ainsworth, no relationship between maternal responsiveness and the amount of crying by infants could be observed for the first three months; for the last three months of the first year, on the other hand, there was a significant but negative relationship, that is, the more responsive the mothers were the less the infants cried. This is a finding contrary to that which principles of operant conditioning would lead one to expect, namely that responsiveness provides a reward in the form of attention and should thus strengthen the infant's tendency to cry. It is therefore not surprising to find that Bell and Ainsworth's conclusions have been hotly disputed and their study criticised on methodological grounds (Gewirtz and Boyd, 1977b); the precise relationship that maternal responsiveness has with amount of infant crying remains at present an open issue.

What accounts for such considerable variation in the responsiveness of different mothers has not been firmly established. The mother's personality, her previous experience of rearing children, the commitment she feels to other family and household tasks and the cultural norms to which she is subjected no doubt all play some part. The latter in particular is worth stressing: reports such as that by Konner (1972) on the child rearing practices of a hunter-gatherer society in Botswana serve to highlight the extent to which responsiveness is shaped by cultural forces. Thus in the society described by Konner there was virtually continuous contact between mother and infant; as a result mothers were readily able to anticipate hunger and other sources of discomfort from bodily or facial cues and could therefore easily prevent crying from taking place at all. Any cry that was heard was regarded as an emergency signal and responded to with an average latency of 6 seconds. One may contrast this with our own society where the question of how readily, and indeed whether, the parent should respond to a baby's non-pain cries has become one of those topics on which child care experts dispense advice to unsure mothers – advice which seems to change from one generation to the next. Thus at one moment of time mothers are urged not to respond for fear of 'spoiling' the child; at another they are admonished to answer immediately in order to avoid the trauma of frustration. Particularly in today's society, through the potent influence of the media, parental practices are inevitably affected by such forces.

Of concern is not only how readily a mother responds to her infant's crying but also the manner of her response, that is, its appropriateness to the cause of distress. According to Korner and Thoman (1970, 1972) the crying of newborn infants can most readily be stopped by those forms of caretaker intervention that combine vestibular stimulation with an upright posture. Thus picking up and putting the child to the shoulder is not only the most effective method of quieting him but also induces a state of alertness during which he can visually attend to his environment and thus be distracted from the original source of distress. Not all children will necessarily benefit from such treatment: highly active infants who dislike any form of physical restraint, including cuddling, require different forms of comfort (Schaffer and Emerson, 1964), and it is part of the mother's sensitivity to take such idiosyncrasies into account and respond accordingly. And for that matter, as children get older other kinds of treatment emerge as the most appropriate, including verbal forms that have little applicability at earlier stages.

Face-to-face interactions

Direct confrontation in face-to-face situations may, of course, occur at any age, but it assumes a particular significance during the period from two to five or six months (Schaffer, 1984). It is then that the child first 'discovers' the social partner; whereas in the earlier weeks the regulation of inner states was the principal theme for infant and caretaker, from about two months onwards there is an increasing turning to the outer world, with particular reference to the most salient feature in it, namely other people. Prolonged periods of face-to-face exchanges with the caretaker now ensue, and it is during these that the child must be helped to learn about the characteristics of other people and the way in which episodes of mutual attention can be managed.

Even the earliest interactive episodes are of a highly complex nature, for they involve the exchange of multiple signals (visual, vocal, bodily) in communicative packages meaningful to the other person. In many respects such early interactions already have the characteristics of proper dialogues, in that the to-and-fro of the exchange is conducted in an apparently orderly fashion to which both partners (infant as well as adult) contribute. On closer examination it becomes clear that the dialogues are really pseudo-dialogues (Schaffer, 1979): there is an initial asymmetry in the parts played by adult and child, in that it is largely up to the adult to ensure that interaction sequences are initiated and maintained with all the precision of temporal patterning and role exchange typical of more mature interactions. How the adult caretaker does so illustrates once again the part which sensitive responsiveness plays.

Data from a considerable number of studies (for example, Brazelton et al., 1974; Stern et al., 1977; Trevarthen, 1977) are now available as to how such early face-to-face interactions are conducted. They draw attention to several features of parental behaviour, three of which in particular are relevant here. The first refers to the great watchfulness adopted by parents when interacting with an infant. Thus gazing periods on the part of mothers observed in the laboratory are prolonged and uninterrupted – especially so in comparison with adult–adult interactions, where a far more rapid alternation of gaze-at and gaze-away is the norm. Mothers may spend as much as 90 per cent of the total observation time looking at the infant, thereby (as Fogel, 1977, has put it) providing a 'frame' within which the infant's gaze at the mother may cycle to and fro. Such diligant monitoring means, of course, that the mother is in a constant ready-state to respond to the infant, interpret his signals correctly and intervene at the appropriate moments in order to maintain the interactive sequences. It is in fact highly unlikely that such a pattern can be maintained for long in environments, like the home, that are not as free from distractions as the laboratory tends to be; nevertheless, these findings do convey something of the asymmetry to be found in the roles of adult and child.

The second feature of parental behaviour refers to the idiosyncratic style with which facial and vocal stimulation is presented to young infants. Stern (1974; 1977; Stern et al., 1977) has described in great detail the highly exaggerated facial expressions which mothers adopt while playing with their infants: the pursed lips or open mouth, the arched eyebrows, the tilted head, as well as the prominent gestures and the voice that covers a much wider range of pitch and loudness than found in conversations among adults. By these means the mother makes her face a highly attention-worthy stimulus; what is more, she can at a moment's notice modify her impact, reducing it if she feels the child is getting too excited or increasing it in order once more to raise his arousal level to some intuitive optimum. In addition, the tempo at which stimulation is presented tends to be slowed down: facial displays are formed gradually and then held for a long time; vocal phrases, particularly with regard to vowel duration, are exaggerated; and also the pauses in between utterances are considerably longer than found in adult-directed speech. It is as though the mother realises the infant's limited capacity for taking in information and hence presents it in chunks that are more easily assimilated and at a pace that gives plenty of time for processing. This is furthered by yet another characteristic of parental style, namely its repetitiveness. As Stern et al. (1977) have shown, both vocal and kinaesthetic acts tend to be repeated over and over again; parental interactive behaviour, that is, contains far more redundancy than is found in behaviour directed at more mature partners. A highly predictable stimulus world is thus presented to the infant; at the same time mothers are continually ready to re-tune their behaviour, varying it as soon as the infant's attention begins to stray. A fine balance between repetition and change should thus be continually struck, aimed at keeping the infant in a state of alertness and continuing involvement with the parent.

The enmeshing of interactive behaviours usually takes place at a split-second level. Take vocal interchange: the majority of speaker-switches in the 'conversations' of 12-month-old infants and their mothers were found by Schaffer et al. (1977) to be accomplished in a time less than one second, with rarely any clashes (in the form of overlapping vocalisations) marring the smoothness of the interchange. Amongst adult interlocutors such smoothness is usually brought about to equal extent by both partners; in mother–infant pairs it is more likely to be primarily accomplished by the skill of the adult in inserting her own contribution in the pauses between the vocal bursts of the child. A very precise sense of timing is required for this purpose – hence no doubt the great watchfulness, which enables the parent to provide the appropriate form of stimulation at the right moments of time.

Thus in order to construct interaction sequences at this early stage of development the sensitive parent must package her social stimulation in a manner appropriate to the abilities of the infant to process it and she must present it at those points within the interactive flow when it will promote rather than interrupt the exchange. This holds equally for maternal stimulation during feeding sessions (Kaye, 1977), free play (Arco and McCluskey, 1981), structured games (Gustafson et al., 1979) and directed task activity (Schaffer and Crook, 1979): under all these circumstances the adult must take the initiative in constructing mutually satisfying social exchanges. Where the child's partner is incompetent to provide such support, as happens in the case of peer interaction (for example, Fogel, 1979), the exchange is likely to be brief and primitive. With increasing age children's social interactions with adults will assume a more symmetrical character (Holmberg, 1980); it seems likely that this can only happen if the child has been previously involved in 'successful' interactions through the initiative of a sensitive adult.

Not all parents are equally adept at supporting the infant in face-to-face interactions. For one thing, experience matters: it is necessary for the parent to learn how to match her interactive style to the natural rhythms of particular infants. Stern (1971), in a study of three-month-old non-identical twins, showed how each infant elicited different behaviour from the mother; as a result of the mother's awareness of each child's individuality the two interactive patterns differed radically in a number of respects. Opportunity to learn about particular infants is clearly required if the manner appropriate to that individual is to be adopted. Trevarthen (1979) interchanged mothers and infants to produce pairs of 'strangers'; when comparing them with the natural pairs he found the communication process to be out of tune in certain respects and to be less elaborate and more cautious. Experience appears to matter more than the sex of the parent: Field (1978), in a study of four-month-old infants and their parents, found few differences between mothers and fathers – if, that is, the parent observed was the child's primary caretaker. However, differences did emerge between primary and secondary caretakers, irrespective of the parent's sex – a finding which points to the differential amount of experience with the child as the crucial factor.

In addition, personality factors may interfere with the interactive process. To take the extreme case, a mother suffering from depression is unlikely to tune in to the child in a sufficiently sensitive manner to be able to construct with him/her a joint interactive sequence. Cohn and Tronick (1983) observed infants under conditions of simulated maternal depression and found their behaviour to be markedly affected, with an increase in protest, wariness and generally negative behaviour being particularly marked. Simulation experiments have indeed proven to be particularly useful in indicating the way in which infants are affected by inappropriate forms and timing of stimulation in face-to-face situaltions. Thus Trevarthen (1977) has shown how infants, confronted by mothers who have been asked to adopt an immobile expression during the interactive session, look puzzled, become unhappy and avoid prolonged eye contact with the mother. The same reactions of puzzlement and strong signs of unhappiness were evident when the mother's attention was experimentally switched from the infant to another adult by means of a change of lighting either side of a partially reflecting window. The mother's response to her child was then no longer contingent on his own behaviour, and though the infants were only eight weeks old they reacted with confusion, distress, inert dejection and withdrawal. Mismatching, through the adult's faulty timing and through over- or under-loading, may well have drastic consequences if continuously experienced as part of the child's daily life.

In face-to-face situations, just as with regard to responsiveness to crying, cultural norms exert an influence. For instance, Brazelton (1977) has described how the Mayan Indians of Mexico provide far less stimulaton for their infants in such a situation: smiling, vocalising and even looking are very much reduced in comparison with Western norms. Mothers' glances at the baby tend to be perfunctory and without apparent expectation of response. When the infant does make a social overture it is rarely responded to; it is as though the mothers want a quiet and undemanding infant and therefore do their best to avert a demand-response pattern from developing. There may be adaptive value in that particular setting to such behaviour; what does become apparent is that watchfulness and a high level of readiness to respond are affected by cultural pressures and are not an inevitable manifestation of parenting.

Language input

In the period beyond infancy the topic which provides most evidence of parents' sensitive responsiveness is children's language development.

Precisely how children acquire language remains conjectural, but it is clear that in some way the linguistic environment plays a crucial part in this process. In particular, it has been proposed that the speech which the child hears from his caretakers must in various ways be carefully adapted to his level of comprehension if it is to help him in acquiring linguistic competence. It follows that adult speech to children differs from speech addressed to other adults, and a large body of data is now in existence which attests to this conclusion (for reviews see dePaulo and Bonvillian, 1978; deVilliers and deVilliers, 1978).

'Motherese' (as the adult-to-child code has come to be known) differs from adult-addressed speech in many ways. It is grammatically simpler, better formed, more repetitive and briefer (Phillips, 1973; Snow, 1972). It contains more imperatives and questions and fewer statements (Gelman and Shatz, 1977; Newport, 1976). Its references are mainly to the here-and-now situation and not to past or future events (Shatz and Gelman, 1973). It is characterised by various phonological differences, such as high pitch and exaggerated intonation (Sachs et al., 1972). It is also closely geared to the child's ongoing activity, in that the adult generally ensures that the child's attention is appropriately focused on the relevant objects before these are referred to (Collis, 1977; Messer, 1981; Schaffer and Crook, 1979). In short, motherese is marked by a great many features that one might expect to make the task of the language learning child an easier one.

It has also been frequently asserted that motherese is fine-tuned to the age (or, to be more precise, to the linguistic competence) of the individual child being addressed. Thus the younger the child the more marked are the features designating the code (Snow, 1972; Fraser and Roberts, 1975). Bellinger (1980), using a composite index of maternal speech derived from a number of variables, found such a relationship with age but noted that the greatest rate of change in the complexity of the mothers' speech occurred when the children were between one year and eight months old and two years three months old. This closely mirrored changes in the children's growing linguistic competence: here, too, the period at about the end of the second year showed the greatest rate of advance. It can be assumed that the mothers were sensitive to the children's achievements and adapted their input accordingly. Such adaptation is not found to equal extent in all aspects of maternal speech; fine tuning, it appears, is more likely to occur in some features that in others (Cross, 1977). The best evidence of adjustment appears to be at the discourse level, that is, in the manner in which mothers use speech to perform such interpersonal functions as linking the topic of their utterances to that of the child's previous utterance, rather than at the syntactic level. This conclusion has not been confirmed by all studies (Ellis and Wells, 1980); nevertheless, it does seem plausible that features of maternal speech describing the *purpose* for which it is used should be particularly sensitive to the characteristics of the listener.

'Motherese' is in fact a misnomer, for the particular style which the term describes is by no means confined to mothers. It has also been found in women who are not mothers (Snow, 1972), in fathers (Golinkoff and Ames, 1979), and even in young children (Shatz and Gelman, 1973). What is more, special ways of talking to children have been found in all speech communities investigated so far in different parts of the world, though the precise nature of the modification may vary according to the characteristics of the language and the socialisation goals of the particular society (Ferguson, 1977). It seems as though almost any human being, confronted with another person, quite automatically adjusts his or her style of communication to the receptive abilities of the addressee. It is assumed that this is done on the basis of feedback information provided by the listener: for example, Snow (1972) asked mothers to speak to an *imaginary* child of a specified age and found the features of motherese not to be as marked as they were to a physically present child. Just what the cues are that provide the necessary information remains as yet unsettled.

However common the fine-tuning of child-addressed speech may be, individual differences in the extent to which it is manifested do occur. As Ellis and Wells (1980), have put it, having found substantial variation in this respect among a sample of parents investigated: '. . . in the population as a whole, the level of adult speech adjustment relating to communicative intent may constitute a continuum with, at one pole, a group of adults highly sensitised to the communicative needs of their children and, at the other, a group of adults who monitor their speech only in coarse syntactic terms.' There have been various suggestions as to how such differences should be considered. One approach is to concentrate on the general style of communication rather than on isolated variables. for example, Nelson (1973) classified mothers according to their directiveness, that is, the extent to which they intruded on the child's activity with commands and requests, imposing their own ideas on his behaviour, as opposed to letting him lead and then responding to his way of formulating his experience. The amount of directiveness was found to have implications for the child's language development: the more intrusive the mother the slower the rate of that development was found to be. Other stylistic classifications (for example, Howe, 1981; Wood et al., 1980) similarly suggest that the more open the adult is to the child's interests and the more she motivates him to talk about these by taking up topics originally raised by him the greater will be the facilitating effect on his developmental progress.

EFFECTS ON CHILDREN

That there are differences between parents in the degree of their sensitive responsiveness is probably not surprising. However, the reason that such differences have been taken seriously and attempts have been made to describe their precise behavioural manifestations is the belief that there are implications for children's development in such variation. Unfortunately any effort to specify cause-and-effect relationships between parental and child characteristics in human beings encounters considerable methodological difficulties. Almost invariably statements are of a correlational nature; there are, that is, no ways of distinguishing between the following alternatives: that the parental behaviour has 'caused' the child characteristic; that, on the contrary, the child's characteristics have 'caused' the parent's behaviour; or that both are due to common factors such as genetic influences. In the absence of experimental interference in the developmental process, evidence for aetiology is bound to be scarce. Efforts to investigate the relationship have, however, been made, with particular reference to two aspects, namely children's emotional security and their developmental progress.

Parental responsiveness and child security

Ainsworth et al. (1978) have put forward the proposition that the quality of children's attachment relationships is a function of the types of interactive experiences they have with their mothers in infancy, and that the sensitive responsiveness which the mothers show in the course of such interactions is likely to have a particularly significant effect in this respect. Hence, a child whose mother responds to his/her signals and communications promptly and appropriately throughout infancy will develop confidence in his/her mother's availability and will have the security to use her as a base from which to explore his/her environment; the insecure child, on the other hand, tends to be anxious regarding the mother's availability and show little confidence in his reactions to the world.

Ainsworth used the 'Strange Situation' to highlight such differences in children's security. This is a standardised laboratory procedure, designed primarily for children around the age of 12 months, that takes place in an unfamiliar setting in which the child is successively introduced to a variety of events designed to put him under stress, thus evoking an intensification of attachment behaviour. These events include introduction of a stranger, being left by the mother with the stranger and being left entirely alone, and on the basis of their reactions to these episodes children can be classified into three main groups: Group A infants, who are judged to be insecurely attached; Group B infants, who are said to show secure attachments; and Group C infants, who are ambivalent towards their mothers but are generally regarded as insecure.

These three patterns are said to have their antecedents in the maternal rearing practices to which the children have been exposed. As Sroufe and Waters (1982) put it: 'Caregiver responsiveness to the infant's signals predicts quality of attachment.' The evidence for this link comes mainly from a study by Blehar et al. (1977), which investigated the interactive experiences of a sample of infants in the period 6 to 15 weeks and related these to their attachment classification at 12 months. Securely attached children (Group B) were found to have mothers who were much more frequently contingent in their pacing during the early weeks, were more encouraging of further interactions and tended to be more playful and lively: anxiously attached children (Groups A and C combined), on the other hand, had mothers who often initiated face-to-face interaction with a silent, impassive face and frequently failed to respond to the babies' attempts to initiate interaction. From these and other related findings Ainsworth et al. conclude that 'babies who differ qualitatively in their attachments to the mother at the end of the first year have mothers whose behaviour and attitudes, as early as the baby's first 3 months, differ in salient ways'. The key variable in such behaviour is said to be the sensitive responsiveness with which the mother responds to infant signals and communications.

Security is such an important quality that attempts to find its antecedents are most praiseworthy. Unfortunately there are problems about the Ainsworth formulation which indicate caution in accepting the particular link proposed. Some of these concern the use of the Strange Situation for diagnostic purposes: for example, the procedure is a highly artificial and constraining one that lacks ecological validity; furthermore, assessment is based almost exclusively on the reunion episodes following separation from the mother and these may last only a few seconds – much too brief a period on which to base such an assessment (Schaffer, 1984). Other problems refer to the suggestion that it is the *earlier* interactions that are responsible for the outcome; as Lamb et al. (1984) have argued, it is more likely to be *concurrent* experiences that have such an influence, for children whose family circumstances undergo drastic change do not necessarily show stability in their attachment classification. Thus, taking into account various other methodological problems to which Lamb et al. also point, it has to be concluded that the link between the child's security of attachments and the mother's sensitive responsiveness has not been firmly established, however plausible it may seem.

Parental responsiveness and developmental progress

The suggestion that parental responsiveness is linked to the rate of children's developmental progress comes mainly from the studies of language acquisition to which we have already referred. The linguistic environment of some children, it is argued, is such that it enables them to become verbally competent more quickly than children experiencing different kinds of input. In particular, the fine-tuning of motherese

Parental Responsiveness and Child Behaviour 301

functions as a teaching aid in helping children to master linguistic skills, and variations in the extent of its usage should therefore result in different rates of acquisition.

There have been a number of attempts to find evidence for such a relationship (Newport et al., 1977; Cross, 1977; Furrow et al., 1979). All have yielded some suggestive evidence that the degree of fine-tuning to which a child is exposed is related to the rate of language development, but all are subject to two reservations. The first is the aforementioned difficulty of inferring cause-and-effect relationships from correlations: it is at least conceivable that adult speech adjustments are child elicited and have no direct influence on developmental rate. In an effort to circumvent this difficulty Furrow et al. used cross-lagged correlations, and found that several characteristics of maternal speech sampled when the children were one year old significantly predicted language development nine months later. They concluded that motherese is indeed an effective teaching device and that the nature of the caretaker's linguistic input to the child does have a definite influence on the language learning process. However, the argument advanced by these authors depends primarily on the ability of the cross-lagged technique to tease apart cause and effect, and, unfortunately, recent criticisms have shown that this is not necessarily the case (Rogasa, 1980).

The second reservation is that even the correlational evidence that has emerged applies only to some aspects of motherese and not to others. For example, Newport et al. (1977) found no consistent relationship of the child's language development to such measures of maternal speech as the complexity or length of utterance or the amount of repetition; on the other hand, extensive use of feedback in respone to the child's own utterance was related to language growth. These authors believe that a particularly important feature is the extent to which mothers ensure that they and their children have matching referents – a conclusion which was borne out by Cross (1978) in a study comparing the mothers of normally developing children with mothers of children showing accelerated language development. No differences were found on a large number of indices of maternal speech; however, the two groups were distinguished by the fact that the mothers of the accelerated children semantically linked their utterances to the child's preceding utterance to a significantly greater extent than the other mothers. By thus following up on topics introduced by the child these women established a joint focus of reference, and moreover one that the child himself or herself had determined in the first place and was thus motivated to attend.

Joint enterprise contexts

The findings just quoted suggest that the degree to which parents achieve success in furthering developmental progress may be related to the extent to which they can set up joint enterprises with the child. There are two aspects to such efforts which can be usefully distinguished: first, the way in which joint enterprises are initially established, and second their maintenance and development. The parent's sensitive responsiveness plays a crucial part in each.

Establishing joint enterprises involves the role of topic sharing. There can be no communication without a shared focus of reference; this must first be established before the interaction can proceed. A number of devices exist for this purpose, enabling one person to communicate to another the topic to which he is attending: verbal reference, indicative gestures such as pointing, gazing at the referent and physical contact with it – these are among the most prominent means of bringing about topic sharing and all play a role even in the earliest interactions (Schaffer, 1984). However, in these early interactions there is an initial asymmetry: almost invariably it is the child who chooses the topic and it is left to the adult to follow that choice. Take the phenomenon of visual coorientation (Collis and Schaffer, 1975): the vast majority of episodes when mother and infant are both simultaneously attending to the same object are brought about by the mother following the infant's visual gaze rather than the infant following the mother's gaze – indeed there are indications that children below the age of one year are not yet capable of appreciating the significance of another person's gaze direction. Or take verbal reference: according to Shugor (1978) most early conversations are based on a repeated pattern where the child chooses the topic and the adult then comments on it; once again, the ability to take up another's topic (verbally expressed in this case) and elaborate upon it is a phenomenon seen only at later stages of development. Before then, if the interaction is to take place at all, it is left to the adult's initiative to respond to the infant's behaviour as though it has communicative intent and to interpret any signs of interest on his part in some environmental feature as an invitation to share that interest. To do so successfully, especially in the case of rapidly shifting gaze direction, the adult must be highly attuned to the infant.

As to the second aspect of joint enterprises – their maintenance and development – there is again an imbalance in the roles of adult and child. This is seen particularly clearly when one compares adult–child interactions with peer interactions, for in the latter the participants are of equal developmental status and lack the support that a more mature partner can provide. Thus Holmberg (1980), in a study of children between 12 and 42 months interacting with peers and with adult caretakers in day care centres, found that 'elaborated interchanges' (that is, interactive sequences involving at least two turns from each partner) occurred rarely between peers in the first two years and only then began to increase in frequency; when the partner was an adult, however, they occurred much more frequently at even the youngest age, maintaining that level of frequency throughout the age-range investigated. One must conclude that it is the adult who is primarily responsible for keeping the interchange going with younger children; as Holmberg noted, adults did so by picking up on the responses of the child (naming what he pointed to, offering another puzzle as he completed the first one, and so on) and so showing sensitivity to his capabilities and response level.

It is this sensitivity to the child's current level of functioning that enables the adult to help the child progress to further developmental levels. Analyses of parental teaching patterns have shown how the parent's awareness of the child's present capabilities and his/her setting of realistic targets for further achievements of the child are basic ingredients of successful tuitional strategies (Wood and Middleton, 1975). Vygotsky (1978), by using the concept of the 'zone of proximal development', attempted to put forward a mechanism to explain the adult's role in situations of joint involvement. Vygotsky defined the zone of proximal development as the distance between the actual developmental level of the child working on his own and the potential level of development when working under the guidance of an adult. By carefully monitoring the child's independent activity the adult is able to judge when to intervene and how to present bits of the task in sequence and in the appropriate amount for the child to absorb. In this way the child is helped gradually to take over patterns of behaviour, internalise them as his own and thus move to progressively more complex modes of functioning. This applies not only to intentional tuition but also to non-deliberate interactive encounters between a child and a more mature partner.

Thus any joint enterprise context provides a sensitive adult with the opportunity of providing facilitative input to the child. Wells (1979) examined the various situations in which conversations take place between mothers and young children and found that the proportion of speech occurring within specifically joint enterprise contexts (doing housework together, play with adult participation, looking at books or just talking together) was significantly related to language development at age 2½. The more rapid development of first-born children which Wells found was similarly ascribed to the greater opportunity that mothers of only children have for engaging in talk in the context of shared activities. Once again one has to be careful in making aetiological statements, but at the very least one can assert that joint enterprise situations provide the *opportunity* for sensitively adapted input to children. As Dunn et al. (1977) note, maternal speech occurring in the context of joint attention to pictures and books is particularly rich in those features thought to be important in rapid language development; in such a situation the mother's speech is likely to be especially closely tied to the child's interests and to what he is trying to communicate, and the extent to which such situations occur in the daily lives of mothers and children may well have considerable implications for the child's mastery of linguistic skills. A finding by White and Watts (1973) is pertinent: from observations of pre-school children and their mothers at home it emerged that a great many of the encounters between mother and child in the course of any one day could be characterised as 'core teaching situations', in that it is in their context that the adult has the opportunity of influencing the child's behaviour in ways that can contribute towards his/her development. However, children developing a relatively high degree of competence generally had mothers who tended to demonstrate and explain things at the *child's* instigation rather than their own, thus providing help and guidance oriented around the child's interests and at times determined by the child himself. Children of lower developmental competence, on the other hand, experienced rather more didactic handling from their mothers, whose respect for the child's own interests was thus more limited and more discouraging. Certain kinds of interactive situations, it appears, lend themselves well to the provision of facilitative input; certain kinds of parents, it also appears, are more likely to make use of these opportunities and may in consequence affect their children in ways regarded as beneficial in our society.

DETERMINANTS OF SENSITIVE RESPONSIVENESS

If it is at all likely that sensitive responsiveness on the part of adults has beneficial effects on the children in their care then it becomes highly desirable to foster this quality. However, one can only do so if one knows something of its origins and the factors that influence it. Unfortunately there is still considerable ignorance in this respect; nevertheless, it is useful to consider some of the likely influences involved. These can be grouped under factors referring to the parent, the child and the context respectively.

Parental factors

As we have already mentioned, to regard sensitive responsiveness as a fixed personality trait that acts as the sole determinant of the parent's behaviour is an unjustified view. However, this is not to deny the existence of predispositions that can be found to varying degree in individual parents, but only to assert that these dispositions act in conjunction with other influences to determine behaviour on any one occasion. To enquire about the origins of individual differences in this sense is therefore a legitimate undertaking.

One possible influence refers to the parent's sex. Women, it has been frequently asserted, are biologically pre-adapted for child care in a psychological as well as physical sense, and their responsiveness is thus greater than that of men. One source of this argument lies in comparative data, in that care of the young is generally undertaken by the mother animal in most species. However, in some species other arrangements do exist, such as shared parental care or even care as primarily a paternal responsibility. In any case the evidence for humans is far from unequivocal. The notion, for example, that maternal care is under the control of hormonal mechanisms that are brought into being by the process of birth (and therefore are not operative in men) receives no support once one also takes into account the effects of differential socialisation pressures on the sexes (Murray, 1979). Indeed, as Berman (1980) concludes from her extensive review, cultural factors are probably more important in determining whatever differences do exist between males and females than are hormonal factors; furthermore, studies such as those by Parke and O'Leary (1976) provide little evidence that mothers are more responsive than fathers during the period following birth. It is therefore unlikely that hormonal factors on their own can in some simple mechanical fashion account for variations in responsiveness.

The same applies to another possible influence, namely extended contact during the neonatal period between mother and child. As the detailed review by Sluckin et al. (1983) shows, the proposal put forward by Klaus and Kennell (1976), that there is a critical period immediately following birth when the amount of contact determines the quality of future maternal care, has not been upheld by other investigators. Though short-term effects of a beneficial nature may well result from extended contact between mother and baby at this time, there has been a consistent failure to replicate the findings by Klaus and Kennell with respect to the long-term effects of such an experience. Thus there is no support for the notion that mothering is affected on a once-and-for-all basis by events occurring during a highly specific period immediately following birth.

A much more likely experimental factor to play a part is the mother's own upbringing and her experience of being mothered as a child. In particular, it has been suggested that deprived children in turn become depriving parents, setting up a cycle that may be difficult to break (Rutter and Madge, 1976). As yet there is little firm evidence to substantiate this claim; however, some preliminary findings by Rutter (1982) about a group of mothers who had been in care during their childhood show that there are indeed many respects in which these women differ in interactions with their children when compared with women who had not been in care. As the differences are nearly all such as to indicate 'inferior' mothering on the part of the in-care group, the possibility arises that prolonged experience of disturbed mothering in the early years may seriously affect the individual's competence to undertake mothering herself as an adult. To what extent this affects the specific aspect of parental behaviour with which we are concerned remains uncertain, though some of Rutter's measures would appear to be direct reflections of this attribute.

Child factors

Parental effectiveness depends as much on the child as on the parent himself or herself. Some children come into the world in a behaviourally disorganised state as a result of neurological deficits, prematurity or other deviant conditions, which cause them to be less well adapted to dyadic interaction than normal children and makes the task harder for their caretakers. Thus, quite apart from the emotional distress experienced by the parents, they have to cope with a child whose disorganised condition makes his signals more difficult to interpret and whose unusual demands cause confusion and inappropriate response patterns in the parents. Pathological crying, disturbed sucking patterns, unusual waking-sleeping rhythms, distractability – these are but some of the infant characteristics with which some parents are confronted. It is hardly surprising that under such conditions the normal course of social interaction is interfered with; it is perhaps more surprising that some parents do manage to adjust to the deviant nature of their child.

The numerous studies of pre-term infants that have recently been undertaken provide some useful examples of these trends. Such children tend to be characterised at birth by lowered responsiveness and poor behavioural organisation; they are restless and distractible and provide the mother with cues that are often ambiguous and difficult to interpret. As Gorski et al. (1980) conclude from their investigation of pre-term infants: 'Our research has shown that the stressed premature may be unable to participate reciprocally with caregivers. The high-risk infant must first develop sufficient physical integrity and internal stability before he is able to use caregiver support.' While this obviously presents considerable problems to parents there are also indications that many are able to meet these difficulties in a constructive way. Beckwith and Cohen (1978) found that the more neonatal complications there were in the infant the more interactive stimulation he received from the parent, and they suggest that this may well indicate the existence of a compensatory mechanism in the caretaker that is triggered by the infant's condition, as a result of which he/she is provided with extra care, thus attenuating the effect of the hazardous events. Eventually many premature infants catch up and group differences between them and full-term infants disappear; at the same time, as Crawford (1982) has shown for the age range 6 to 14 months, differences between the two groups of mothers similarly disappear. At six months the mothers of the prematures showed more caretaking and affectionate behaviour; with increasing age the gap between the two sets of mothers steadily diminished.

Compensation by the parent for the child's deficiencies has been decribed by others too. Sorce and Emde (1982), for example, noting that Down's Syndrome infants tend to be considerably less emotionally expressive than normal children, found that their mothers were much more likely to react even to low-intensity emotional signals than mothers generally do – as though they had recalibrated their responsiveness threshold to ensure that the child was not deprived of interactional experience as a result of an inherently depressed level of signalling. Thus the parent-child system as a whole may continue to function satisfactorily despite some deficiency on one part, thanks to adjustment made on another part. Not that such adjustment will occur inevitably; thus the difficulty of providing phased stimulation, whether during feeding (Field, 1977) or during vocal exchanges (Jones, 1977), may be a recurrent one in the case of many deviant infants; indeed Berger and Cunningham (1983) found that the incidence of vocal clashing, indicating a breakdown of the vocal turn-taking pattern of mothers and Down's Syndrome infants, tended to increase with age during the first half-year. Such children slur their volcalizations and do not produce the clear-cut burst-pause patterns that characterize normal children, making it much more difficult for the mothers to know when the child has come to the end of his contribution to the exchange and when to insert hers.

Contextual factors

Interactions always occur in particular physical and social settings, the nature of which may have considerable implications for the kind of interactive behaviour observed. In the study of parental responsiveness, as with so many other aspects of human behaviour, psychologists have until quite recently tended to neglect such contextual factors; thus mother and infant have frequently been observed in a specially constructed laboratory situation and findings obtained there were then generalised to all other settings. However, in a dyadic situation where the adult is shielded from all other commitments and can thus pay undivided attention to the child sensitive responsiveness can be relatively easily achieved. Constant monitoring and fine tuning may be much more difficult in real-life settings where the adult is often subjected to various other competing demands.

This arises, for example, in a group care situation, where the adult is simultaneously responsible for several children rather than for just one child. Schaffer and Liddell (1984) compared the behaviour of nursery nurses when confronted by either one child or by four children. Marked differences were obtained in several respects, such as the amount of attention paid by the adult to any one individual child, the amount of speech addressed to him and the nature of the interactive style adopted towards the child. Of particular significance, however, was the marked reduction in responsiveness in the polyadic situation: whereas in the dyad nearly every bid by the child for the adult's attention was promptly answered, almost half of all children's bids failed to elicit a response from the adult in the larger group. What is more, when the adult did respond the resulting bout of joint involvement tended to be far briefer than in the dyadic setting. Having to divide his/her attention among several children therefore meant not only that the adult was often unavailable to any one child but that he/she also showed none of the concentrated interest in the child's activity that he/she showed in the dyadic situation.

There are no doubt various environmental conditions which militate against sensitive responsiveness. It seems likely that the greater the number of children a mother has and the more closely spaced they are, the more difficult will be her task to meet their demands in a consistently sensitive and contingent fashion. In this connection it is relevant to note that, according to home observations by White et al. (1979), mothers' responsiveness to first-born (and at the time only) children was at a consistently higher level than to subsequent children. For that matter physical conditions may also affect the adult's behaviour: thus Neill (1982) found children in open-plan nurseries to receive rather less attention from staff than in smaller, more intimately designed nursery rooms where they tended to be in closer proximity to the adults. Responsiveness, in short, is not merely a function of the adult's personality and of the make-up of the particular child to whom he/she is responding; it is also affected by a variety of setting conditions relating both to the other people present in that situation and the physical characteristics of the environment in which the interaction occurs.

CONCLUSIONS

There are many questions that remain to be answered about sensitive responsiveness. To what extent is it a unitary characteristic as opposed to an umbrella term for a number of only loosely associated aspects of parenting? What are its (no doubt multiple) determinants? What are its effects on children and by what sort of processes are these brought about? Further work must address itself to these issues, but in the meantime it appears justified to conclude that sensitive responsiveness is indeed an important aspect of parenting, with implications not only for attempts to understand the nature of the parent-child relationship but also for practical efforts made to bring about improvements in that relationship in individual cases. One point should be borne in mind, however, whether one approaches the topic from a theoretical or a practical point of view: sensitive responsiveness characterizes not so much parents as *individuals* as the parent's *interaction* with the child; it is basically a statement about that interaction and hence is meaningless without reference to *both* partners.

One further matter must also be raised. It refers to the current vogue for 'intervention' in parent-child relationships and the models on which such efforts are based. If sensitive responsiveness is to be valued as a desirable characteristic that is to be fostered, then a 'the more the better' philosophy may well prevail. As part of the model for the ideal relationship a 100 per cent level of responsiveness could thus easily be adopted, with relationships falling below that level being regarded as sub-optimal and intervention being geared to raising it to its maximum. Let us note, however, that many of the studies of parent-infant interaction from which this model is derived have been conducted in laboratory settings where the two partners were completely sheltered from all distractions and where it was therefore very easy for the parent to attend to the child virtually non-stop. Observations in real-life settings such as the home indicate a rather different picture. Ainsworth and Bell (1969), for instance, found that the mother with the *highest* responsiveness level in their sample responded to her baby's crying on only 55 per cent of occasions during the first quarter year. White and Watts (1973) noted that everyday interactions tended to be of an episodic nature: during such episodes there was generally great watchfulness on the part of the parent, but these were incorporated in the rest of the parent's duties and activities, in the course of which attentiveness to the child might well be sharply reduced. The norm, it seems, is other than that suggested by the laboratory studies (though it is also as well to remember that this norm varies from one culture to another: see the observations by Konner and by Brazelton previously referred to). What is more, there is nothing to indicate that the level of responsiveness required for optimal development must be 100 per cent. Might there not even be some benefits in a certain amount of ignoring of the child's signals – benefits such as developing self-reliance and learning to share the parent with others? We do not know the answer to this question, for interactive studies investigating the exchange of signals have by and large focused on those instances where a response did occur rather than analyse instances where it failed to be provided. A fairly high level of responsiveness may well be beneficial, indeed essential, to the child's development; it does not follow that 'the higher the better' must be adopted as a guideline and that all intervention efforts should be geared to such an end. The importance of responsiveness is apparent; ignoring might, however, well prove to be a most informative research topic. Until we redress the balance and learn more about that side of the coin we are in no position to state what the 'ideal' relationship involves.

REFERENCES

Ainsworth, M. D. S. and Bell, S. M. 1969: Some contemporary patterns of mother-infant interaction in the feeding situation. In A. Ambrose (ed.), *Stimulation in Early Infancy*, London: Academic Press.

Ainsworth, M. D. S., Bell, S. M. and Stayton, D. J. 1971: Individual differences in strange-situation behavior of one-year-olds. In H. R. Schaffer (ed.), *The Origins of Human Social Relations*, London: Academic Press.
—— 1974: Infant–mother attachment and social development: 'socialization' as a product of reciprocal responsiveness to signals. In M. P. M. Richards (ed.), *The Integration of a Child Into a Social World*, Cambridge: Cambridge University Press.
Ainsworth, M. D. S., Blehar, M. C., Waters, E. and Wall, S. 1978: *Patterns of Attachment*. Hillsdale, New Jersey: Lawrence Erlbaum.
Arco, C. M. B. and McCluskey, K. A. 1981: 'A change of pace': an investigation of the salience of maternal temporal style in mother–child play. *Child Development*, 52, 941–9.
Beckwith, L. and Cohen, S. E. 1978: Preterm birth: hazardous obstetrical and post-natal events as related to caregiver–infant behavior. *Infant Behavior and Development*, 1. 403–11.
Bell, R. Q. 1977: Socialisation findings re-examined. In R. Q. Bell and C. V. Harper (eds), *Child Effects on Adults*, Hillsdale, New Jersey: Erlbaum.
Bell, S. M. and Ainsworth, M. D. S. 1972: Infant crying and maternal responsiveness. *Child Development*, 43, 1171–90.
Bellinger, D. 1980: Consistency in the pattern of change in mothers' speech: some discriminant analyses. *Journal of Child Language*, 7, 469–87.
Berger, J. and Cunningham, C. C. 1983: Development of early vocal behaviors and interactions in Down's Syndrome and non-handicapped infant–mother pairs. *Developmental Psychology*, 19, 322–31.
Berman, P. W. 1980: Are women more responsive than men to the young? A review of developmental and situational variables. *Psychological Bulletin*, 88, 668–95.
Blehar, M. C., Lieberman, A. F. and Ainsworth, M. D. S. 1977: Early face-to-face interaction and its relation to later infant–mother attachment. *Child Development*, 48, 182–94.
Bowlby, J. 1969: *Attachment and Loss. vol. 1: Attachment*. London: Hogarth Press.
Brazelton, T. B. 1977: Implications of infant development among the Mayan Indians of Mexico. In P. H. Liederman, S. R. Tulkin and A Rosenfeld (eds), *Culture and Infancy*, New York: Academic Press.
Brazelton, T. B., Koslowski, B. and Main, M. 1974: The origins of reciprocity: the early mother–infant interaction. In M. Lewis and L. A. Rosenblum (eds), *The Effect of the Infant on its Caregiver*, New York: Wiley.
Caldwell, B. M. 1964: The effects of infant care. In M. L. Hoffman and L. W. Hoffman (eds), *Review of Child Development Research*, New York: Russel Sage Foundation.
Caudill, W. A. and Schooler, C. 1973: Child behavior and child rearing in Japan and the United States: An interim report. *Journal of Nervous and Mental Disease*, 157, 323–38.
Caudill, W. and Weinstein, H. 1969: Maternal care and infant behavior in Japan and America. *Psychiatry*, 32, 12–43.
Cheney, D. L. 1977: The acquisition of rank and the development of reciprocal alliances among free-ranging immature baboons. *Behavioural Ecology and Sociobiology*, 2, 303–18.

Clarke, A. M. and Clarke, A. D. B. 1976: *Early Experience: Myth and Evidence.* London Open Books.

Cohn, J. F. and Tronick, E. Z. 1983: Three-month-old infants' reaction to simulated maternal depression. *Child Development*, 54, 185-93.

Collis, G. M. 1977: Visual coorientation and maternal speech. In H. R. Schaffer (ed.), *Studies in Mother-Infant Interaction*, London: Academic Press.

Collis, G. M. and Schaffer, H. R. 1975: Synchronisation of visual attention in mother-infant pairs. *Journal of Child Psychology & Psychiatry*, 16, 315-20.

Crawford, J. W. 1982: Mother-infant interaction in premature and full-term infants. *Child Development*, 53, 957-62.

Cross, T. 1977: Mothers' speech adjustment: the contribution of selected child listener variables. In C. E. Snow and C. A. Ferguson (eds), *Talking to Children: Language Input and Acquisition*, Cambridge: Cambridge University Press.

—— 1978: Mothers' speech and its association with rate of linguistic development in young children. In N. Waterson and C. Snow (eds), *The Development of Communication*, Chichester: John Wiley.

Cummins, M. S. and Suomi, S. J. 1976: Long-term effects of social rehabilitation in rhesus monkeys. *Primates*, 17, 43-51.

dePaulo, B. M. and Bonvillian, J. D. 1978: The effect on language development of the special characteristics of speech addressed to children. *Journal of Psycholinguistic Research*, 7, 189-211.

deVilliers, J. and deVilliers, P. 1978: *Language Acquisition.* Cambridge. Massachusetts: Harvard University Press.

Dunn, J., Wooding, C. and Hermann, J. 1977: Mothers' speech to young children: variation in context. *Developmental Medicine & Child Neurology*, 19, 629-38.

Ellis, R. and Wells, C. G. 1980: Enabling factors in adult-child discourse. *First Language*, 1, 46-62.

Ferguson, C. A. 1977: Baby talk as a simplified register. In C. E. Snow and C. A. Ferguson (eds), *Talking to Children*, Cambridge: Cambridge University Press.

Field, T. M. 1978: Interaction behaviors of primary versus secondary caretaker fathers. *Developmental Psychology*, 14, 183-4.

—— 1977: Maternal stimulation during infant feeding. *Developmental Psychology*, 14, 539-40.

Fogel, A. 1977: Temporal organization in mother-infant face-to-face interaction. In H. R. Schaffer (ed.), *Studies in Mother-Infant Interaction*, London, Academic Press.

—— 1979: Peer vs. mother-directed behavior in one- to three-month-old infants. *Infant Behavior and Development*, 2, 215-26.

Fraser, C. and Roberts, N. 1975: Mothers' speech to children of four different ages. *Journal of Psycholinguistic Research*, 4, 9-16.

Frodi, A. M., Lamb, M. E., Leavit, L. A. and Donovan, W. L. 1978: Fathers' and mothers' responses to infant smiles and cries. *Infant Behavior & Development*, 1, 187-98.

Furrow, D., Nelson, K. and Benedict, H. 1979: Mothers' speech to children and syntactic development: some simple relationships. *Journal of Child Language*, 6, 423-42.

Gelman, R. and Shatz, M. 1977: Appropriate speech adjustments: the operation of conversational constraints on talk to two-year-olds. In M. Lewis and

L. Rosenblum (eds), *Interaction, Conversation, and the Development of Language*, New York: Wiley.
Gewirtz, J. L. 1968: Mechanisms of Social Learning: some roles of stimulation and behaviour in early human development. In D. A. Goslin (ed.), *Handbook of Socialization Theory and Research*, Chicago: Rand McNally.
Gewirtz, J. L. and Boyd, E. F. 1977a: Experiments on mother–infant interaction underlying mutual attachment acquisition: the infant conditions the mother. In T. Alloway, P. Pliner and L. Krames (eds), *Attachment Behaviour*, New York: Plenum.
—— 1977b: Does maternal responding imply reduced infant crying? A critique of the 1972 Bell and Ainsworth report. *Child Development*, 48, 1200–7.
Golinkoff, R. M. and Ames, G. J. 1979: A comparison of fathers' and mothers' speech with their young children. *Child Development*, 50, 28–32.
Gorski, P., Davidson, M. F. and Brazelton, T. B. 1980: Stages of behavioral organzization in the high-risk neonate: theoretical and clinical considerations. In P. M. Taylor (ed.), *Parent–Infant Relationships*. New York: Grune & Stratton.
Gustafson, G. E., Green, J. A. and West, M. J. 1979: The infant's changing role in mother–infant games: the growth of social skills. *Infant Behavior & Development*, 2, 301–8.
Hinde, R. A. 1979: *Towards Understanding Relationships*. London: Academic Press.
Hinde, R. A. and Davies, L. 1972: Removing infant rhesus from mothers for 13 days compared with removing mother from infant. *Journal of Child Psychology & Psychiatry*, 13, 227–37.
Holmberg, M. C. 1980: The development of social interchange patterns from 12 to 42 months. *Child Development*, 51, 448–56.
Howe, C. 1981: *Acquiring Language in a Conversational Context*. London: Academic Press.
Immelmann, K. 1972: Sexual and other long-term aspects of imprinting in birds and other species. *Advances in the Study of Behaviour*, 4, 147–74.
Jones, O. H. M. 1977: Mother–child communication with pre-linguistic Down's Syndrome and normal infants. In H. R. Schaffer (ed.), *Studies in Mother–Infant Interaction*. London: Academic Press.
Kaye, K. 1977: Toward the origin of dialogue. In H. R. Schaffer (ed.), *Studies in Mother–Infant Interaction*. London: Academic Press.
Klaus, M. H. and Kennell, J. H. 1976: *Parent–Infant Bonding*. St Louis: Mosby.
Konner, A. F. and Thoman, E. B. 1970: Visual alertness in neonates as evoked by maternal care. *Journal of Experimental Child Psychology*, 10, 67–78.
Konner, M. J. 1972: Aspects of the developmental ethology of a foraging people. In N. Blurton Jones (ed.), *Ethological Studies in Child Behaviour*, Cambridge: Cambridge University Press.
Kroodsman, D. 1978: Aspects of learning in the ontogeny of bird song: where, from whom, when, how many, which, and how accurately? In G. M. Burghardt and M. Bekoff (eds), *The Development of Behaviour: Comparative and Evolutionary Aspects*, New York: Garland.
Lamb, M. E. and Easterbrooks, M. A. 1981: Individual differences in parental sensitivity: origins, components, and consequences. In M. E. Lamb and L. R.

Sherrod (eds), *Infant Social Cognition: Empirical and Theoretical Considerations*, Hillsdale, New Jersey: Lawrence Erlbaum.
Lamb, M. E., Thompson, R. M., Gardner, W., Charnov, E. L. and Estes, D. 1984: Security of infantile attachment as assessed in the 'Strange Situation': its study and biological interpretation. *Behavioral and Brain Sciences*, 7, 127-71.
Lenneberg, E., Rebelsky, F. and Nichols, I. 1965: The vocalizations of infants born to deaf and hearing parents. *Human Development*, 8, 23-37.
Lester, B. M. and Zeskind, P. S. 1979: The organisation and assessment of crying in the infant at risk. In T. M. Field (ed.), *Infants Born at Risk*, In T. M. Field New York: Spectrum Publications.
Lewis, M. and Goldberg, S. 1969: Perceptual-cognitive development in infancy: a generalised expectancy model as a function of the mother-infant interaction. *Merrill-Palmer Quarterly*, 15, 81-100.
McGinnis, L. M. 1980: Maternal separation studies in children and nonhuman primates. In R. W. Ball and W. P. Smotherman (eds), *Maternal Influences and Early Behaviour*, Lancaster, MTP Press.
McGrew, W. C. 1977: Socialization and object manipulation of wild chimpanzees. In S. Chevalier-Skolnikoff and F. E. Poirier (eds), *Primate Bio-social Development*, New York: Garland.
Messer, D. J. 1981: Non-linguistic information facilitating the young child's comprehension of adult speech. In W. P. Robinson (ed.), *Communication in Development*, London: Adademic Press.
Millar, W. S. 1972: A study of operant conditioning under delayed reinforcement in early infancy. *Monographs of the Society for Research in Child Development*, 37, 2 (whole no. 147).
Mischel, W. 1977: The interaction of person and situation. In D. Magnusson and N. S. Endler (eds), *Personality at the Crossroads*, Hillsdale, New Jersey: Erlbaum.
Murray, A. D. 1979: Infant crying as an elicitor of parental behavior: an examination of two models. *Psychological Bulletin*, 86, 191-215.
Neill, S. R. St. J. 1982: Preschool design and child behaviour. *Journal of Child Psychology and Psychiatry*, 23, 309-18.
Nelson, K. 1973: Structure and strategy in learning to talk. *Monographs of the Society for Research in Child Development*, 38, serial no. 149.
Newport, E. 1976: Motherese: the speech of mothers to young children. In N. Castellan, D. Pisoni and G. Potts (eds), *Cognitive Theory: vol. II*. Hillsdale, New Jersey: Erlbaum.
Newport, E. L., Gleitman, H. and Gleitman, L. 1977: Mother, I'd rather do it myself: some effects and non-effects of maternal speech style. In C. E. Snow and C. A. Ferguson (eds), *Talking to Children: Language Input and Acquisition*, Cambridge: Cambridge University Press.
Parke, R. D. and O'Leary, S. E. 1976: Father-mother-infant interaction in the newborn period. In K. Riegel and J. Mcadam (eds), *The Developing Individual in a Changing World, vol. II, Social and Environmental Issues*, The Hague: Mouton.
Phillips, J. R. 1973: Syntax and vocabulary of mothers' speech to young children: age and sex comparisons. *Child Development*, 44, 182-5.
Prechtl, H. F. R. and O'Brien, M. J. 1982: Behavioural states of the full-term

newborn. The emergence of a concept. In P. Stratton (ed.), *Psychobiology of the Human Newborn*, Chichester: Wiley.

Prechtl, H. F. R., Theorell, K., Grausbergen, A. and Lind, J. 1969: A statistical analysis of cry patterns in normal and abnormal newborn infants. *Developmental Medicine and Child Neurology*, 11, 142-52.

Provence, S. and Lipton, R. C. 1962: *Infants in Institutions.* New York: International Universities Press.

Rogosa, D. 1980: A critique of cross-lagged correlation. *Psychological Bulletin*, 88, 245-58.

Rutter, M. 1982: Parenting qualities: origins and effects. *Paper presented to International Association of Child Psychiatry*, Dublin.

Rutter, M. and Madge, N. 1976: *Cycles of Disadvantage.* London: Heinemann.

Sachs, J., Brown, R. and Salerno, R. 1972: Adults' speech to children. In W. von Raffler Enger and Y. Lebrun (eds), *Baby Talk and Infant Speech*, Lisse, Swets & Zeitlinger.

Schaffer, H. R. 1977: *Mothering.* London: Fontana.

—— 1979: Acquiring the concept of the dialogue. In M. H. Bornstein and W. Kessen (eds), *Psychological Development from Infancy: Image and Intention*, Hillsdale, New Jersey: Lawrence Erlbaum Associates.

—— 1984: *The Child's Entry Into a Social World.* London: Academic Press.

Schaffer, H. R., Collis, G. M. and Parsons, G. 1977: Vocal interchange and visual regard in verbal and pre-verbal children. In H. R. Schaffer (ed.), *Studies in Mother-Infant Interaction.* London: Academic Press.

Schaffer, H. R. and Crook, C. K. 1979: Maternal control techniques in a directed play situation. *Child Development*, 50, 989-98.

Schaffer, H. R. and Emerson, P. E. 1964: Patterns of response to physical contact in early human development. *Journal of Child Psychology and Psychiatry*, 5, 1-13.

Schaffer, H. R. and Liddell, C. 1984: Adult-child interaction under dyadic and polyadic conditions. *British Journal of Developmental Psychology*, 2, 33-42.

Shatz, M. and Gelman, R. 1973: The development of communication skills: modifications in the speech of young children as a function of the listener. *Monographs of the Society for Research in Child Development*, 38, (serial no. 152).

Shugar, G. W. 1978: Text analysis as an approach to the study of early linguistic operations. In N. Waterson and C. Snow (eds), *The Development of Communication*, Chichester: John Wiley.

Sluckin, W., Herbert, M. and Sluckin, A. 1983: *Maternal Bonding.* Oxford: Basil Blackwell.

Snow, C. E. 1972: Mothers' speech to children learning language. *Child Development*, 43, 549-65.

Sorce, J. F. and Emde, R. N. 1982: The meaning of infant emotional expressions: regularities in caregiving responses in normal and Down's Syndrome infants. *Journal of Child Psychology & Psychiatry*, 23, 145-58.

Sroufe, L. A. and Waters, E. 1982: Issues of temperament and attachment. *American Journal of Orthopsychiatry*, 52, 743-6.

Stern, D. N. 1971: A micro-analysis of mother-infant interaction: behavior regulating social contact between a mother and her 3-month-old twins. *Journal of the American Academy of Child Psychiatry*, 10, 501-17.

—— 1974: Mother and infant at play: the dyadic interaction involving facial, vocal, and gaze behaviour. In M. Lewis and L. A. Rosenblum (eds), *The Effect of the Infant on its Caregiver.* New York: Wiley.
—— 1977: *The First Relationship.* London: Open Books/Fontana; Cambridge Massachusetts: Harvard University Press.
Stern, D. N., Beebe, B., Jaffe, J. and Bennett, S. J. 1977: The infant's stimulus world during social interaction. In H. R. Schaffer (ed.), *Studies in Mother-Infant Interaction.* London: Academic Press.
Suomi, S. J., Harlow, H. F. and Novak, M. A. 1974: Reversal of social deficits produced by isolation rearing in monkeys. *Journal of Human Evolution,* 3, 527–34.
Tizard, B. 1978: *Adoption: A Second Chance.* London: Open Books.
Trevarthen, C. 1977: Descriptive analyses of infant communicative behaviour. In H. R. Schaffer (ed.), *Studies in Mother-Infant Interaction,* London: Academic Press.
—— 1979: Communication and cooperation in early infancy: a description of primary intersubjectivity. In M. Bullowa (ed.), *Before Speech: The Beginning of Interpersonal Communication,* Cambirdge: Cambridge University Press.
Vygotsky, L. S. 1978: *Mind in Society.* Cambridge, Massachusetts: MIT Press.
Wachs, T. D. and Gruen, G. E. 1982: *Early Experience and Human Development* New York: Plenum.
Wasz-Hockert, O., Lind, J., Vuorenkoski, V., Parteuen, T. and Valanne, E. 1968: *The Infant Cry: A Spectographic and Auditory Analysis. Clinics in Developmental Medicine (Report no. 29).* London: Spastics International Medical Publications.
Wells, C. G. 1979: Variations in child language. In P. Fletcher and M. Garman (eds), *Language Acquisition,* Cambridge: Cambridge University Press.
White, B. L., Kaban, B. T., and Attanucci, J. S. 1979: *The Origins of Human Competence.* Lexington, Massachusetts: D. C. Heath.
White, B. L. and Watts, J. C. 1973: *Experience and Environment: Major Influences on the Development of the Young Child.* Englewood Cliffs, New Jersey: Prentice-Hall.
Wolff, P. H. 1969: The natural history of crying and other vocalizations in early infancy. In B.M. Foss (ed.), *Determinants of Infant Behaviour,* vol. 4, London: Methuen.
Wood, D., McMahon, L. and Cranstoun, Y. 1980: *Working with Under Fives.* London: Grant McIntyre.
Wood, D. and Middleton, D. 1975: A study of assisted problem solving. *British Journal of Psychology,* 66, 181–92.
Yarrow, M. R. Campbell, J. D. and Burton, R. U. 1968: *Child Rearing: An Enquiry into Research and Methods.* San Francisco: Jossey Bass.

11 The Pathology of Human Parental Behaviour

M. Herbert

INTRODUCTION

The birth of a first child to a couple makes parents of them and transforms their partnership into a family. Living together in small groups, or what might broadly be thought of as 'families' has been the universal pattern for *homo sapiens* and his forebears for some 500,000 years. It would seem reasonable to claim that an institution that has endured from the middle of the Pleistocene period to the present day must have significant survival value for its individual members, and their species.

However, the family, led by adults (parents), is not only crucial for the care and protection it offers the young during their relatively prolonged period of dependence; it plays a major role in the introduction of infants into the ways of their social environment – that is, their culture. Our cultural heritage does not and cannot depend upon any biological device such as a genetic code, as is the case with our physical equipment for living.

The transmission of cultures cannot be inflexible and yet it cannot be left to chance. As Lidz puts it:

The welfare of the individual and the continuity of the culture which is essential to man depend upon his having satisfactory means of indoctrinating the members of the new generation into the mores, sentiments and instrumentalities of the society, and to assure that they, in turn, will become satisfactory carriers of the culture. The major task of indoctrination devolves upon the family, even though the members of the family often do not know that they have the task. (Lidz, 1968)

This is not to say that the family as an institution is a static entity, stereotyped in its forms, or unchanging in its functions. It is a dynamic system, susceptible to change; it is influenced, in the short term, by the personality, development and relationships of its members, and in the longer term by the pressures of economic events and historical processes.

PARENTAL CHOICE

Biological evolution in human beings has resulted in a degree of choice and a variety of motives in the assumption of parenthood which is absent in lower species. Different motives lead to a variety of parental attitudes toward the parental role, and, indeed, to different ways in which parents formulate goals for their children's future. In the past (and even today, in societies remote from industrialization) the child was likely to be groomed along very rigid lines for a predetermined role in the kinship system. Unlike the modern industrial society, the subsistence economies could not afford the luxury of choice for their offspring.

The family is less a mirror image of the wider community than it was; it is no longer the indispensable 'cell' in the economic body of society. It can branch out individualistically, as can its members, and while it still serves an economic function, it is now primarily concerned with socializing the members of society and providing care for them. The idea of privacy in the family household is a relatively modern conception; such privacy dilutes somewhat the pressures on parents to conform; and this degree of flexibility results in a greater range of individualism in the children produced by the society. In addition, there is a greater risk that many parents *and* their children will be out of step with their society – or so it is claimed. The very freedom which families and individuals have won for themselves – the loosening of the social fabric – may have brought in its wake the problems of non-conformist, idiosyncratic, even alienated youth (see Bronfenbrenner, 1970; Seligman, 1975).

Contemporary family patterns vary a great deal. Unlike children in more traditional Eastern families who are instructed unambiguously in the rules, conventions, laws, ethics and religious practices of the community and the importance of these traditional beliefs, these matters are conveyed rather haphazardly to many Western children (Morgan, 1975). Children in Western society lack such clear-cut terms of reference. Class differences, occupational and kinship groupings cut across each other, often resulting in a confusion of roles, expectations and tasks for the individual child.

Many commentators are concerned about what they would see as the adverse features inherent in the nuclear family as a context for rearing children. But this is not the place to debate such value judgements; in any event it would prove inconclusive and probably unrewarding. However, it is worth noting that although society delegates its most crucial functions to the family, there is little formal education or training offered to would-be parents; even the informal learning and experience once offered to older children caring for younger siblings in large families, or the help from the experienced members of the extended family and from relatives living nearby, may not be available to the relatively isolated nuclear family.

PARENTAL FUNCTIONS

Specific family functions are usually divided between *survival functions* such as provision of shelter, space, food, income, physical care, health and safety; and *psycho-social functions* such as:

a Socialization: encouraging, guiding, supporting and rewarding
Parents endeavour within the family setting to transform a broadly asocial biological organism into a social being by the process called 'socialization'. Socialization has, as one of its objectives, the preparation of children for their future; yet today's parents are less sure than their ancestors (perhaps because of the rapidly changing nature of society) of the life their young should be made competent to deal with. Notions of right and wrong, a code of behaviour, a set of attitudes and values, the ability to see the other person's point of view – all of these basic qualities which make an individual a socialized being – are nurtured in the first instance by parents. They induct the infant into his culture by determining which physical and social stimuli will mainly be presented to the infant, what he will be taught, which opinions and behaviours will be reinforced and consolidated, and which will be discouraged by various means.

b The provision of a support system for parents and their offspring
The family is linked with outside social institutions and one of its functions is to provide parents with a variety of social, emotional and economic support systems (for example, through links with extended family, community, workplace). There have been several investigations showing that parents who abuse their children have fewer associations outside the home, receive less help in caring for their offspring and perceive their neighbours more negatively, than other parents (Orford, 1980). This emphasis on parents' needs is a useful corrective to the emphasis later in this chapter on the effects on offspring of specific parental and other inter-personal behaviours.

PARENTAL INFLUENCE

Parents, as 'agents' of society, exerting social influence and control on an impressionable, malleable child, bear an awesome responsibility. The very processes which help the child adapt to social life can, under certain circumstances, contribute to the development of deviant, dysfunctional modes of behaviour. An immature child who learns by imitating an adult is not necessarily able to comprehend when it is undesirable behaviour that is being modelled. Parents who teach their child (adaptively) on the basis of classical conditioning and instrumental learning processes, to avoid dangerous situations can also by mismanagement, condition him (maladaptively) to avoid school or social gatherings.

The family can be conceptualized as a 'system' with its individual members as elements or sub-units within it. Whatever happens to one or

The Pathology of Human Parental Behaviour 319

more of its elements – say, mental illness in the mother or father, or serious marital disharmony – can affect the entire system. The privatization of the family (that is, the heightened emotional intimacy and interdependence of the members) is thought to place a great burden on parents, most particularly the mother, and also the children (see Gavron, 1966; Madge, 1983; Mishler and Waxler, 1968). Parents are the crucial and therefore (potentially) the weak link in the chain of socialization. Taking care of young children is likely to be more stressful for some parents than others, especially in unfavourable circumstances (poverty or poor housing for instance). The identification and specification of 'pathological' parenting should, in theory, lead to the remediation and better still, prevention of problems; or, rather, this is the hope. Indeed, a popular aphorism of the 1950s (and the ideology underlying it persists today) stated that 'there are no problem children, only problem parents'.

When one considers the intimate, protracted and highly influential nature of parents' relationships with their children, it seems self-evident that the quality of such relationships must have a vital bearing on the development of the child's personality and general adaptation. The scientific inquiry into human parental behaviour arises, in part, from a conviction about the power and reach of the early experiences the parents provide for their offspring.

This quest has produced disappointingly meagre results. The reasons are not difficult to find: the sheer complexity of the subject, a daunting number of methodological problems, and some quirky biases in scientists' approach to this area of research. Leaving aside the complexity issue, which is self-evident, there are particular doubts about many of the studies of human parenting due largely to flawed research designs, biases in sampling, and a tendency for social-class and ethnocentric values to determine the questions asked and the assumptions made, in various investigations. These matters will be commented upon where appropriate. Of more immediate concern is a theoretical issue. Ever since the publication of Mischel's critique of trait psychology (Mischel, 1968) there has been a controversy as to whether or not social behaviours demonstrate stability across situations and across time. One result of this debate has been the emergence of the 'interactionist perspective' (Bell and Harper, 1977) which has had relatively little impact on research into parental behaviour, the emphasis of which is still on intrinsic dispositions (motives, traits, attitudes, and so on) in the individual parent. The finding that children *initiate* approximately 50 per cent of interactions with their parents (Bell and Harper, 1977), helps to explain the paucity of generalizations about parent–child transactions: the traditionally unidirectional (parent-to-child) account of what occurs. Interpretations of socialization in terms of social reinforcement have shared a common model of the child as a 'tabula rasa' – an essentially passive organism under the control of a socializing agent (in this case, a parent) who dispenses rewards and punishments. This preconception has resulted in the neglect of those factors which are not under the control of external agents: for example, maturational processes, as well as hereditary and congenital conditions. Among the latter are the causes of temperamental attributes found in certain 'difficult' childen, which make them highly resistant to socialization. Even the simplest training requirements may involve an uphill struggle (Thomas, Chess and Birch, 1968). In many ways a child's behaviour can have as much effect on his parents' actions as their behaviour has on his (Bell and Harper, 1977). So when parents meet these temperamental or inborn factors – extreme moodiness, over-activity, defiance, over-sensitivity, physical problems, general irregularity, and unadaptability (that is, a noisy unwillingness to adapt to changes in routine) – they can be overwhelmed.

OPERATIONAL DEFINITIONS OF PARENTAL BEHAVIOUR

Whatever the deficiencies of research methodology (an excessive reliance on interviews with parents rather than on direct observations or information from children; on field studies rather than experimental work), studies consistently produce evidence of a reasonably predictable structure underlying parental and other interpersonal transactions (Benjamin, 1974; Schaefer, 1959). A wide range of parental behaviours reduce to two major dimensions, with orthogonal axes described as warm–hostile and control–autonomy (see figure 1a). Thus we have: transactions which are 'loving' (or warm) at one extreme, and 'rejecting' (hostile) at the other; transactions which are restrictive (controlling) at one extreme, and permissive (encouraging autonomy) at the other.

Schaefer (1959) denotes parental behaviour in terms of the interactions of the two main attributes, thus a 'democratic' mother is one who is both loving and permissive; an 'antagonistic' mother combines hostility and restrictiveness; a 'protective' mother is one who is both loving and restrictive, and so on. There is evidence (Orford, 1980) that dominance may be demarcated into restrictiveness and firmness with their opposites of autonomy-giving and laxness respectively. Having identified some key parental behaviours we are still left with the issue of what constitute pathological variants.

CRITERIA OF PATHOLOGICAL PARENTAL BEHAVIOUR

It is no simple task to specify precisely what is 'pathological' or 'abnormal' parental behaviour. One would be hard pressed to find absolute distinctions between the characteristics of parents who come to be labelled

The Pathology of Human Parental Behaviour 321

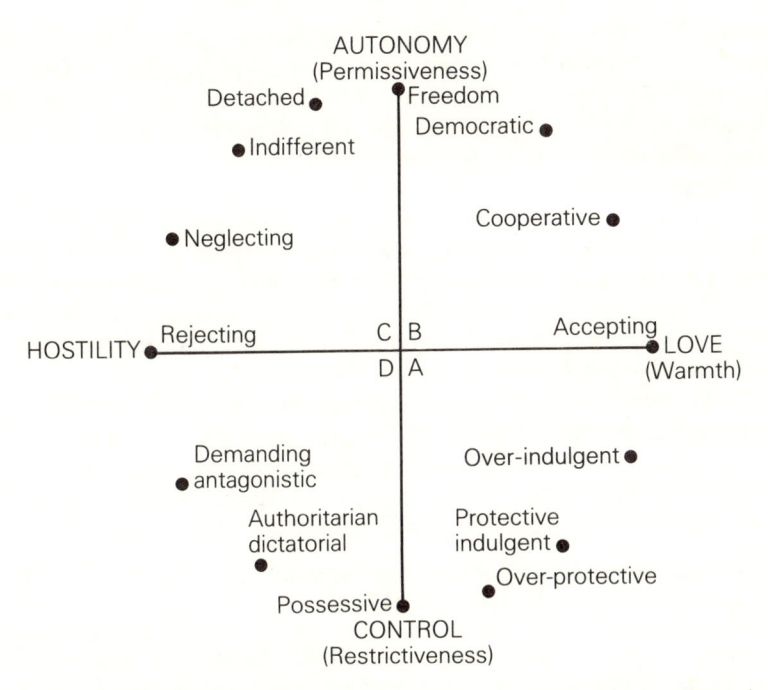

```
                    AUTONOMY
                  (Permissiveness)
        Detached •    • Freedom
                       Democratic •
           • Indifferent


                              Cooperative •
         • Neglecting


                    C | B              Accepting
HOSTILITY • Rejecting ─────────────────── • LOVE
                    D | A              (Warmth)


        Demanding           Over-indulgent •
        • antagonistic

          Authoritarian      Protective
          dictatorial        indulgent •
          •                   • Over-protective
              Possessive •
                    CONTROL
                  (Restrictiveness)
```

Figure 1a The range of parental behaviour-types within two major dimensions: autonomy-control/hostility-warmth

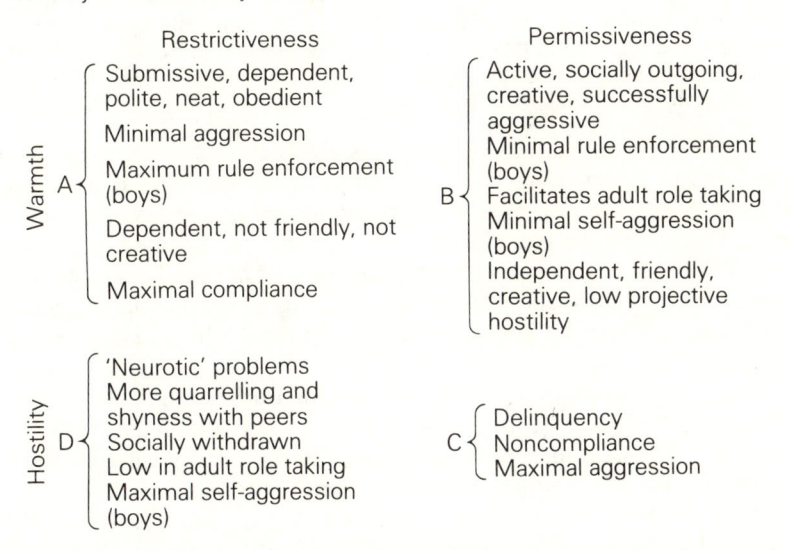

	Restrictiveness	Permissiveness
Warmth A	Submissive, dependent, polite, neat, obedient Minimal aggression Maximum rule enforcement (boys) Dependent, not friendly, not creative Maximal compliance	B — Active, socially outgoing, creative, successfully aggressive Minimal rule enforcement (boys) Facilitates adult role taking Minimal self-aggression (boys) Independent, friendly, creative, low projective hostility
Hostility D	'Neurotic' problems More quarrelling and shyness with peers Socially withdrawn Low in adult role taking Maximal self-aggression (boys)	C — Delinquency Noncompliance Maximal aggression

Figure 1b Details of children's behaviour within the parameters of two major dimensions: restrictive parenting and permissive parenting

problematic by psychiatrists, social workers and other professionals, and those of other unselected parents. The judgement of what is pathological or 'abnormal' (that is, a *deviation* from a norm or standard) is largely a social one; the parent fails to meet certain of society's expectations of what constitutes appropriate behaviour. Unfortunately, terms like 'normal' and 'abnormal' are commonly applied to parents (more often mothers than fathers) as if they are mutually exclusive concepts.

These 'pathological mothers', have appeared in the clinical and social work literature over many years: schizophrenogenic mothers, asthmagenic mothers, mothers accused of figuratively 'suffocating' their offspring, or in some other way overprotecting, rejecting, or double-binding them into abnormality.

Our matricentric culture has provided a refinement in the inculpation of parents as causes of their children's difficulties, by pointing the finger mainly in the direction of the mother. A distinguished psychiatrist and theoretician, Stella Chess, refers to this phenomenon as 'mal de mère'. She wrote in an editorial for the American Journal of Orthopsychiatry as follows:

> The standard procedure is to assume that the child's problem is reactive to maternal handling in a one-to-one relationship. Having come to this conclusion, the diagnostician turns his further investigations unidirectionally toward negative maternal attitudes and the conflicts presumed to underlie these. Investigation in other directions is done in a most cursory fashion or not at all. At the diagnostic conference, speculations are made concerning the mother's relationship with her own parents, her degree of immaturity, her presumed rejection of this child and her over-compensations for this rejection. Single bits of data fitting in with these speculations are quoted as typical of the child's feelings and the mother's attitudes and are taken as proof of the thesis of anxious maternal attitudes as universal causation. (Chess, 1964)

Certain parental actions or attributes are more or less abnormal, and they tend to be associated with particular circumstances and situations. Clearly, actions which negate the fundamental caregiving and growth-enhancing components of parenting (for example, gross neglect) are abnormal. They deviate from the cultural standards and norms of actual parental conduct in the statistical sense, and also from universal human values as to what constitute appropriate responsibilities of parents toward their helpless offspring.

Problems of definition arise in that area that lies between the extremes of palpable neglect (child abuse for instance) and nurturance (for example, persistent infantalizing of the child) because of the rich variety of approaches to parenthood that has manifested itself at different times and in different societies (see Sluckin, Herbert and Sluckin, 1983; Rapoport, Rapoport and Strelitz, 1977). Signs of abnormal parental behaviour involve exaggerations, deficits or harmful combinations of behaviour patterns common to all parents. Expressions of emotional feelings and behavioural acts have certain allowable intensity levels. Very 'high' intensities – emotional responses of excessive magnitude – which have unpleasant consequences for children, might be indicative of pathological parenting. Thus, when dutiful concern for a child's wellbeing becomes a morbidly intense preoccupation with his health, or when caring attitudes are transformed into endless fussing, the child is likely to suffer and the parent to be 'diagnosed' as overprotective (Levy, 1943). There is an opposite extreme ranging from attitudes to the child which are casual, *laissez-faire,* lax or even indifferent – in other words deficits in caring attitudes.

A high rate of emission of certain behaviours may also be taken as a criterion of the harmfulness of certain parental behaviours and therefore the appropriateness of the description 'abnormal'. Most parents interfere in the choices and activities of their children up to a point; indeed they would object to their offspring's use of the word 'interference'. However, there are parents whose interference in their children's lives amounts to 'gross intrusion', something postulated to be inimical to their healthy development (Laing and Esterson, 1970).

CHILDREN'S NEEDS

If a major defining criterion of pathological parental behaviour is the *consequence* of particular parental practices for the child's well-being, then it would seem fruitful to identify children's 'needs'. Several lists have been constructed along the lines of Maslow's 'hierarchy of needs' formulations (Maslow, 1954) and, not surprisingly, they tend to differ markedly despite areas of overlap (see Kellmer Pringle, 1980; Rapoport, Rapoport and Strelitz, 1977).

The lists tend to produce a dichotomy between *survival* functions (such as the need for food, shelter, physical care, and so on) and *psycho-social* functions (such as requirement of love, security, attention, protection, new experiences, acceptance, praise and recognition, responsibility, education, belongingness, and so on).

Theoretically, such lists should allow, say, a social worker to analyse which actions on the part of mothers and fathers facilitate or are antipathetic to the meeting of crucial needs. In practice these lists have proved to be too abstract and general to be of much use as social workers have found to their cost; they become the source of fruitless quibbling over which needs are most important and how, when, where and by whom, they can best be satisfied. They arguably tend to divert attention (because of their global and therefore sometimes unmanageable nature) from more modest and precise objectives in work with parents and children (Iwaniec and Herbert, 1982; Herbert, Suckin and Sluckin, 1982).

SURVIVAL FUNCTIONS

Obviously, the child's survival may be put at risk if parents repudiate their offspring. For some children rejection means callous and indifferent neglect or positive hostility from the parents, and this form of rejection does not always take the form of physical cruelty or negligence; it may be emotional and subtle. The child comes to believe that he is worthless, that his very existence makes his parents unhappy.

Parental rejection

A rejecting parent tends not only to ignore his child's requests for nurturance, but also to punish his dependent behaviours. One would expect, therefore, that more severe forms of rejection would lead children to suppress such behaviour; this is reflected in the finding that aggressive boys who have suffered a lot from parental rejection show much less dependent behaviour than non-aggressive boys who have been accepted by their parents (Herbert, 1974). There is an exception – if the parents withold, or are sparing with, their attention and care, but do not actually punish dependent behaviour, they are likely to intensify the child's needs for attention and care. The more a child is 'pushed away' (figuratively speaking) the more he clings.

Clearly, rejection of a child is seen as unnatural and pathological, not only because of its tragic effects, but also because it flies in the face of our expectations that parental (and particularly, maternal) love is something natural, indeed instinctive. Maternal attachment (as we have seen in chapters 7, 8 and 9) has been studied more closely than the paternal bond, yet our knowledge of the nature of maternal affection and rejection is limited. This lack of understanding is particularly regrettable when we are faced with the enigma of the mother who is hostile to her child, neglects, abuses or abandons it – fortunately, a relatively rare phenomenon. The same actions on the part of the father are condemned by society but not with quite the same intensity of feeling or incomprehension. The woman who feels little or nothing for her offspring offends against the cherished myths about 'mother-love', its sanctity and universality. Nowhere is this more evident that in the treatment of the subject of child abuse in the mass media.

Violence towards children in the home

Just as historians have neglected evidence of infanticide and the brutalization of children to be found in old documents, the public (and the media that serve it) find it difficult to accept fully the disturbing reports of child abuse (Gil, 1971).

The Pathology of Human Parental Behaviour 325

Women, according to official statistics, are more likely than men to be the perpetrators of violence towards their offspring. However, these statistics – and speculations about causation – must be treated with caution: they are likely to be biased as they are collected in social and health agencies, and sex ratios vary widely depending upon the particular sample analysed. We do not know how many children are physically abused, why, how, and by whom, in the wider community outside these agencies. There is another bias to consider: the fact that women spend more time in contact with children than men. Several reviews of the literature make it clear that descriptions of abusive parents' characteristics are inconsistent (Allan, 1978; Belsky, 1978).

Social workers have been very influenced by the work of Kempe and Kempe (1978), who report that the most consistent features of the histories of abusive families is the repetition, from one generation to the next, of a pattern of abuse, neglect and parent loss or deprivation. The Kempes admit that 'no one knows quite how the ability to be a parent is passed on from one generation to the next. Probably the most significant channel is the experience of having been sympathetically parented, of having experienced what it feels like to be an infant, helpless but cherished and nurtured into childhood.' The evidence for an association between violence against children from one generation and the next within the same family is, in fact, slender (Belsky, 1978) and, in any event, only accounts for a fraction of known cases.

Whatever the antecedents of parental hostility, the issue of rejection becomes crucial when one considers that range of parental behaviours that come under the general rubric of discipline, and more particularly, the heading of 'punishment'. It leads us on to a consideration of the psycho-social consequences of hostile parenting.

PSYCHO-SOCIAL FUNCTIONS

All societies are remarkably successful, within broad limits, at inculcating in their members the approved skills, codes of conduct, prosocial norms, values and personal attributes, despite the wide variety of parental practices referred to above. According to Fromm (1955), for any society to function well, its members 'must acquire the kind of character which makes them want to act in the way they have to act as members of their society'. To put it another way, they have to desire to do what it is objectively necessary for them to do. Certain parental attitudes and actions facilitate such developments; others, such as rejection, produce adverse outcomes.

An empirical approach to the issue of rejection and its 'effects' and, indeed, outcomes of other parental practices, have provided the findings

(Caldwell, 1964; Becker, 1964; Frank, 1965; Yarrow, Campbell and Burton, 1968) summarized in figure 1b. Schaefer's dimensions in figure 1a have been combined with evidence (summarized by Becker, 1964) of the sort of behaviour problems associated with different combinations of parental attitudes and behaviours. The 'outcomes' of these combinations (trends, of course) appear below the figure.

Parental hostility

The most serious consequences (see Herbert, 1978) result from punitive methods persistently used against a background of rejecting, hostile parental attitudes. These methods are often referred to as power-assertive; the adult asserts dominant and authoritarian control through physical punishment, harsh verbal abuse, angry threats and deprivation of privileges. There is a positive relationship between the extensive use of physical punishment in the home by parents and high levels of aggression in their offspring outside the home (see Becker, 1964). It would seem that physical violence is the least effective form of discipline or training when it comes to moulding a child's behaviour. All the evidence to date (Johnson and Medinnus, 1965) shows that physical methods of punishment (the deliberate infliction of pain on the child) may for the time being suppress the behaviour that it is meant to inhibit, but the long-term effects are less impressive. Violence begets violence. What the child appears to learn is that might is right. Delinquents have more commonly been the victims of adult assaults – often of a vicious, persistent and even calculated nature – than non-delinquents (Herbert, 1978).

Parental permissiveness

According to Baumrind (1971) the technical meaning of the designation 'permissive parent' is that the mother (or father) attempts to behave in a non-punitive, accepting, and affirmative manner toward the child's impulses, desires, and actions; consults with him about policy decisions and gives explanations for family rules; makes few demands for household responsibility and orderly behaviour; presents himself or herself to the child as someone to call upon for help and company as he wishes, not simply as an active agent responsible only for shaping or altering his ongoing or future behaviour; allows the child to regulate his own activities as much as possible; avoids the excessive exercise of control; does not encourage him to obey absolute, externally-defined standards; and attempts to use reason but not overt power to accomplish his ends.

This empirical analysis is very different from the popular usage of the word 'permissiveness', which tends to be restricted to the extreme end of the phenomenon with its connotations of *laissez-faire* parenting. The emotional background to this extreme pattern of permissiveness or 'lax discipline' is, only too often, outright indifference.

Reviews of research (Becker, 1964; Baumrind, 1971) into child-rearing techniques suggest an empirical basis for the notion that there is a happy medium, and that the extremes of permissiveness and restrictiveness entail risks. A blend of permissiveness and a warm, encouraging and accepting attitude fits the recommendations of child-rearing specialists who are concerned with fostering the sort of children who are socially outgoing, friendly, creative and reasonably independent and self-assertive.

Parental warmth

Bandura and Walters (1963) suggest the variables of maternal warmth and demonstrativeness as possible antecedents of dependency behaviour. Clinicians, generally, have been concerned about the effects on children of an excess of mothering. Maternal nurturance, in other words, may be a vital ingredient for the child's healthy development, but there can be 'too much of a good thing'. (We return to this theme later in the chapter).

It is tempting to ask why rather vague notions of maternal wrongdoing have been raised to the status of sufficient and necessary causes of a wide range of childhood problems. So much of the clinical and casework evidence is based on *ex post facto* studies generating correlations between various variables; this can lead to questionable conclusions. Correlations are often, without further evidence, assumed to imply a cause–effect relationship. This fallacy has been particularly pervasive (along with *post hoc, ergo propter hoc* reasoning) in the 'maternal separation' and 'maternal rejection' literature.

It has to be said by way of a caveat that there are many inconsistencies, even contradictions, in the literature on parent–child attitudes and relationships and parental child-rearing practices. The isomorphism between parental *attitudes,* as measured, and specific *behaviours* is limited. Global assessments of such independent variables as parental warmth, hostility, rejection and the like, are too abstract and too coarse to capture many of the subtle nuances of parental behaviour. They also lack sufficient contextual anchorage; that is to say, they do not specify the variations in behavioural interactions between parents and children which occur in particular situations and which are necessary to define precise relationships between independent and dependent variables. This applies especially to the fund of dependent variables psychologists are interested in, namely, moral behaviour, delinquent reactions, and so on. Behavioural conformity to specific norms in particular situations is more likely to depend on situation-specific sanctions than on the general pattern of parent–child relationships (Danziger, 1971). Nevertheless, there is a tendency for hostile and rejecting parents to produce a disproportionate number of delinquent children, although individual predictions are rarely possible. The complex interactions of person and social context have conspired to make a nonsense of the simple classical linear causality which is so pervasive in the literature on parent education, psychopathy and preventive work. Frank (1965) reminds us that:

> ... we may observe the same or similar dynamic process operating in different individuals at different times and in different contexts to produce different personalities. Moreover, we may also observe different processes producing similar or equivalent personalities when operating at different times, in different life situations, in different individuals.

ANTECEDENTS OF PATHOLOGICAL PARENTAL BEHAVIOUR

Although we have some understanding, at a descriptive level, of parental behaviours and also of some of the likely effects of different types of child care on the offspring, there is little to say about the antecedents of parenting – be it normal or abnormal. There is a growing interest in the development of prosocial behaviour generally, and, as it covers a wide range of phenomena such as helping, sharing, self-sacrifice and norm observance, it is of relevance to the study of parenting. After all, these phenomena have one common characteristic – namely, that an individual's actions are orientated toward protection, maintenance, or enhancement of the well-being of an external social object: a specific person (Reykowski, 1982). There are, however, immature and self-centred individuals who cannot cope with the demands for the selflessness, altruism and hard work implicit in parenthood.

There is a sense in which parenthood can be conceptualized as a new stage of development for the couple who have given birth to their first child. Lidz (1968) makes the point that the tasks of the marital partners, the roles they occupy, their orientation toward the future, all change profoundly when this happens. However, 'this simple step into parenthood, so often taken as an inadvertent mis-slip, provides a severe test of all preceding developmental stages and consequent integration of the individual parent. The inevitable changes in the husband and wife will, in turn, alter the marital relationship and place stress upon it until a new equilibrium can be established.'

This notion of developmental stages is put forward (*inter alia*) to explain the link between poor parenting and the experience by the 'pathological' parent (during his or her own childhood) of inadequate care. Many of these ideas can be traced to Erikson's work. Erikson (1965) regards early personality development as one stage in the ongoing developmnent of a pattern of reciprocity between the self and others. At each stage a conflict between opposite poles of the relationship has to be resolved; there is a series of crises to which Erikson gave the names of 'trust–mistrust', 'confidence–doubt', 'initiative–guilt', and so on. There appears, prima

The Pathology of Human Parental Behaviour

facie, to be a connection between these bi-polar pairs and Piaget's concepts of assimilation and accommodation. Assimilation occurs when the person alters the environment to meet his own needs; accommodation occurs when the person adapts his own behaviour in response to the demands of his milieu. Play and imitation are examples of cognitive behaviour marked by a lack of balance, in one direction or the other, between assimilatory and accommodatory processes.

It is possible to understand Erikson's bi-polar pairs as involving a similar lack of balance or conflict in the development of social reciprocity – an imbalance between ego and alter in one direction or the other. A mature resolution of these opposing forces is critical when the individual reaches adulthood. Erikson describes the 'triggering' during the adult years of an interest in producing, guiding and laying the foundations for the next generation. This desire to nurture and teach he called 'the capacity for generativity' with stagnation as the negative outcome. His concept of stages has a certain face validity but is not particularly susceptible, in its details, to proof or disproof. Rather more profitable, from an empirical standpoint, have been the attempts to elucidate the origins of the various kinds of prosocial behaviour referred to above.

This area of research is very much in its infancy, but it is possible to summarize the major factors which foster prosocial behaviour as: parental affection and nurturance; parental control; induction – the use of reasoning in disciplinary encounters; modelling; and assigning responsibility (Staub, 1975). The balancing of these components is perhaps best illustrated in the philosophy of what (on the basis of her investigations) Baumrind (1971) calls the 'authoritative' mother. This kind of mother attempts to direct her child's activities in a rational manner determined by the issues involved in particular disciplinary situations; she encourages verbal give-and-take and shares with the child the reasoning behind her policy; she values both the child's self-expression and his so-called 'instrumental attributes' (respect for authority, work and the like); and she appreciates both independent self-will and disciplined conformity. Therefore, she exerts firm control at points where she and the child diverge in viewpoint, but does not hem in the child with restrictions. She recognizes her own special rights as an adult, but also the child's individual interests and special ways.

DISRUPTION OF SOCIALIZATION

Although it is not usually made explicit, substantial deviations from this ideal as represented by authoritarian parenting at one 'extreme' and excessively permissive (*laissez-faire*) parenting at the other, are regarded, if not as pathological, at least as undesirable. In an area of value judgements such as this, it is unwise to draw firm conclusions. Nevertheless, available evidence (Herbert, 1974) suggests that what is important in child-rearing (from the cultural perspective) is the general social climate in the home – the attitudes and feelings of the parents which form a background to the application of specific methods and the manifestation of particular interactions during child-rearing.

If the authoritative parent does indeed represent an ideal (granted an ethnocentric Western, middle-class desideratum) there are many circumstances which might disrupt parental attachments and activities. Some whould say that the seeds of disorder lie within the structure and function of contemporary family life. In the industrialized societies of Western Europe and the United States, a striking feature of life is the diversity of family life, parenting practices and expectations of children. Also, in an urbanized society there remain few rituals or ceremonies which are generally recognized by the community as special marks of the child's status and role or the various stages in a child's development. On the other hand, in isolated rural or peasant communities and isolated tribal societies, formal initiation rites are a principal means of demonstrating to the community and to the child himself that he is socially as well as physically acceptable as an adult; these rites have a traditional order. They may have gone on for generations with little change; their meaning is clear to all. This is in complete contrast to the blurring of age- and sex-roles which are stated by some commentators to be a feature of contemporary urban family life.

Whatever the truth of such claims, it would appear that, notwithstanding variations in family pattern and style of parenting, all societies seem to be broadly successful in the task of transforming helpless, self-centred infants into more or less self-supporting, responsible members of their particular form of community. Indeed, there is a basic 'preparedness' on the part of most infants to be trained – that is, an inbuilt bias toward all things social. The baby *responds* to the mother's characteristic infant-orientated overtures (see Stern, 1977) in a sociable manner that produces in her (in her turn) a happy and sociable reaction. He also initiates social encounters with vocalizations or smiles directed to the mother which cause her in turn to smile back and to talk, tickle or touch him. In this way she elicits further responses from the baby. A chain of mutually rewarding interactions is thus initiated on many occasions. Parents and child learn about each other in the course of these interactions.

TEMPERAMENTAL INDIVIDUALITY

Some babies get off to a bad start, as we saw earlier. Not all infants are as rewarding or as co-operative as expected in this process of socializing and socialization. Parents are sometimes taken by surprise by the 'difficult' temperament of their new-born baby, and by this resistance to changes of routine and other demands of the socializing process.

Although it might seem far-fetched to propose that the stimulus characteristics of the child can constitute a sufficient or necessary provocation to maltreatment, factor analytic studies indicate that abnormal attributes in the child are at least as substantial a factor in explaining incidents of abuse as deviance in the parents (Gil, 1971). It has been reported (Gil, 1971) that abused children have been ill-treated in a foster home, transferred, and then abused in another foster home. Episodes of persistent, intense crying or screaming by the child, bouts of bladder or bowel incontinence, and incorrigible defiance are frequently cited as immediate antecedents of parental violence.

There is evidence (reviewed by Rutter, 1977) that temperamental characteristics have a significant association with the later development of behaviour disorders of various kinds. Children who are hostile, restless, impulsive and manifest poor concentration and who, in addition, display insensitivity to the feelings of others are predisposed to delinquent tendencies. It is not, of course, a simple matter of some children having 'adverse temperamental attributes' – for example, poor adaptability – which make them difficult to rear. Children's characteristics interact with parental attributes; a mismatch of temperament can result in an extended series of mutually unrewarding interactions. They can also lead to faulty or incomplete socialization (Herbert, 1978).

Predisposing factors

Of all the factors cited by Staub (1975), and listed above, the establishment of an affectional bond between parent and child is perhaps the most critical – the foundation on which all social training is based (Hoffman, 1970). The factors which make for a rejecting parent are likely to be multiple and additive in their influence. They may be, in part, situational; for instance a baby might be born at a particularly bad time for the parents, a time of emotional vulnerability and/or financial hardship. Parents identified as abusers of children are disproportionately represented among the under-twenties at the birth of their first child. They have produced relatively large families, and they report much marital disharmony (Brown, Harris and Peto, 1973; Ounsted, Oppenheimer and Lindsay, 1975). From studies of child abuse incidents (Gil, 1971) it is suggested that the battered child is frequently born at a time in the history of the family when there was unusual stress originating in problems such as subsistence or housing. (We shall return to a consideration of this type of family).

PATHOLOGICAL MATERNAL BEHAVIOUR

The literature on child abuse and childhood psychopathology reveals a preoccupation with the background, psychopathology or personality of the individual (usually the mother) who ill-treats her child or provides inappropriate rearing. In terms of formal studies of childhood psychopathology it was in the forties that it became explicit that problem children frequently had problem parents, and it was inferred from this that they had become problematic mainly because of this fact. There followed a plethora of studies of parental attitudes and actions and many simplistic schemata were devised to demonstrate the relationship between parental attitudes and childhood behaviour disorders.

Maternal overprotection

Mothering can be 'overdone' according to some theorists (for example, Levy, 1943) resulting in excessive mother–child contact and attachment. Levy (1943) investigated 15 cases of over-protective mothers in depth and found that their child-rearing methods ensured excessive mother–child contact. In a typical example, the child slept in the same room as his mother for years; she tended to fondle him excessively, watch over him constantly, and prevented him from taking risks or acting in an independent manner; she fussed a lot about his health, over-medicating and over-dressing him; and she made many of his decisions for him; she tended to indulge his every whim in return for absolute obedience. Levy interpreted this transaction as an attempt on the part of the mother to prolong his childhood and keep him 'tied to her apron strings'. Over-protective mothers frequently alternated between dominating the child and submitting to him.

According to Levy there are two types of maternal over-protection. The dominating form of over-protection may lead to excessively dependent behaviour on the part of the child – certainly, the children in his study tended to be dependent, passive and submissive. It is thought that if the child is discouraged from acting independently, exploring, and experimenting, he acquires timid, awkward and generally apprehensive behaviours. It is important to remember that what is being referred to here is really extreme maternal behaviour which involves maternal domination (restrictiveness and nagging.)

The postulated outcome of the second kind of over-protective parenting (that is to say, the behaviour profile of children with over-protective, but indulgent rather than dominant mothers) fits the popular stereotype of the feckless, spoilt child. His behaviour is characterized by disobedience, impudence, tantrums, excessive demands, and varying degrees of tyrannical behaviour – all thought to represent accelerated growth of the aggressive components of personality and to be related to maternal indulgence. Levy refers to this type of child, in the fully developed form, as 'an infant-monster or egocentric psychopath' and points out that, in addition to his low frustration tolerance, the child gradually evolves into an exploitive character using every device ranging from charm, wheedling and coaxing to bullying, in order to get his own way. Levy predicted that unless this was stemmed by reality experiences, the child would con-

The Pathology of Human Parental Behaviour

tinue into adult life to play the part of the beloved tyrant of an ever-responding mother. The incessant infantilizing of his life-style might leave the 'enfant terrible' with a permanent illusion of omnipotence.

It has to be stressed again that sampling and the generalisation of findings from clinic-attending children, has led to many erroneous conclusions. It is not surprising that when special-risk groups of children are examined, relationships between specific parental patterns and subsequent behaviour disturbances in the offspring are found which then disappear when they are looked for in the more general population outside the clinics. What we do not know from aetiological studies of restricted clinical samples is how many run-of-the-mill children have suffered the allegedly 'pathogenic' influence without any particular ill-effect. What it is possible to affirm, on the basis of more broadly based surveys, is that in all the risk categories invoked in the literature on psychopathology (be the categories prenatal or perinatal pathology, parental characteristics, family problems, and so on) it is possible to identify significant numbers of children who, although subjected to these influences, nevertheless developed normally (Thomas & Chess, 1977).

The precursors of maternal attachment are several, and they are also speculative. It may be that a child is infantilized and 'held on to' because he perhaps is particularly 'precious'; he may have followed a prolonged period of infertility, a series of miscarriages, or the loss of a sibling. Some mothers are anxious by temperament; others by circumstances, as when an infant is ill or disabled; or simply because he is the first-born, arriving when the parents are apprehensive 'learners ' (Herbert, 1985).

Maternal rejection

At the opposite extreme, mothers have become rejecting when their babies are unresponsive, or indeed, in some of their attributes, quite unlovable. The infant may lead to disharmony between the mother and a pathologically jealous husband, or even to the eventual disintegration of a marriage. Doubtless there are temperamental (for example, narcissistic) aspects in the make-up of some mothers which make it difficult for them to love anyone but themselves; there may be little room left over for babies. But rejection is not a fixed characteristic – feelings can and do change.

An understanding of 'affectionless' mothers requires a more general understanding of so-called affectionless personalities in people who are sometimes said to suffer from psychopathic disorders. It is quite likely that adults who are capable only of relating to other adults (spouses for example) in a superficial, exploitive and hostile manner, show similar relationships with their children. There is certainly a relationship between child abuse and psychological disorder, notably depression (Elmer, 1967) and alcohol abuse (Belsky, 1978). The latter is significant on account of the facilitating effect of alcohol upon violence of all varieties, but in the case of a drunken parent, particularly the risk of attacks directed toward a convenient child victim.

Several theorists (see Sluckin et al., 1983) have proposed that an early disruption of mother-to-child attachment – distortion or failure of maternal bonding – is brought about by the deliberate or fortuitous separation of mother and infant in the postpartum hours or days. (This subject is dealt with in chapter 6.)

Several investigations have looked at the effects of a woman's childhood experiences on her own subsequent maternal attitudes; they suggest a relationship between disrupted childhood and family life (perhaps of separation or rejection) and negative attitudes to parenting.

The theory of maternal rejection most frequently encountered relates the predisposition in a hostile parent to neglect or abuse her offspring to experiences of a similar nature (or the absence of critical learning opportunities) in her own childhood (for example, Kempe and Kempe, 1976; Rutter, Quinton and Liddle, 1983). Variants of three main explanatory themes are put forward to account for this supposed intergenerational transmission, sometimes referred to as the 'cycle of disadvantage' (Rutter and Madge, 1976): the modelling of violent behaviour by a parent or parents; the reinforcement by parents of the child's (and thus parent-to-be's) aggressive actions (on the basis of operant learning principles); and the deprivation of the child's basic needs. This might be the repudiation and frustration of her dependency needs, a lack which leads, hypothetically, to an inability later in life to empathize with her own children's needs for nurturance.

The last factor has been the subject of much psychodynamically-orientated speculation. However, in the light of empirical evidence (for example, Robson, 1981) we know that there are many factors in the mother's background that can influence how she will relate to her infant. Unfortunately they do not allow us to predict a particular mother's feelings for her child from known circumstances, especially those extreme cases of pathological indifference or loathing.

Theorists have had to admit that it is not only the mother's *presence* that can be pathogenic; her *absence* can also have effects for children. But therein lies the dilemma: even if 'maternal separation' were a reasonable and precise explanatory concept it would not necessarily follow from the discovery of a positive correlation between maternal separation and (say) delinquent activities in children that the one caused the other. The factor being investigated may have no causal role at all, or may play only a minor contributory part in producing the behaviour disturbance or, for that matter, may only influence developments in very special instances. Like many other would-be explanatory concepts in psychiatry, 'maternal deprivation' (or 'maternal rejection') is over-inclusive and too imprecise to be of any predictive value. They have been used in a manner which suggests they describe a *unitary* phenomenon. In fact, as the evidence suggests, there are many moderating influences which determine the seriousness of the consequences for the child.

Fathering

There are not many references to 'paternal rejection' *per se,* in the literature concerning parental rejection, which is surprising, since fathers are much more likely to abandon their families than mothers. However, much of the present interest in discovering how fathers contribute to children's development has been stimulated by studies of children's development when the father is absent from the home environment. This deficit model of development is similar to the approach used when researchers first became interested in the effects of maternal deprivation. Studies of children who lack the fathering experience indicate that their adjustment is difficult in the areas of personality and social development (see review by Bigner, 1979).

Despite the fact that many fathers spend the majority of their time away from their families, they are expected to participate in child-rearing responsibilities as much as possible. The control function of fathering has received more attention from researchers than the nurturant function because social control is considered to be an integral by-product of the power and authority associated with fatherhood. Research generally indicates that the father rather than the mother is responsible for reinforcing the dependent nature of the child's behaviour, and this effect may be more visible in a girl than in a boy. Sears, Rau and Alpert, (1965) found that a number of dependent-type behaviours are shown by children whose fathers place intense pressures on them to be neat and orderly, punish or fail to reward independent behaviour, reward traditional sex-typed behaviour of both boys and girls, and discourage children from expressing their affection for them physically.

Many studies indicate that fathers, as opposed to mothers, rely on power-assertive techniques to control children's behaviour. One of the more consistent findings about the effects of power assertion by a father is that children so-treated show a higher level of aggression toward others (Becker, 1964). Aggression toward the child by the parent is thought to breed aggression of the child toward others. The parent may, perhaps, act as a model of aggression for the child, who, in seeing the father behave in this manner, transfers the aggression to his or her relationships with siblings and peers.

PSYCHIATRIC DISORDER IN PARENTS

When adverse circumstances make parenting very difficult, the range, flexibility and robustness of the mother's and/or father's coping repertoires are of particular significance. In a study by Rutter (1966) it was found that, of children with neurotic and behaviour problems attending a child psychiatric clinic, one in five had a psychiatrically ill parent. This was three times the incidence found in the parents of a control group consisting of children attending a dental clinic. Also, the children under psychiatric treatment whose parents had a history of psychiatric disorder had more severe and extensive problems than other problem children at the same clinic. However, not all, by any means, of the children of mentally ill parents develop emotional problems; many grow up to be healthy adults – nevertheless, children in this situation are at risk. The children of psychotic parents are less vulnerable than those with psychopathic parents.

Childhood problems are more likely to be associated with psychiatric illness in the mother than in the father. When both parents are psychiatrically disturbed, the child is particularly vulnerable as there is no one, except possibly brothers and sisters, to act as a buffer between himself and his pathological parents. Rutter's study of psychiatrically ill parents showed that there was little relationship between the specific type of mental illness in the parent and the type of problem displayed by the child. What does seem to make a child most susceptible is his direct involvement in the mother's or father's symptoms. He is particularly at risk, for example, if the parent has delusions about the child, paranoid feelings, obsessional fears of harming him, morbid anxiety concerning his development, or hostile feelings towards him. Affective symptoms (that is to say, disturbed emotions such as hostility and depression) are particularly associated with the development of problems in the offspring.

The problems of those members of families who are genetically predisposed to some psychiatric or psychosomatic disorder, or those who are recovering from illness or trauma, are mitigated when they can look to a cohesive family for support.

Psychiatric illness and maternal bonding

Particular concern has been shown in relation to the effects of psychiatric disorder on the mother's attachment to her infant. Does it lead to rejection of the baby or older child? Can it seriously distort the mother's relationship with her offspring thus putting the child at risk? We have already touched on some of these issues. Sluckin, Herbert and Sluckin (1983) have reviewed the conceptual complexities of the term 'maternal bonding' and the evidence for a sensitive period for the formation of attachments (see chapter 6), but little detailed work has been done on the mother–baby relationship during the mother's psychiatric disorder. Practical competence is often relatively intact, at least at the crude level at which it has been measured, but social activities such as play and touching are more frequently disrupted (see Game et al., 1976; Margison, 1982). Sadly, research designs are not usually of a standard to allow confident generalizations or conclusions in this area.

Margison (1982) is of the opinion that different aspects of maternal psychopathology may interfere with the mother–baby relationship in different ways. An ill mother's rejection of her baby is a common concern among nursing staff and doctors. Rejection of the infant might occur in several ways: negative thoughts about the child; a wish that the child had never been born or could be changed in some fundamental way (for example by being of the opposite sex); denial that the child is hers or even that she has given birth at all (Margison, 1982).

Brew and Seidenberg (1950) in a review of 83 postpartum psychoses admitted to the Syracuse Psychopathic Hospital between 1933 and 1946 stated that in both the schizophrenic and manic-depressive postpartum reactions, one point in common was the rejection of the newborn infant. The instances of actual violence were high in this series; three mothers attempted infanticide by strangulation and another tried to throw the child out of the window. However, the other patients under examination showed a wide variety of symptoms which were also interpreted as rejection – delusions, hallucinations, amnesia and over-solicitude to the baby. The delusions were commonly in the form of fears that the baby was dead or disfigured.

Manic patients commonly show a marked disorganization of practical care, relating warmly to the infant but only for short periods and then inconsistently (Margison, 1982). Mothers with chronic schizophrenia also show deficits in their organization of daily routines and poor practical competence, but to an even more marked degree, and consequently their babies are often placed in the care of the Local Authority Social Services (Margison, 1981). Mothers with depression frequently report a lack of feelings which may involve the infant. They may express concern about lack of love for the baby and, commonly, an inability to cope.

Margison (1982) concludes a detailed and comprehensive review on the pathology of mother–child relationships by stating that as yet there has been insufficient research to disentangle the effects of psychiatric illness, maternal attitudes and behaviour, and the various aspects of treatment. He is of the opinion that an important prerequisite for future progress is that researchers stop considering normal and abnormal bonding as two discrete categories and instead consider separately the components of the bonding process and their relation to psychiatric disorders. He believes that the mother's problems with her baby are best viewed individually and the contingencies and triggers for each problem be kept separate. Bonding problems are complex phenomena involving different levels of analysis: family, marital, intrapersonal, interpersonal, and behavioural (see Herbert, 1981).

Family cohesion

Evidence reviewed by Orford (1980) indicates that family cohesion (see below) is a key factor linking work in otherwise separated problem areas. Orford classified the various conditions which give rise to the syndrome of cohesion in families, hypothesizing that the syndrome is associated with the psychological well-being of groups members:

1 More time spent in shared activity.
2 Less withdrawal, avoidance and segregated activity.
3 A higher rate of warm interactions, and a lower rate of critical or hostile interactions amongst members.
4 Fuller and more accurate communication between members.
5 A more favourable evaluation of other members; a lower level of criticism of other members.
6 More favourable meta-perceptions; that is, members more likely to assume that other members have a favourable view of them.
7 A higher level of perceived affection between members.
8 A higher level of satisfaction and morale, and greater optimism about the future stability of the family group.

Families which lack the characteristics listed above make their members vulnerable to experiencing psychological distress. Particularly at risk are those family members who are already vulnerable for other reasons – the young, the elderly, those suffering from other types of stress (for example, hospitalization, alcohol dependence, or coping with a large number of children).

Lidz (1965) maintains that a coalition between the parents is necessary not only to give unity of direction to the children but also to provide each parent with the emotional support essential for carrying out his or her cardinal functions. In one type of family situation, which is frequently found (Lidz et al., 1957; Wynne et al., 1958) to be the background to the development of schizophrenic withdrawal in children, the members are torn by a 'schismatic conflict' between husband and wife, so that the family becomes divided into two hostile factions and the children become involved in an emotional tug-of-war. This destructive situation – a complete negation of a parental coalition – may go on and on. Although a divorce would probably end everybody's misery, it does not occur, and a state of what is called 'emotional divorce' persists – a corrosive situation pervaded by continual bickering, mutual recriminations and venomous hatred. If the child shows affection to one parent, this is regarded as betrayal by the other.

In another disturbed familial pattern, called 'marital skew', one parent dominates the roost, and his (or her) psychopathology – abnormal thinking, bizarre style of living and abnormal manner of child-rearing – is

The Pathology of Human Parental Behaviour

passively accepted by the spouse. Suspiciousness and distrust of outsiders amounting to paranoia may prevail and these deluded attitudes, together with other irrational interpretations of life, may be conveyed to the children. Such parents provide faulty models for their offspring to emulate, transmitting faulty modes of thinking to their children. Both these types of environment immerse the child in an irrational family atmosphere which is thought to negate the development of a healthy ego.

Danzinger (1971) comments on some of the psycho-social factors which have a bearing on the psychological disorganisation (in particular the lack of a clear and autonomous sense of ego-identity) in schizophrenics. As he puts it:

One may speculate that the primary effect of patterns of parental demand and support lies in the establishment of ego boundaries, that is, in the creation of a zone of self-expression bounded by a clearly perceived social reality. In the absence of such boundaries there is confusion between self and non-self; impaired appreciation of external reality and impaired autonomy of the self are mutually complementary. The development of effective ego boundaries depends on both the demands and the supports for which socialization agents are the source. A boundary is, of course, a relationship. To establish the kind of relationship between 'inside' and 'outside' which we associate with normal personality functioning, parental demands must establish wide but firm limits within which generous support is extended to the growth of autonomous ego functions.

Wynne and Singer (1963) regard a stable and coherent environment –that is, one which provides opportunities for the child to test reality in a variety of roles during development – as a prerequisite for the formation of a healthy ego. However, it is postulated that the families of schizophrenics lack these attributes of stability and coherence. According to Lidz, Fleck and Cornelison (1965) the age- and sex-roles in these families are ambiguous. Add to that extreme instability at home, and all the ingredients are present to inhibit the development of appropriate forms of behaviour and a stable sense of identity in the child.

The findings from several studies appear to demonstrate the disruptive effect produced by parents of schizophrenic children in crucial areas of their psychological development (for example, Mishler and Waxler, 1968; Laing and Esterson, 1970). There is a fairly consistent tendency for the parents of schizophrenic children to deny communicative support to their offspring. Verbal interactions between parents and child tend to be stereotyped with almost no outlet for spontaneous expression. Frequently they fail to respond to their child's communications or to his demands for a recognition of his own point of view. Their own statements tend to be intrusive and take the form of interventions rather than replies to the child. The replies they do make tend to be selective, being responses to those of the child's expressions which have been initiated by themselves rather than to any expressions originated by the child. The child's spontaneous utterances and self-expression are restricted, as if he were being denied the right to an independent point of view.

There are theorists who contend that the complexities of language and logic are such that a condition as serious as schizophrenic thought disorder may be a consequence of the child having received a faulty grounding in the consensual (linguistic) meanings as well as other instrumentalities of society. These deficiencies limit his adaptive capacities and permit him to escape from insoluble contradictions by abandoning the 'meaning system' of his culture: he takes refuge in irrationality and withdrawn behaviour.

These theories have indeed been highly influential. However, the conditions described in the home situation are by no means unique to the parents of psychotic children (see Frank, 1965) and Bender and Grugett (1956) point out the diversity of personalities among parents of schizophrenic children. While some of the investigations mentioned above are concurrent or longitudinal studies, much of the work on the psychosocial background to psychosis is retrospective. Life histories are among the most popular methods used in clinical research, but data based on life histories retrospectively obtained can be fraught with error and difficulties of interpretation (Schofield and Balian, 1959).

One of the key concepts at the basis of theories of conflicting, confusing communications – the 'double-bind' – has been subjected to heavy criticism (for example, Sluki and Ransom, 1976) because of its vagueness, the disagreements about the criteria thought necessary to produce double-bind situations and the lack of empirical support for the far-reaching claims that have been made for it.

SOME IMPLICATIONS

We have explored, in this chapter, the available findings relevant to human adults who adopt a deviant parental role. The literature on parent–child relationships encompasses a belief (and one with a long ancestry) that optimal care and training of children during the impressionable years of life will 'inoculate' them against present and future problems of life – as such, it has all the pathos of the researches of those early alchemists, seeking a formula to transform base metal into pure gold. Sadly the base metal today is our meagre supply of knowledge –mainly ambiguous correlations and fragile generalizations hedged around with a multitude of qualifications – of the nature of parental practices and their influence on child development.

All this is not to claim that an interest in parenting (pathological or normal) is misplaced. It is the failure empirically to provide convincing explanatory links between the cause–effect ends of the 'problem parent–

problem child' equation that is worrying, especially when social and clinicians so often appear to find no such difficulty (wi dent 'causal' linking of 'failure of maternal bonding' with ch early maternal deprivation with the 'affectionless' personality).

An intolerable burden of anxiety and guilt is foisted on parents. The bland assumption that parent pathology underlies all childhood psychopathology implies that parents are all-powerful, all-responsible 'and must assume the role of playing preventive Fate for their children' (Bruch, 1954). Mothers frequently ask their doctors whether they have done the wrong thing or have made an awful mistake by doing X or Y with the child, as if a one-off allegedly 'traumatic' event would leave an indelible mark on the child.

Traditionally, the case history in social work and clinical practice has taken a 'vertical' form; it is an attempt to relate present troubles to past experiences so as to see how they have evolved. Not surprisingly several difficulties attach to this approach and they necessitate caution in interpreting retrospective data. Parental recall of the significant events in their child's life is notoriously unreliable (Robbins, 1963). Then again, it is only too easy for the parent to unwittingly reconstruct the past, to rewrite history, so to speak, so as to make sense of it. It is much easier to be wise *after* the event than to make good predictions (and therefore prescriptions) about a child's future behaviour. Freud observed that:

So long as we trace the development (of a mental process) backwards, the connection appears continuous, and we feel we have obtained an insight which is completely satisfactory and even exhaustive. But if we proceed the reverse way, if we start from the premise inferred from the analysis and try to follow up the final result, then we no longer get the impression of an inevitable sequence of events, which could not have been otherwise determined. We notice at once that there might be another result, and that we might have been just as well able to understand and explain the latter. (Freud, 1956)

Clarke and Clarke (1976), on the basis of a painstaking review of the evidence, conclude that early experience is no more than a link in the development chain, shaping behaviour less and less powerfully as age increases. What is probably crucial is that early (maladaptive) learning is continually reinforced and it is in this way that long-term effects appear. There is also the possibility that some of the later deviance is the result of later reinforcement as well as original learning experiences. Human young are malleable, capable of learning and highly susceptible to change – it is not surprising, then, that we find no hard evidence of a relationship between *specific* early child-rearing practices and the child's adjustment and personality development.

Yet parent education is still based on a false assumption that there is a body of knowledge about the best techniques of child care. If it is accepted that this body of knowledge exists, it follows that these can be taught to parents. Much has been made (particularly in the psychoanalytic literature) of the allegedly disruptive effects on infant adjustment of different child-care practices (notably of feeding, weaning and toilet-training procedures). Yet what is really important in childrearing is the general social climate in the home. For example, the mother does best who does what she and the community to which she belongs believe is right for the child, and a mother acting in this way implies that she is a relaxed and confident mother (Behrens, 1954).

As Marian Radke Yarrow and her co-authors of the book *Child Rearing* regretfully reflect: questions of child-rearing have not yielded easily to scientific study. They conclude from their own investigations and their extensive review of available evidence that we are still searching for the specific conditions in the child's cumulative experience with his parents that evoke, strengthen or modify his behaviour.

REFERENCES

Allan, L. J. 1978: Child abuse: a critical review of the research and the theory. In J. P. Martin (ed.), *Violence and the Family*. Chichester: Wiley.

Bandura, A., and Walters, R. H. 1963: *Social Learning and Personality Development*. New York: Holt, Rinehart and Winston.

Baumrind, D. 1971: Current patterns of parental authority. *Development Psychology Monographs*, 4, (1), part 2, 1–103.

Becker, W. C. 1964: Consequences of different kinds of parental discipline. In M. L. Hoffman and L. W. Hoffman (eds), *Review of Child Development Research, Vol. I*, New York: Russell Sage Foundation.

Behrens, M. L. 1954: Child rearing and the character structure of the mother. *Child Development*, 25, 225–38.

Bell, R., and Harper, L. 1977: *Child Effects on Adults*. Hillsdale, New Jersey: Erlbaum.

Belsky, J. 1978: Three theoretical models of child abuse: a critical review. *Child Abuse and Neglect*, 2, 37–50.

Bender, L., and Grugett, A. E. 1956: A study of certain epidemiological factors in a group of children with childhood schizophrenia. *American Journal of Orthopsychiatry*, 26, 131–43.

Benjamin, L. S. 1974: Structural analysis of social behaviour. *Psychological Review*, 81, 392–425.

Bigner, J. J. 1979: *Parent–Child Relations: An Introduction to Parenting*. New York: Macmillan.

Brew, M. F., and Seidenberg, R. 1950: Psychotic reactions associated with pregancy and childbirth. *Journal of Nervous & Mental Diseases*, 111, 408–23.

Bronfenbrenner, U. 1970: *The Two Worlds of Childhood: USA and USSR*. London: Allen and Unwin.

Brown, G. W., Harris, T. O., and Peto, J. 1973: Life events and psychiatric disorder, 2: Nature of causal links. *Psychological Medicine*, 3, 158–76.

Bruch, H. 1954: Parent education, or the illusion of omnipotence. *American Journal of Orthopsychiatry,* 24, 723–6.
Caldwell, B. M. 1964: The effects of infant care. In M. L. Hoffman and L. W. Hoffman (eds), *Review of Child Development Research,* New York: Russell Sage Foundation.
Chess, S. 1964: Editorial: 'Mal-de-mère'. *American Journal of Orthopsychiatry,* 34, 613.
Clarke, A. and Clarke, A. D. B. 1976: *Early Experience: Myth and Reality,* London: Academic Press.
Danziger, K. 1971: *Socialization.* Harmondsworth: Penguin.
Elmer, E. 1967: *Children in Jeopardy: A Study of Abused Minors and their Families.* Pittsburgh: University of Pittsburgh Press.
Erikson, E. H. 1965: *Childhood and Society.* (revised edn.), Harmondsworth: Penguin.
Frank, G. H. 1965: The role of the family in the development of psycho pathology. *Psychological Bulletin,* 64, 191–203.
Freud, S. 1956: Homosexuality in a woman. In E. Jones (ed.), *Collected Papers, Vol. 2.* London: Hogarth Press.
Fromm, E. 1955: *The Sane Society.* London: Holt, Rinehart and Winston.
Gamer, E., Gallant, D., and Grunebaum, M. 1976: Children of psychotic mothers: an evaluation of one year olds on a test of object permanence. *Archives of General Psychiatry,* 33, 311–17.
Gil, D. 1971: Violence against children. *Journal of Marriage and the Family,* 33, 639–48.
Herbert, M. 1974: *Emotional Problems of Development in Children.* London: Academic Press.
—— 1978: *Conduct Disorders of Childhood and Adolescence.* Chichester: John Wiley and Sons.
—— 1981: *Behavioural Treatment of Problem Children: A Practice Manual.* London: Academic Press.
—— 1985: Triadic work with children. In F. Watts (ed.), *Recent Developments in Clinical Psychology.* Chichester: John Wiley and Sons.
Herbert, M., Sluckin, W., and Sluckin, A. 1982: Mother-to-infant 'bonding'. *Journal of Child Psychology & Psychiatry,* 23, 205–21.
Hoffman, M. L. 1970: Moral development, in P. H. Mussen (ed.), *Carmichael's Manual of Child Psychology, Vol. II.* (3rd edn), London: John Wiley and sons.
Iwaniec, D., and Herbert, M. 1982: The assessment and treatment of children who fail to thrive. *Social Work Today,* 13, 8–12.
Johnson, R. C., and Medinnus, G. R. 1968: *Child Psychology: Behaviour and Development.* New York: John Wiley and Sons.
Kellmer Pringle, M. 1980: *The Needs of Children.* London: Burnett Books (with André Deutsch).
Kempe, R. S., and Kempe, C. H. 1978: *Child Abuse.* London: Fontana.
Laing, R. D., and Esterson, A. 1970: *Sanity, Madness and the Family.* Harmondsworth, Penguin.
Levy, D. M. 1943: *Maternal Overprotection.* New York: Columbia University Press.

Lidz, T. 1968: *The Person: His Development Throughout the Life Cycle.* London: Basic Books.
Lidz, T., Cornelison, A., Fleck, S., and Terry, D. 1957: The intrafamily environment of schizophrenic patients: II, Marital schism and marital skew. *Americal Journal of Psychiatry,* 114, 241–8.
Lidz, T., Fleck, S., and Cornelison, A. R. 1965: *Schizophrenia and the Family.* New York: International University Press.
Madge, N. (ed.) 1983: *Families at Risk.* London: Heinemann.
Margison, F. R. 1981: Assessing the use of a pyschiatric unit for mothers with their babies: Risk to the babies. M.Sc. thesis, University of Manchester.
—— 1982: The pathology of the mother–child relationship. In I. F. Brockington and R. Kumer (eds), *Motherhood and Mental Illness,* London: Academic Press.
Maslow, A. H. 1954: *Motivation and Personality.* New York: Harper and Row.
Mischel, W. 1968: *Personality and Assessment.* New York: John Wiley and Sons.
Mishler, E. G., and Waxler, N. E. 1968: *Interaction in Families,* New York: John Wiley and Sons.
Morgan, P. 1975: *Child Care: Sense and Fable.* London: Temple Smith.
Orford, J. 1980: The domestic context. In P. Feldman and J. Orford (eds), *Psychological Problems: The Social Context,* Chichester: John Wiley and Sons.
Ounsted, C., Oppenheimer, R., and Lindsay, J. 1975: The psychopathology and psychotherapy of the families: aspects of bonding failure. In A. Franklin (ed.), *Concerning Child Abuse,* Edinburgh: Churchill Livingstone.
Rapoport, R., Rapoport, R. N., and Strelitz, Z. 1977: *Fathers, Mothers and Others.* London: Routledge and Kegal Paul.
Reykowski, J. 1982: Development of prosocial motivation: a dialectic process. In N. Eisenberg (ed.), *The Development of Prosocial Behaviour,* London: Academic Press.
Robbins, L. C. 1963: The accuracy of parental recall of aspects of child development and of child rearing practices. *Journal of Abnormal Social Psychology,* 66, 261–70.
Robson, K. M. 1981: A study of mothers' emotional reactions to their newborn babies. Ph.D. thesis, University of London.
Rutter, M. 1966: *Children of Sick Parents.* Oxford: Oxford University Press.
—— 1977: Prospective studies to investigate behavioural change. In J. S. Strauss, H. M. Babigian and M. Roff (eds), *The Origins and Course of Psychopathology,* New York: Plenum.
Rutter, M., and Madge, N. 1976: *Cycles of Disadvantage.* London: Heinemann.
Rutter, M., Quinton, D., and Liddle, C. 1983: Parenting in two generations: Looking backwards and looking forwards. In N. Madge, (ed.), *Families at Risk,* London: Heinemann.
Schaefer, E. S. 1959: A circumplex model for maternal behaviour. *Journal of Abnormal Social Psychology,* 59, 226–35.
Schofield, W., and Balian, L. 1959: A comparative study of the personal histories of schizophrenic and nonpsychiatric patients. *Journal of Abnormal Social Psychology,* 59, 216–25.
Schuham, A. I. 1967: The double-bind hypothesis: a decade later. *Psychological Bulletin,* 68, 409–16.

Sears, R. R., Rau, L., and Alpert, R. 1965: *Identification and Child Rearing.* Stanford: Stanford University Press.
Seligman, M. E. P. 1975: *Helplessness: On Depression, Development, and Death.* San Fransisco: Freeman.
Sluckin, W., Herbert, M., and Sluckin, A. 1983: *Maternal Bonding.* Oxford: Basil Blackwell.
Sluki, C. E., and Ransom, D. C. (eds) 1976: *Double Bind: The Foundation of the Communicational Approach to the Family.* New York: Grune & Stratton.
Staub, E. 1975: *The Development of Prosocial Behaviour in Children*, Morrison, New Jersey: General Learning Press.
Stern, D. 1977: *The First Relationship: Infant and Mother.* London: Fontana/Open Books.
Stuart, R. B. 1969: Operant interpersonal treatment for marital discord. *Journal of Counselling and Clinical Psychology,* 33, 675–82.
Talbot, N. (ed.) 1974: *Raising Children in Modern America.* Boston: Little, Brown.
Thomas, A., and Chess, S. 1977: *Temperament and Development.* New York: Brunner/Mazel.
Thomas, A., Chess, S., and Birch, H. G. 1968: *Temperament and Behaviour Disorder in Children.* London: University of London Press.
Wynn, L. C., Rycoff, I. M., Day, J., and Hirsch, S. I. 1958: Pseudomutuality in the family relations of schizophrenia. *Psychiatry,* 21, 205–11.
Wynn, L. C., and Singer, M. T. 1963: Thought disorder and family relations of schizophrenics, II: Classification of forms of thinking. *Archives of General Psychiatry,* 9, 199–203.
Yarrow, M. R., Campbell, J. D., and Burton, R. V. 1968: *Child Rearing: An Inquiry into Research and Methods.* San Francisco: Jossey-Bass.

12 Methods and Approaches to the Study of Parenting

K. Browne

It is obvious from the preceding chapters that parental behaviour is multifactorial and not amenable to general explanation in terms of one or even a few causal factors. The purpose of this chapter is to assess the ways in which comparative or ethological methods and approaches can contribute to our understanding of parental care-giving and attachment. The limits of this contribution for the study of human parenting will also be discussed with reference to sociological and psychological approaches.

THE ETHOLOGICAL APPROACH

As outlined by Sluckin and Herbert in chapter 1, ethology focuses upon the reasons why an animal behaves in the way it does, and studies the function, causation and development of observed behaviour patterns in order to learn what selection pressures have shaped their evolution. Classical ethology considered that there was an absolute dichotomy between instinctive units of behaviour and those acquired through learning, and that gross behavioural patterns represented a composition of both learned and innate elements (Lorenz, 1965). Much controversy arose between this viewpoint and the 'behaviourist' position that almost all observable behaviour was a product of learning (for example, Skinner, 1953). Modern ethology, however, considers that environmental influences on behaviour can be much more diverse than is implied by the nature–nurture controversy.

Wilson (1978) supports the view that behavioural characteristics which are genetically determined may develop rather than emerge perfectly on the first occasion, and that suitably large deviations from normal experience may lead to the modification of these behaviour patterns or their non-appearance. A view which deviates even further from the classical one is that espoused by Moltz (1965). He holds that the genetic influences upon behaviour should be seen as interacting with the environment and that every behavioural system is the end result of a combination of both environmental and genetic factors. Therefore, rather than trying to decide whether a particular behaviour is innate or acquired, we should consider a continuum ranging from environmentally stable characteristics, which are relatively impervious to environmental influence, to environmentally labile characteristics which develop largely under the control of the environment.

It is generally accepted that genetically determined aspects of behaviour become less evident as we ascend the phylogenetic scale. Nevertheless, merely because behaviours are more varied in appearance due to the effects of learning, this does not mean that their biological bases are unimportant, or their comparative nature invalid. Hinde (1974) believes that studies in animal behaviour, when used for the development of methods and the formation of general principles can contribute to an understanding of human behaviour. But with the reservation that genetic and environmental factors which influence the behaviour of animals, can only be generalized to man with caution and possibly in conjunction with more direct evidence. Thus it is desirable to maintain a distinction between the adoption of ethological methods and the acceptance of associated theoretical perspectives. For example, the sociobiological notions of 'parental investment' (Trivers, 1972) and 'parent–offspring conflict' (Trivers, 1974) are based on observations of animal behaviour and evolutionary theory. They suggest that an infant will tend to maximize the chances of passing-on its own genes by demanding more parental investment than the parents, in terms of their own 'genetic fitness', are willing to give. However, this 'selfish gene' approach to behaviour (Dawkins, 1976a) is regarded by many behavioural scientists as too simplistic an explanation for activity, since it neglects the role played by the environment and cultural evolution. Therefore it should be emphasized that methods and theories of ethology may be separated. We are free to adopt one without the other.

Ethological methods are characterized by an insistence upon: (1) the prolonged observation and description of the animal to be investigated; (2) the study of behaviours that are meaningful in the natural habitat, preferably by direct observation within the animal's own environment; (3) description of behaviour in anatomical terms rather than the use of broad categories; (4) the comparative study of a wide range of species and their related behaviours. During the past decade comparative approaches and ethological methods have become commonplace in the study of the human as well as animal relationships (for example, Hinde, 1979). In particular, they have been applied to the examination of parent–child interactions (for example, Lewis and Rosenblum, 1974; Schaffer, 1977; Lytton, 1980; Kaye, 1982) and its theoretical interpretation in terms of survival and evolution (Bowlby, 1969; Ainsworth et al., 1978). This has encouraged a change in emphasis regarding the methods used by developmental psychologists from descriptive and psychometric techniques toward those involving direct observation of behaviour, such as the techniques described by Hutt and Hutt (1970) and Sackett (1978). There has also been a drift away from the traditional experimental setting of a structured laboratory environment, with more use made of naturalistic observations in the home (for example, Bronfenbrenner, 1979).

These methodological issues have accompanied a conceptual change in the study of human parenting. As mentioned by Schaffer and Collis in chapter 10 and by Herbert in chapter 11, attention has shifted away from an interest in parental attitudes and disciplinary techniques (a unidirectional approach: parent → child), to patterns of family interaction, parental responsivity, the development of relationships and the quality of attachments (a bidirectional/interactive approach: parent ⇄ child).

This conceptual change has not been limited to human studies, as exemplified by Higley and Suomi's review of primates (chapter 6). Thus the comparative or ethological model emphasizes the study of parent–offspring relationships, which Rosenblum and Moltz (1983) consider to have symbiotic properties that are essential for the survival and development of both participants.

THE DIRECT OBSERVATION OF BEHAVIOUR

Behaviour can be defined as the observable movements of part or whole of an individual's body in response to internal or external environmental factors (Alexander, 1975). Thus following Hinde (1974), behaviour patterns may be described by their 'physical characteristics' which outline the pattern of body movements (for example, postures), or by 'consequence' which refers to an activity that produces some result (for example, approach, evade). Although these two systems are different they are not mutually exclusive and both have advantages and disadvantages which have been reviewed by Bekoff (1979). Reference should also be made to Slater (1978) who provides a comprehensive guideline to their categorization.

Recently, Drummond (1981) in his review of 'the nature and description of behaviour patterns' has claimed that few authors concern themselves with how behaviour units are selected and described, while considerable emphasis is placed on techniques of analysis. This is surprising since the outcome of any behavioural study will be heavily influenced by the initial descriptive formulations.

A stream of observed behaviour is usually split into meaningful patterns of instantaneous 'events' and/or 'states' of appreciable duration.

Methods and Approaches to the Study of Parenting 349

Patterns, but not necessarily actual behaviours, must be operationally defined and described by their regularities in appearance, and then labelled. A behavioural catalogue is then constructed, based on the above descriptors. This can yield a large checklist of behavioural items, which may lack practical application. In such cases behaviour items are generally classified and clustered into categories according to their function and shared causal factors (Hinde, 1975). This makes comparisons between samples or cases more manageable.

Social behaviour such as parenting, which leads to the proximity of and interaction with another, has further complications for the observer. In addition to the behaviour catalogue, a list of possible respondents within the subject's environment must be identified.

The reliability and validity of observations

In any study in which data are generated by direct observation, it is necessary to find a way of assessing the reliability and validity of the observer's sampling and to what extent the behavioural catalogue is effective in standardizing the description of behaviour. As a general rule, the more complex the sampling method, the more difficult it is to achieve high reliability.

Intra-observer reliability. An intra or 'within individual' observer check is conducted in order to highlight inconsistencies and weaknesses in the description and coding of behavioural acts (Hollenbeck, 1978). This is usually carried out by a 'test-retest' comparison of a permanent record of behaviour on film or videotape.

Inter-observer reliability. The reliability of observations between observers can again be carried out by a 'test-retest' comparison of a permanent record of behaviour. Alternatively, two or more observers record behaviour directly in the same context and at the same moment in time. For both techniques the percentage agreement (Hartmann, 1977) or Kendall's coefficient of concordance (see Siegel, 1956) between (N) observers may be calculated. For measuring agreement between just two observers, product moment correlation coefficients can be calculated, although the most favoured method at present is 'Cohen's Kappa' (Conger, 1985).

Complete sample coverage. To determine whether an observation is valid and has adequately sampled all the items in a behaviour catalogue, a 'coverage indicator' can be established. Following Fagen and Goldman (1977), 'complete sample coverage' is defined as the probability of an independent act, observed at a random point in the recording, being of a type already represented within the total sample of behaviour. Hence, to measure incompleteness, the number of behaviour types represented once or not at all (N) is found for each recording, and this figure divided by the total number of acts observed (I). This calculation is then taken away from one, to determine the probability of a rare act occurring within an observation. Thus a useful behaviour catalogue 'coverage indicator' (g) is established with the application of the following formula:

$$g = 1 - \frac{N}{I}$$

Table 1 Analysis of videotape recording of behavioural acts to show coverage index at a 1 per cent level of significance

Film sequence	Total number of behavioural acts observed	Behavioural acts represented once or not at all	Coverage index
A	535	1	0.998
B	424	2	0.995
C	555	2	0.996

The methods of sampling behaviour

Whether behaviour is permanently recorded on film or directly observed using checklists and event recorders, it is necessary to decide how to sample the data. Altmann (1974) has classified seven sampling techniques that may be adopted singularly or in combination. With reference to Hinde (1982), they may be summarized as follows:

1 *Ad libitum sampling* (*events or states*) The observer records as much behaviour as possible, or notes those activities that are of interest. The problem with this type of sampling is that it assumes all types of behaviour have an equal probability of occurring and of being recorded, whereas, in fact, some behaviours and individuals are more readily observed than others. Hence this technique is usually employed in preliminary studies of the behaviour under investigation and is superseded by the following techniques, which are more precise.

2 *Sociometric matrix completion (events)* In addition to sampling behaviour *ad libitum,* the observer records particular individuals or pairs of individuals. The data are cast into matrices in which rows and columns represent two interactants (for example, 'groomer' and 'groomee'). However, no meaning can be attached to the row and column totals, so the uses of these data are generally restricted. The technique is designed to give information on directionality and the degree of one-sidedness in relations between pairs of animals, but gives no indication of duration.

Methods and Approaches to the Study of Parenting 351

3 *Focal individual sampling (events or states)* All occurrences of specified types of behaviour are recorded for a selected individual or group of individuals. This method is carried out for a preset sampling period. For example, the observer concentrates on the father within a family group and records only the father's actions and responses to others. For interaction studies it is also important to record the behaviour of other members of the family that relates to the father's actions. This method is designed to give information about frequency, duration and spatial relations of behaviour. However, unless the interactive behaviours of others are also recorded the method does not provide information about the synchrony of individuals within a group.

4 *Sampling all occurrences of some behaviours (usually event)* The observer attempts to record all occurrences of a certain type of behaviour within a group of individuals. This technique may be limited unless the behavioural events are 'attention attracting' and never occur too rapidly to be recorded. The method is designed to give information on frequency, duration and synchrony for a particular type of behaviour. For example, the occurrence of aggressive acts within a family group.

5 *Sequence sampling (events or states)* The observer concentrates on recording sequences of interaction between individuals. Recording begins when one of a series of specified interactive initiatives is observed (for example, the parent starts to feed the offspring). The observation continues until the sequence is concluded in one of a number of specified ways. The problem with this technique is the difficulty in defining the beginning and the end of a sequence. Nevertheless it does enable the observer to obtain large samples of social behaviour.

6 *One-zero sampling (usually state)* Interval recording and time sampling account for 41 per cent of data collection methods (Kelly, 1977). Using a one-zero convention, as opposed to other forms of time-sampling (see Tyler, 1979), the occurrence or non-occurrence of particular behavioural acts are scored for each successive time interval (often a period between 10 and 30 seconds in length). An act is scored once, irrespective of the number of onsets within the interval or the amount of the interval that an activity occupied. From this a score, or Hansen frequency (Hansen, 1966), is determined, which is equivalent to the total number of intervals in which a particular act has occurred. Therefore what is recorded is not the 'true' frequency of each type of behaviour but the frequency of intervals that include any amount of time spent in that behaviour.

In fact, one-zero sampling has been the subject of much criticism (Altmann, 1974; Dunbar, 1976; Powell et al., 1977; Kraemer, 1979) as the convention tends to underestimate 'true' frequency of acts and overestimate time spent in an activity, although in certain circumstances this may give a better guide to the relative representation of a behaviour (Slater, 1978). It has been suggested that Hansen frequencies can be related to rates of performance and duration of an activity with a correction factor (Fienberg, 1972; Simpson and Simpson, 1977) but these conventions are based on unrealistic assumptions about the organization of the behaviour.

The advantage of using the one–zero sampling method is its simplicity and ease of use with a checklist (Hutt and Hutt, 1970) and its high reliability of measurement compared with other observational techniques (Sackett, 1978). Provided that the time interval is short compared to the duration of each behaviour, the score will approximate the 'true' frequency.

7 *Instantaneous and scan sampling (usually state)* The observer records an individual's current activity at pre-selected moments in time. When this instantaneous sampling method is used on groups, it is termed 'scan' sampling. Each individual is watched for a specified time and the observer focuses on each individual in turn. Once this 'scan' has been completed (for example, the members of a family), then the sequence is repeated. This technique can be used to obtain data from a large number of group members and information on synchrony may be obtained. The percentage of time that individuals devote to various activities can be estimated from the percentage of samples in which a given activity is recorded (Bekoff, 1979).

Associated equipment. During direct observation, some form of time marker is essential for many of the above sampling methods to indicate the beginning and end of pre-selected time periods or intervals. Many studies report the use of a stopwatch for this purpose, but this distracts the observer's attention away from the behaviour under investigation. Some authors accommodate this problem by recording every other time interval – a highly unsatisfactory compromise! A 'behavioural timer' should be used to mark the beginning and end of the time periods. This consists of a pocket-sized electronic device with an earpiece attachment. The electronic timing mechanism is pre-set for the interval required. After each successive time interval has elapsed the machine bleeps to indicate the end of the previous time period and the start of the next. This sound is only made audible to the observer through the small earpiece.

New developments in data collection

The direct observation of behaviour has always been seen to be labour intensive, both in data collection and analysis. However, with the recent development in battery-powered portable microcomputers there is the potential to automate the direct recording of behavioural events both for laboratory and field work. They can provide an effortless way of obtaining a complete record of behaviour, which was previously possible only by the use of cine-film or video. They may be equipped with a built-in printer, microcassette drive, LCD screen and a full-size keyboard, yet still be compact enough to rest easily on the observer's lap.

For example, Browne and Madeley (1985) have developed a software package for use with the portable Epson HX20, which affords the ability to record rapidly pre-defined behaviours with minimum keyboard input. For each behaviour key pressed the sequence and elapsed time from start are automatically recorded up to seven times per second. This can be done for single subject observations, for multiple subject observations and for interactions between subjects observed in their natural environment. The recorded data may be immediately printed out and stored on microcassette for later transfer to a larger computer. The package also provides facilities to create, edit and store permanent libraries of behaviours and subjects for use in subsequent observations.

For some time software packages have been available for the collection and analysis of observation data with laboratory-based microcomputers, such as the Apple II (Flowers, 1982). In addition, programs have been written for the analysis of verbal interaction (Hargrove and Martin, 1982), and the calculation of inter-observer reliability statistics (Burns and Cavallaro, 1982). These new developments make traditional and comparatively expensive 'event recorders' (for example, Datamite) obsolete.

ANALYSIS AND TREATMENT OF OBSERVATIONAL DATA

Several types of analysis can be applied to observational data based on a behavioural catalogue of items.

Frequency and bout analysis

Proportional frequency scores. These are necessary to take account of varying lengths of observation sample time. To compare every parent, each behaviour item can be calculated as a proportion of the total number of time interval units per episode e.g. 'feeding offspring' might equal 18 units of a total of 36 observed units, and would receive a score of 0.50 for that sample.

Interaction frequencies. For every parent and offspring the total number of time units spent in interaction can also be calculated as a proportion of the total number of observed time units in the sample. The number of (1) initiating and (2) responsive interactive behaviours should then be determined and calculated as a proportion of the interaction score for the observation time.

Types of interaction may be identified in the following manner. If a general interactive initiative is shown (that is, behaviour directed toward another), it can have one of three possible outcomes: first, it may result in the respondent reacting with another interactive initiative (mutual interaction); second, it may result in the respondent reacting with a non-interactive behaviour (causal interaction); third, it may receive no reaction at all (failed interaction). For example, if the parent is eating and the offspring reaches for the food (interactive initiative), the parent may give the offspring some food (mutual interaction); stop eating and attend to the offspring (causal interaction); or continue to eat (failed interaction).

The interaction frequencies may be calculated globally and in relation to specific behaviour items. For example, vocal – visual; tactile – visual; tactile – vocal.

Bout analysis. The length of an 'interaction bout' can be established by summing the number of time-interval units that occur in sequence, containing an ongoing interactive behaviour between parent and offspring. The start and finish of such a sequence can be determined by periods of non-interactive behaviour. It is usually best to choose bout criteria which lead to few within and between bout intervals being assigned to the wrong category (Slater and Lester, 1982).

Evidence shows that infants play an important role in initiating, monitoring, differentiating and terminating bouts of interaction. Indeed, Harper (1977) claims that few significant exchanges between mammalian care-givers and their young will fail to involve reciprocal stimulation. Furthermore, analysis of mother–infant interaction data by 'time-series analysis' (Gottman et al., 1982) has demonstrated that the interaction becomes more complex with age by adding faster cycles of reciprocal behaviour as the infant becomes more responsive.

The analysis of interaction can also indicate pathology, for example, in humans, abusing mothers have significantly fewer mutual interactions with their infants compared to non-abusing mothers who show less failed interaction (Hyman et al., 1979). In addition, abusing mothers show shorter sequences of uninterrupted interaction (interaction bout) with their infants than non-abusers (Browne, 1982).

Sequential analysis

Recent studies on human parent–child relationships (for example, Martin et al., 1981; Cohn and Tronick, 1982; Dowdney et al., 1984; Phelps and Slater, 1985) can be distinguished by the emphasis on temporal relationships and contingencies of behaviour, in interactive situations. This novel approach to human study is a direct result of its success in animal investigations. Dyadic interactions are based on the interweaving of the participants' behavioural flow as time passes, and it

Methods and Approaches to the Study of Parenting

has been stressed that sequential analysis is a powerful method for analysing this behavioural flow.

Slater (1973) states that the simplest type of sequencing of events is a 'deterministic sequence', in which events always follow each other in a fixed order, so that the nature of the preceding act defines precisely the nature of that which will follow. But most behavioural sequences are 'probabilistic' rather than 'deterministic' in form, meaning that while the probability of a given act depends on the sequence of those preceding it, it is not possible to predict at a particular point exactly which behaviour will follow. If the sequences are highly ordered they are usually referred to as 'chained responses'. In these cases, the probability of a particular event is markedly altered by the event immediately before it. If the sequences are not so highly ordered, some transitions may be observed between almost every behaviour and every other, and only those transitions which have a high probability of occurrence are then useful.

In the parent–offspring case, sequential analysis can be applied initially by determining sequences within the individual. For example, if a parent shows act A it could be interpreted as an indication that activity B may follow, if the parent shows B after A more frequently than any other behaviour. However, the sequences within the parent may also depend on what the infant does, and activity B might follow A, only if the infant shows activity C. Therefore, the determination of inter-individual sequences is important.

Thus, sequential analysis can yield three types of data: (a) sequential flow of behaviour for parent; (b) sequential flow of behaviour for infant; and (c) sequential flow in interactive behaviour between parent and infant (that is, interactive sequences, Castellan, 1979). For example:

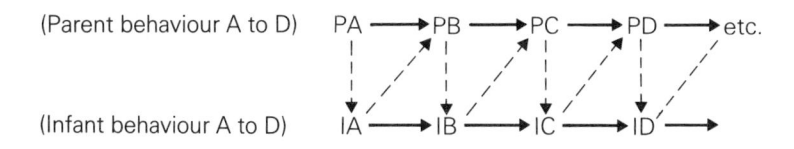

(Parent behaviour A to D)   PA ⟶ PB ⟶ PC ⟶ PD ⟶ etc.

(Infant behaviour A to D)   IA ⟶ IB ⟶ IC ⟶ ID' ⟶

The first stage in such an analysis is the cross-tabulation of the number of times each behaviour item precedes or follows every other behaviour item, which produces a 'transition matrix' or 'contingency table' (see figure 1).

The sequential analysis of each transition matrix, appropriate to describe the sequence of behaviours A → B → C → D, can be based on conditional probabilities where the sequence of each behaviour depends only on the immediately preceding act and not any earlier one, and this is termed a 'first order chain' of behaviour (Cane, 1978). But if

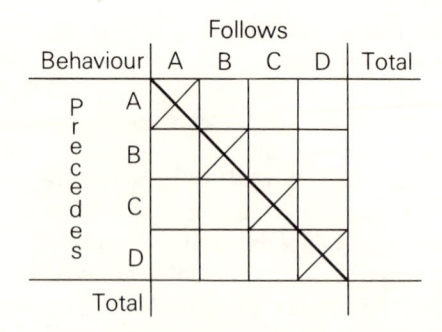

Figure 1 Transition matrix cross-tabulating sequential behaviour items A to D

the 'first order' model is inadequate to account for the structure of the behaviour and the probability of C depends on A as well as B in a given sequence A → B → C, then second and third order analysis can be applied. This is achieved by the use of Markov chain analysis (for example, Altmann, 1965). However, because of the underlying assumptions, behavioural data rarely meet the requirements of this type of analysis. One of its main limitations is the requirement of stationarity, such that when behaviours are not governed by a 'steady state' due to changing causal factors, they will be unsuitable for vigorous Markov chain analysis (Slater, 1973). In addition, there is the problem of 'repertoire size', that is, the number of behaviour items being observed. Fagen and Young (1978) estimated that the minimum sample size needed for a first order sequential analysis is $5R^2$, where R is the repertoire size within the matrix.

There are a number of ways in which single cells in a transition matrix can be analysed to detect significant relationships without the same rigid statistical prerequisites (Bakeman, 1978). For example, one analysis which can be applied to parent-infant data is derived from the chi-squared and binomial test (Siegel, 1956) and has been previously used by Stevenson and Poole (1976) to determine first order relationships between behaviours in the common marmoset.

To identify the 'first order' dependencies, a matrix of observed transition frequencies can be compared with that which would be expected if all acts were independent of one another. The expected values are calculated as for a contingency table using chi-squared. (The use of a χ^2 test on such data has been validated by Bartlett, 1951.) The null hypothesis is that the expected value for each transition (E_{ab}) is based on the chance association of the two behaviour patterns calculated from their own frequency of occurrence, weighted to take account of the impossibility of a behaviour pattern following itself. That is:

$$E_{ab} = \frac{f_a \times f_b}{N_2 - f_a}$$

where

f_a = frequency of bouts of behaviour A
f_b = frequency of bouts of behaviour B
N_2 = total number of pairs of bouts.

A chi-squared value[1] is calculated from the observed and expected matrices, according to the following formula:

$$\text{chi-squared} = \chi^2 = \sum_{\substack{ab \\ a \neq b}} \frac{(O_{ab} - E_{ab})^2}{E_{ab}}$$

where

E_{ab} = expected transition frequency for behaviour A to behaviour B
O_{ab} = observed transition frequency for A to B
a = preceding behaviour
b = following behaviour
$a \neq b$ = diagonals (indicating the number of times a behaviour followed itself) are excluded.

If the difference between the observed and expected matrices is found to be significant, the null hypothesis that the behaviour consists of a sequence of independent acts can be rejected.

The discrepancy between observed and expected values in individual cells of the transition matrix can then be serially examined, by a condensation of the whole matrix into a 2×2 table about the cell of interest followed by a binomial test to detect whether that particular transition is more frequent than expected. Following Poole's (1978) methods for the study of polecats, sequences of behaviour can be analysed by calculating expected values for transition frequencies based upon the frequency of occurrence of the two behaviour patterns being considered. The binomial test (see Siegel, 1956) can be applied to these data using the following formula to calculate the normal deviate (z). For behaviour B following A:

$$z = \frac{(x \pm 0.5) - \frac{AB}{T-A}}{\sqrt{\left[\left(\frac{AB}{T-A}\right)\left(1 - \frac{AB}{T(T-A)}\right)\right]}}$$

---

[1] The chi-squared value should be regarded as a goodness-of-fit test statistic which will be approximately χ^2 with $(t^2 - 3t + 1)$ degrees of freedom, where t is the total number of different behaviours (Chatfield and Lemon 1970).

where

z = normal deviate
x = observed frequency with which behaviour B follows A
A = number of occurrences of A
B = number of occurrences of B
T = total number of transitions between behaviour patterns.

Therefore, associations between acts which are commoner than expected by chance can be extracted, and a flow diagram constructed of 'chained' sequences of behaviour.

A non-statistical approach to the sequential analysis of behaviour is the 'Patterns of Pattern' algorithm method (described in Dawkins, 1976b). The rows of raw data are scanned for patterns of behaviour that actually occur in sequence, and for patterns of patterns. The frequency of various sequential patterns that occur can then be noted. This method can be used to check the flow diagrams produced by the binomial analysis of transition matrix data. The limitations of this approach as a primary form of analysis have been discussed by Douglas and Tweed (1979).

Cluster analysis

Eibl-Eibesfeldt (1975) gives evidence for the fact that behaviour patterns tend to occur clustered in time, the clusters constituting functionally related groups. Formally, a cluster analysis begins with a matrix of preconceived distances between elements and proceeds to group elements together with high similarity (Everitt, 1974). None the less, a more interesting approach has been developed, which highlights the hierarchical organization of behaviour. Dawkins (1967b) outlines a hierarchical cluster analysis based on the 'mutual replaceability' of behaviour patterns rather than any form of temporal proximity. It is not expected that clusters based on sequential or temporal analysis would give similar results, as two acts mutually replaceable would probably not be sequentially close.

Mutual replaceability cluster analysis. This analysis examines in turn all possible pairs of behaviour acts within a transition matrix (see p. 354). The pair with the highest correlation, in terms of their mutual replaceability, is found and designated as members of the same cluster. The matrix table is then collapsed so that no further distinction can be made between the two behaviour items and the entries are lumped by simple addition. The whole operation is then repeated on the condensed table, and continues until only two entries are left or no good correlations can be found. At each stage the pair of elements with the highest correlation is determined, sometimes as single behaviour items; sometimes as already lumped clusters of behaviour items. This produces

Methods and Approaches to the Study of Parenting 359

a branched hierarchy of behavioural acts, which forms a tree-like diagram termed a 'dendrogram'. Dawkins (1976b) argues that by producing a 'dendrogram' based on mutual replaceability, a hierarchy of decisions made to reach a certain behaviour can be established. This has obvious advantages for the study of parental behaviour.

For sociomatrices, where the rows and columns of a $N \times N$ matrix represent individuals rather than behavioural acts, hierarchical cluster analysis can be used to identify close relationships within a social group. This may be based on observations of a particular behaviour or interaction (for example, care-giving) or of time spent together. Thus, pairs of individuals most closely associated in a chosen activity are indentified (for example, mother and infant). Then further individuals associated closely with either of the pair are added to the cluster (for example, father), and so on. Dendrograms, can again be produced to represent diagramatically, for example, parental relationships within a large social group. An application of this method, to show the relationships between 39 chimpanzees in an African reserve, may be found in Hinde (1974, p. 374).

CLINICAL APPLICATIONS OF ETHOLOGICAL METHODS

The above analytical models, when applied to parent–infant interaction, may provide an insight into the processes involved in the formation of different interactive styles shown between normal and pathological groups. For example, by comparing the sequential behaviour of abusing and non-abusing mother–infant pairs it might be possible to detect exactly where the communication between an abusing mother and her infant is breaking down (Browne and Parr, 1980). To test this hypothesis, Browne (1982) used video data from 23 abusing and 23 'matched' non-abusing mother–infant pairs to analyse sequentially 3 minute recordings of mother–infant interaction, observed while they were alone in a strange room.

For the mothers' sequential behaviour, flow diagrams were produced to illustrate the main temporal relationships between different behaviour items at a 1 per cent level of significance (see figure 2). It is evident that the order of occurrence of certain behaviour items are extremely similar and, indeed, no significant differences in the frequency of these acts could be found between the two groups. However, there seems to be a difference in alternatives for the maternal sequence. The non-abusing mothers seem to be more aware of the infant's interests in toys and less concerned with the infants themselves. This suggests that they have a less coercive style of interaction in comparison to abusing mothers who act intrusively as indicated by previous work (Hyman et al., 1979).

For the infant's sequential behaviour, flow diagrams were also produced to illustrate the main temporal relationships between different

a Abusing mother

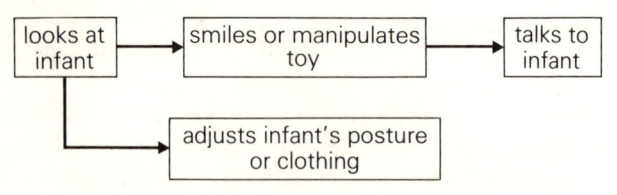

```
looks at  →  smiles or manipulates  →  talks to
infant       toy                        infant
    │
    └──→  adjusts infant's posture
          or clothing
```

b Non-abusing mother

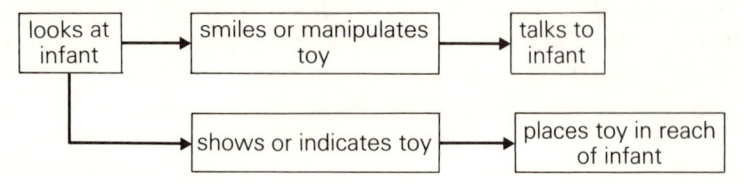

```
looks at  →  smiles or manipulates  →  talks to
infant       toy                        infant
    │
    └──→  shows or indicates toy  →  places toy in reach
                                     of infant
```

Figure 2 Main temporal relationships between different behaviour items of mothers ($z > 2.34$, $p < 0.01$) (Browne and Saqi, forthcoming)

behaviour items, again at a one per cent level of significance (see figure 3). It is evident from these results that the abused infants are ambivalent – as predicted by previous work (Hyman et al., 1979; Crittenden, 1985). They mix distress behaviours, such as crying or fretting with toy-related actions, while non-abused infants have two distinct chains of behaviour which do not significantly link up: one chain consisting of toy-related behaviours and the other exclusively dealing with distress, thus causing less confusion during mother–infant interaction. The non-abused infant's behaviour sequences also contain more interactive initiatives.

Interactive sequences. Finally, Browne (1982) carried out an analysis of data which highlighted interactive sequences between the infant and the mother in abusing and non-abusing dyads. Figure 4 demonstrates the lack of sensitivity shown by abusing mothers. The author asked the question 'What leads a mother to obtrusively adjust her infant's clothing or posture?' Non-abusing mothers significantly responded only to the signal of frets from their infants, which resulted in a change of behaviour if they successfully adjusted the right thing. In contrast, abusing mothers significantly responded to a wide range of behaviours with adjustment, none of which indicated the infant's distress. Ironically, the abusing mothers sometimes induced distress in the forms of struggle and frets from their infants, as a result of their untimely interventions. Indeed, on the majority of occasions the abusing mothers interrupted exploratory behaviour in their infant. If this became a habit, it would have serious effects for the infant's development.

Methods and Approaches to the Study of Parenting 361

a Abused infant

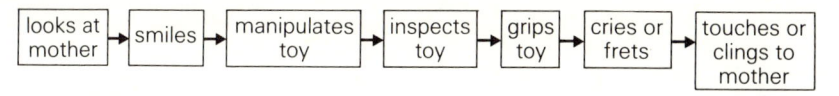

looks at mother → smiles → manipulates toy → inspects toy → grips toy → cries or frets → touches or clings to mother

b Non-abused infant

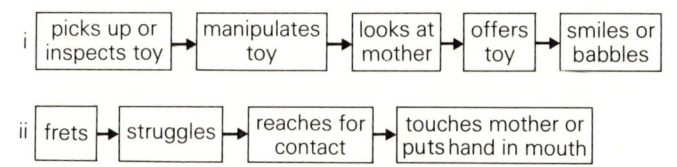

i picks up or inspects toy → manipulates toy → looks at mother → offers toy → smiles or babbles

ii frets → struggles → reaches for contact → touches mother or puts hand in mouth

Figure 3 Main temporal relationships between different behaviour items of infants ($z > 2.34$, $p < 0.01$) (Brown and Saqi, forthcoming)

ABUSING (N = 23)   NON-ABUSING (N=23)

Infant   Mother   Infant   Infant   Mother   Infant

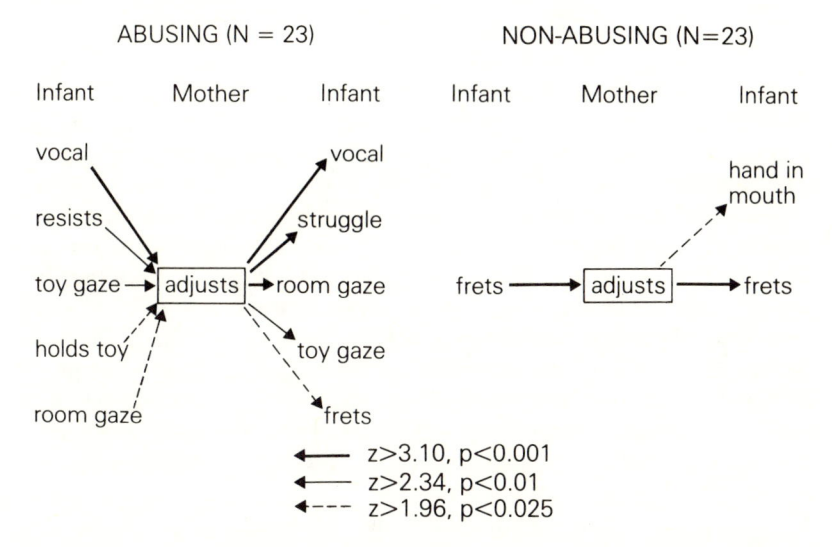

vocal   ↗ vocal   hand in
resists ↘   mouth
toy gaze → adjusts → room gaze   frets → adjusts → frets
holds toy ↗   ↘ toy gaze
room gaze ↗   ↘ frets

← z>3.10, p<0.001
← z>2.34, p<0.01
←--- z>1.96, p<0.025

Figure 4 Infant-to-mother interaction (1) (Browne and Saqi, forthcoming)

Figure 5 demonstrates the ambivalence of abused infants, while in their mother's arms. Whereas normal infants visually explore the environment in such a secure situation, abused infants cry and fret and passively hold on to a toy or their mother. Overall, these findings are in agreement with the predictions made from 'attachment theory' (Bowlby, 1984), and support the conditioning model put forward by Sluckin in chapter 7.

The infant, given the pre-adaptations and cognitive means for interactive development, requires the opportunity for interactions. Therefore, the importance of a turn-taking pattern of parent–infant interaction has

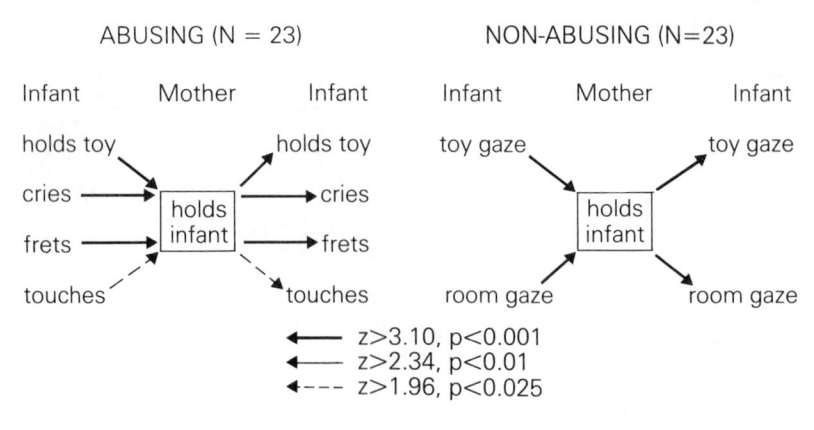

```
                ABUSING (N = 23)                    NON-ABUSING (N=23)

     Infant      Mother      Infant          Infant      Mother      Infant
     holds toy            holds toy          toy gaze               toy gaze
     cries    →  ┌──────┐  → cries                     ┌──────┐
                 │holds │                              │holds │
     frets    →  │infant│  → frets                     │infant│
                 └──────┘                              └──────┘
     touches              touches            room gaze              room gaze
                      ←──── z>3.10, p<0.001
                      ←──── z>2.34, p<0.01
                      ←---- z>1.96, p<0.025
```

Figure 5 Infant-to-mother interaction (2) (Browne and Saqi, forthcoming)

been stressed by Schaffer and Collis in chapter 10. The parents must allow themselves to be paced by the infant and build up routine sequences of a predictable nature to maximize the opportunity for the infant to learn.

The development lag of abused infants and the fewer interactive responses shown by abusing mothers suggests that this process has been distorted (Hyman et al. 1979). This is evident as suggested from the findings of the sequential analysis, which shows that abusing mothers interrupt their routine sequences with coercive, non-interactive behaviour, such as adjusting the infant's posture or clothing. This inhibits reciprocal interactive behaviour and cuts the 'interaction bout' short. Non-abusing mothers interrupt their routine sequences with toy-related behaviours that facilitate further interaction and this produces a secure interactive relationship. Therefore, when a non-abused infant is distressed, he does not change his style of interaction but continues to use behaviours signalling distress until a response is forthcoming. In contrast, the abused infant switches from toy-related action to behaviours indicating distress in an ambivalent fashion, being unsure of a response.

Thus, clinical recommendations can be made on the basis of ethological methodology. From the foregoing it may be suggested that those involved in the treatment of abusive families should be concerned with the development of an interactive relationship between parent and infant. It is not sufficient to evaluate treatment programmes on the basis of the occurrence or non-occurrence of subsequent abuse. Training parents to inhibit aggressive behaviour towards their offspring may still leave the harmful context in which the initial abuse occurred quite unchanged. It should also be recognized that in some cases treatment with the infant remaining in the family may be ineffective and may, in fact, serve to perpetuate the 'abused becoming abuser' cycle (Browne and Parr, 1980).

I consider that ethology can make its most significant contribution to the study of parenting and its problems by providing methods of research and analytical techniques for the investigation of social communication. As Herbert (chapter 11) points out, much work remains to be done in the field of pathological parenting and the application of ethological theory and methods will give help by complementing, rather than displacing, the more established psychological and sociological approaches.

ESTABLISHED APPROACHES TO THE STUDY OF HUMAN PARENTING

The ethological approach emphasizes the importance of description as a basis for the analysis of behaviour. Indeed, a considerable amount of attention has already been paid to such techniques. Nevertheless, this purely behavioural approach is unlikely to be adequate for studies of human parenting; other techniques are often necessary to elucidate the meaning of behaviour. There are four main approaches to data collection concerned with parent–child behaviour and interaction (Hinde, 1982) of which two have already been considered from an ethological perspective.

1. Direct permanent recording – A permanent and complete record of behaviour obtained on film or videotape.
2. Direct observation – An observer records pre-selected aspects of behaviour by use of a checklist, event recorder or commentary spoken into a tape-recorder.
3. Indirect reports – Data obtained from interviews and/or questionnaires given to individuals that spend a great deal of time with the family.
4. Self reports – Interviews and/or questionnaires given to the actual parents or children in the study, or given to an individual (case study). Indeed, one of the commonest methods of investigating child-rearing styles has been to interview the parents.

Interviews, questionnaires and surveys

There are four issues to be considered with the interview method of data collection: interview structure; questionnaire construction; interview strategy; and technique. Nachmias (1976) has reviewed the structure of questionnaires in some detail. In sum, he classifies interviews into three basic forms: schedule-structured, nonschedule-structured ('focused') and non-structured ('nondirective' or 'in depth') interviews. The construction of the questionnaire concerns the wording of the questions, the choice between open-ended and fixed alternative questions, how narrow or broad the questions are and what their sequence is. These points have been considered by Kahn and Cannell (1963) and Gordon (1969). Interview strategy and technique is dictated by the circumstances in which the interviewer is operating. For example, how to open, control and pace the interview to elicit the most information (Gordon, 1969).

The alternative is to use a questionnaire approach without an interview, but this has its drawbacks. There is often a poor return (typically 35 to 40 per cent) when questionnaires are sent by post, even if arrangements are made to collect them. The survey method is usually inflexible with respect to the question format and this can lead to misleading and false answers. Therefore the interview method is mostly favoured for studies of parenting (for example, Newson and Newson, 1980).

Participant observation and subject reactivity

Living with a family in order to observe parental behaviour is another possible approach, derived from anthropology. However, this method, including the use of interviews, is widely criticized on the grounds that it is unscientific, that is, intuitive, subjective and not verifiable.

To be more specific, when carrying out research and giving explanations to subjects the interviewer should be blind to the status of the family in the research design (that is, experimental or control). The interviewer should not be involved in any way with the family being interviewed or observed. Non-involvement essentially allows the respondents to feel that they can impart a confidence without any consequences, and behave naturally. Confidentiality and, if possible, anonymity has to be convincingly assured. The interviewer should not have a perceived role in relation to judgement or to the use of information to be obtained. The parent should be regarded as someone who has useful experience to share. Thus, the interviewer should adopt the impartial role of listener/observer and not offer advice or suggestions. The parent must feel free to talk and behave however he or she wishes and this comes partly from the knowledge that she is not being judged and need not be defensive.

Lytton (1973) and Moss (1964) have both pointed out in their home studies that the introduction of an observer into the home produces some distortion of the 'normal' mother–infant interaction, and both stressed the need to under-emphasize the role of the mother as a subject. Moss (1964) commented that mothers showed few signs of nervousness and distraction when told that the primary interest was in the child. He also suggested that the interviewer/observer should avoid being cast as an 'expert in child-rearing'. A full review and critical analysis of subject reactivity during observational assessment has been published quite recently (Harris and Lahey, 1982a) together with a review of factors that cause bias and inaccuracy in recording and coding of behaviour (Harris and Lahey, 1982b).

Multi-method designs

For both indirect and self-reports, above, it should be noted that reported events may not correspond to actual ones. Nevertheless, how a respondent thinks or feels may be crucial. For example, Browne and Saqi (1986) have used a questionnaire, for abusing and non-abusing parents to rate their child's pattern of sleeping, eating, activity, controlability and interaction. The responses to the questionnaire demonstrated that abusing parents have more negative conceptions about their child's behaviour than non-abusing parents. It was concluded that abusing parents have unrealistic expectations of their children, which may partly explain their observed intrusive and insensitive style of interaction.

Thus, when time and finances permit, a project ideally should contain data collected using all four of the approaches previously outlined. This multi-method technique allows interview/questionnaire data to be validated against directly observed behaviour and vice versa. Indeed, it is essential for the complete assessment of dimensions in the parent–child relationship, especially if one considers that a substantial amount of parental behaviour and care-giving responses are under minimal conscious control (Papousek and Papousek, 1983). Hence, being unaware of their actions, the parent cannot report on them in a questionnaire.

THE DESCRIPTION OF RELATIONSHIPS

Hinde (1979) suggests that a relationship can be described in the following terms:

1 *Content and diversity of interactions* What the partners do together can distinguish the different types of relationship (for example, father–son, mother–daughter etc.), and emphasize behaviour related to sex roles.

2 *Quality of interactions* A parent may handle a child roughly or tenderly – hence how the partners interact is important. The quality of interaction may be measured by the intensity of movement (for example, speed, amplitude), the associated verbal and non-verbal communications, the degree of co-ordination (meshing) between partners and the relative frequency and sequencing of their interaction. The type, promptness and consistency of reward and punishment within their relationship is also important (see chapter 10).

3 *Reciprocity of interactions* Partners doing the same thing simultaneously or in turn.

4 *Complementary interactions* Partners doing different but related things. The appropriateness of their actions.

5 *Intimate interactions* Participants feeling at ease and able to reveal themselves to each other in an uninhibited manner, especially in their display of love and affection.

6 *Inter-personal perception during interactions* This is concerned with the perceptions participants in a relationship have of each other. For example, whether the respondent is seen to resemble, understand, empathize, be satisfied with and be close to their partner.

7 *Commitment* To what extent the participants accept their relationship and wish to continue their involvement. Perceived commitment may be just as important as commitment itself, especially in terms of paternal support. Commitment is also related to permissiveness and control of the partner's behaviour, and governs the 'rules' of the relationship.

8 *The context of the interactions* The importance of context for communication is concerned with the idea that *message plus context equals meaning* (Smith, 1977). Thus the same behavioural act or signal may be perceived differently in various contexts and receive a variety of responses.

Characteristics of affectionate relationships and emotional attachments

Hinde (1979) considers the characteristics of an 'affectionate relationship' as having the following features: (1) The partners engage in different types of interactions; (2) in the absence of the other, the partner attempts to restore proximity, real or imagined; (3) actions conducive to the welfare of the other are likely to be repeated, especially with respect to the parent; (4) the behaviour of each partner is organized in relation to the ongoing behaviour of the other; (5) the presence of the partner alleviates anxiety induced by strange objects, persons or situations; and (6) the partners will be uninhibited by each other and will wish to display love and affection to one another.

As mentioned by Schaffer and Collis in chapter 10, human infants are presumed by Bowlby (1969) to be predisposed to form specific emotional attachments during a relatively limited period of their infancy. It is now generally agreed that the period of attachment formation for the majority of children extends from about the age of six months to one year. As the principal caretaker, the mother is most frequently the primary object of the child's attachment, although this is by no means always the case. This attachment is presumed to develop as a result of attachment behaviours (crying, smiling, tracking, following, clinging) which promote proximity to, or contact with the mother. Exploratory behaviours are seen as reciprocal to attachment behaviours in that they involve loss of contact with the reduction of proximity to the mother. Bowlby's theoretical perspective has been adopted by Ainsworth and her associates who have taken an experimental approach to the study of attachment and exploratory behaviour.

Experimental approaches to the study of attachment

Schaffer and Collis in chapter 10 Lewis in chapter 8 noted that Ainsworth and Witting (1969) have developed a standardized strange-situation procedure for directly observing individual differences in attachment and exploratory behaviours shown by one-year old infants. This involves the infants being placed in a strange environment and confronted with a stranger in both the presence and absence of the parent. This experimental approach has become the most frequently cited method for the assessment of infant to parent attachment, although concern has been expressed for its diagnostic use (see chapter 10).

The strange-situation procedure. This procedure involves two brief separations between a mother and child, in order to assess the different types of their attachment. Anxiety may be increased in the child by the presence of an adult female stranger in three of the seven episodes. The episodes are each three minutes in duration and are continuously recorded on videotape through a one-way mirror in the following sequential sequence:

Introduction	The observer introduces mother and child to the experimental room and then leaves.
Episode 1	Mother and child present. The child explores while the mother sits in her chair.
Episode 2	Mother, child and stranger present. The stranger speaks with the mother and then interacts with the child.
Episode 3	Child and stranger present. The mother leaves and the stranger interacts with the child, comforting him/her if necessary.
Episode 4	Mother and child present. The mother re-enters and the stranger leaves, the mother responds to her child's reunion behaviour and re-interests the child in toys, she then returns to her chair.
Episode 5	Child present. The mother leaves and the child is left alone.
Episode 6	Child and stranger present. The stranger re-enters and interacts with the child, comforting and consoling him/her if necessary.
Episode 7	Mother and child present. The mother re-enters and the stranger leaves, the mother responds to her child's reunion behaviour, comforting him/her if necessary.

Using this procedure Ainsworth et al. (1978) have described three broad categories of response to the strange situation. The normative infants, Group B, may show distress particularly when left alone, but show positive responses to reunion with the mother and generally exhibit a desire for interaction with her in preference to the stranger. Group A infants on the other hand tend to show high levels of exploratory behaviour throughout the strange-situation procedure and tend not to seek interaction with or proximity to the mother, frequently avoiding her contact on reunion. The third category of response involves ambivalent behaviour, the infants in Group C show elements of both approach and avoidance behaviour towards the mother on her return. Group C children also show low levels of exploratory behaviour throughout and are more prone to crying.

In describing early relationships, Ainsworth and her colleagues warn against the use of a simple distinction between attached and non-attached infants and stress that we should consider the quality of attachment in terms of security.

Ainsworth et al. (1978) have examined the relationship between the infant's response to the strange situation procedure and the behaviour of both mother and child in the home environment. Their findings suggest a dimension of maternal sensitivity as being most influential in affecting the child's reactions. In the homes of the securely attached infants (Group B) sensitive mothering was exhibited to the infant's behaviour. While anxiously attached, avoidant infants (Group A) were found to be rejected by the mothers in terms of interaction and it was suggested that the enhanced exploratory behaviours shown by these infants were an attempt to block attachment behaviours which had been rejected in the past. In the home environments of the insecurely attached ambivalent infants (Group C) a disharmonious mother–infant relationship was evident and the ambivalent behaviours shown were seen as a result of inconsistent parenting. A summary of their conclusions with regard to parenting and infant attachment is presented in figure 6.

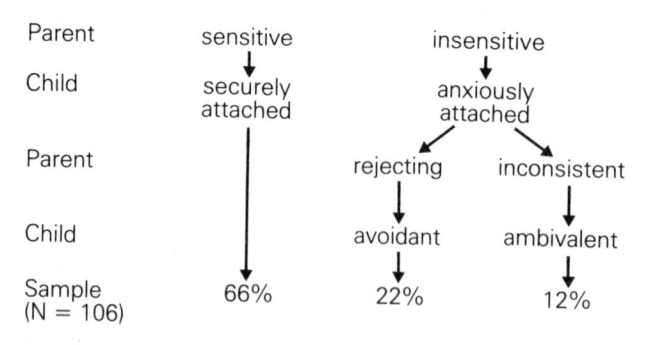

Parent	sensitive	insensitive	
	↓	↓	
Child	securely attached	anxiously attached	
Parent		rejecting ↙	↘ inconsistent
		↓	↓
Child		avoidant	ambivalent
		↓	↓
Sample (N = 106)	66%	22%	12%

Figure 6 Patterns of attachment (Ainsworth et al., 1978)

Maccoby (1980) concludes from the above findings that the parents' contribution to attachment can be identified within four dimensions of caretaking syle:

1 *Sensitivity/Insensitivity* The sensitive parent 'meshes' her responses to the infant's signals and communications to form a cyclic turn-taking pattern of interaction. In contrast, the insensitive parent intervenes arbitrarily, and these intrusions reflect her own wishes and mood.

2 *Acceptance/Rejection* The accepting parent accepts in general the responsibility of child care. She shows few signs of irritation with the child. However, the rejecting parent has feelings of anger and resentment that eclipse her affection for the child. She often finds the child irritating and resorts to punitive control.

3 *Co-operation/Interference* The co-operative parent respects the child's autonomy and rarely exerts direct control. The interfering parent imposes her wishes on the child with little concern for the child's current mood or activity.

4 *Accessibility/Ignoring* The accessible parent is familiar with her child's communications and notices them at some distance, hence she is easily distracted by her child. The ignoring parent is preoccupied with her own activities and thoughts. She often fails to notice the child's communications unless they are obvious through intensification. She may even forget about the child outside the scheduled times for caretaking.

As Maccoby (1980) points out, these four dimensions may be interrelated with individuals showing some aspects of all the dimensions, and this itself may be inconsistent. In a similar way Raphael-Heff (1986) on the basis of her 'clinical experience, mother–child observations and survey data', classifies parental practices into two orientations: The *facilitator* who responds spontaneously to her infant's needs as they arise, and the *regulator* who establishes a routine to foster predictability in her infant. Raphael-Leff claims that these predispositions are established in pregnancy and early parenthood.

Some studies of human parenting have taken a purely naturalistic approach and only observed behaviour in an unstructured home setting (for example, Clark-Stewart, 1973; Lytton, 1980). The findings support the conclusions of Ainsworth's experimental approach and to a certain extent validate her descriptions of infant to parent attachments and its associated interactive processes.

THE BROADER CONTEXT OF HUMAN PARENTING

In chapter 9 Shaw commented on the fact that parenting is not an unvarying phenomenon conducted in a social vacuum. Even if one considers the parental competence to be 'intuitive' with a large biological component (Papousek and Papousek, 1983), certain stress factors in the environment may serve to elicit and maintain pathological patterns of parenting. Societal factors such as unemployment, social isolation and poor housing, can have strong influences on parental behaviour. With respect to these broader issues, Belsky (1984) has developed an ecological perspective to parental functioning. He identifies three determinants of parenting: the personal psychological resources of the parents; the individual characteristics of the child; and the contextual sources of stress and support. His 'process model' presumes that parents' personal resources are most effective in buffering the parent–child relationship from stress, followed by social support systems and then by the child's characteristics. Belsky's (1984) essay on socialization and individual differences in parenting, partly explains why it is only the minority of parents under stress who develop pathological patterns of caretaking.

A parent's personal resources obviously depend on their own personality development and previous experiences. Some authors (Rapoport and Rapoport, 1980, for example) emphasize this fact and take a lifespan approach to the study of parenting. They draw a distinction between individuals who postpone parenthood and people who have their children very young. Furthermore, supporters of the life-span approach hold that parental behaviour will alter during early, mid- and late establishment phases of the family. Certainly parents need to adopt different coping strategies, for stress factors and social support systems change during parenthood as a result of the children growing older.

Recently, a parent questionnaire has been developed to identify parent–child relationships currently under stress and 'at-risk' of dysfunction. This 'parenting stress index' (Abidin, 1983), is a screening instrument that provides scores related to the parent's sense of attachment, competence, social isolation, relationship with spouse, mental and physical health. In addition to an assessment of life stress events, it also provides scores on the child's demands, mood, activity, adaptability and acceptability, as perceived by the parent. In future, it will be of interest to see whether observational research can establish an association between styles of parent–child interaction and high or low ratings on the 'parenting stress index'.

CONCLUSIONS

With reference to the questions posed by Sluckin and Herbert in chapter 1, as to the relevance of parental behaviour in animals for the study of human parenting, it should be clear from the foregoing that a view of the comparative approach which holds that its only contribution to the study of human behaviour lies in naive generalizations from animals to man is inappropriate. For many years psychologists have used animals in studies where ethics prevents the use of human subjects. For example, Suomi's (1976) work demonstrates that motherless rhesus monkeys abuse and neglect their offspring, especially after long periods of social isolation. At first sight one might be tempted to refer to such studies as indicating a possible basis for the reported fact that battering parents often experience unsatisfactory or abusive relationships as children (Kempe and Kempe, 1978). However, severe deprivation, such as that imposed upon these rhesus monkeys, occurs in only the most extreme cases in humans. The importance of this work does not lie in its potentiality for supplying an experimental analogue of human child abuse, but in the effects that it has had upon our ideas about and approaches to the study of early social development in human infants. Animal studies do not supply simple models of human behaviour from which we may generalize with impunity, but may assist in the understanding of basic processes. Theoretical contributions serve to focus attention upon particular periods of development in the relationship of parent and child which should result in more effective prevention, identification and treatment of the problems of parenthood. Finally, the methods derived from ethology provide ways in which a fuller understanding of parent-child interaction may be developed, thereby allowing a more precise appreciation of the difficulties involved in a pathological relationship.

REFERENCES

Abidin, R. 1983: *Manual of the Parenting Stress Index (PSI)*. Charlottesville, Virginia: Pediatric Psychology Press.

Ainsworth, M. D. S., Blehar, M. C., Waters, E. and Wall, S. 1978: *Patterns of Attachment: A Psychological Study of the Strange Situation*. New Jersey: Lawrence Erlbaum Associates.

Ainsworth, M. D. S. and Wittig, B. A. 1969: Attachment and exploratory behaviour in one-year olds in a strange situation. In B. M. Foss (ed.) *Determinants of Infant Behaviour*, vol 4., London: Methuen.

Alexander, R. D. 1975: The search for a general theory of behaviour. *Behavioural Sciences*, 20, 77-100.

Altmann, J. 1974: Observational study of behaviour: sampling methods. *Behaviour*, 49, 227-67.

Altmann, S. A. 1965: Sociobiology of rhesus monkeys II. Stochastics of social communication. *Journal of Theoretical Biology*, 8, 490-522.

Bakeman, R. 1978: Untangling streams of behaviour. In G. P. Sackett (ed.) *Observing Behaviour Vol: II. Data Collection and Analysis Methods*, Baltimore: University Park Press.

Bartlett, M. S. 1951: The frequency goodness of fit test for probability chains. *Proceedings of the Cambridge Philosophical Society*, 47, 87-95.

Bekoff, M. 1979: Behavioural acts: description, classification, ethogram analysis and measurement. In R. B. Cairns (ed.), *The Analysis of Social Interactions: Methods, Issues and Illustrations*, New Jersey: Lawrence Erlbaum Associates, 67-80.

Belsky, J. 1984: The determinants of parenting: a process model. *Child Development*, 55, 83-96.

Bowlby, J. 1969: *Attachment and Loss. Vol. I 'Attachment'*. London: Hogarth.

—— 1984: Violence in the family as a disorder of the attachment and caregiving systems. *American Journal of Psychoanalysis*, 44, 1, 9-31.

Bronfenbrenner, U. 1979: *The Ecology of Human Development: Experiments by Nature and Design*. Cambridge, Mass.: Harvard University Press.

Browne, K. D. 1982: The sequential analysis of social behaviour patterns shown by mother and infant in abusing and non-abusing families. *Proceedings of the 10th International Congress of the Association of Child and Adolescent Psychiatry*, Dublin, 25-30 July, 1982.

Browne, K. D. and Parr, R. 1980: Contributions of an ethological approach to the study of abuse. In N. Frude, (ed.), *Psychological Approaches to Child Abuse*, London: Batsford Press, 83-99.

Browne, K. D. and Madeley, R. 1985: 'Ethogram' - an event recorder software package. *Journal of Child Psychology and Psychiatry*, 26, 6, Software Survey Section, P. III.

Browne, K. D. and Saqi, S. 1986: Maternal perceptions of child behaviour in abusing and non-abusing (high risk and low risk) families. *Proceedings of the British Psychological Society*, Annual Conference, Sheffield University, April 1986.

—— forthcoming: Parent-child interaction in abusing families and its possible causes and consequences. In P. Maher (ed.), *Child Abuse: The Educational Perspective*, Oxford: Basil Blackwell.

Burns, E. and Cavallaro, C. 1982: A computer program to determine inter-observer reliability statistics. *Behavioural Research Methods and Instruments*, 14, 1, 42.

Cane, V. R. 1978: On fitting low-order Markov chains to behaviour sequences. *Animal Behaviour*, 26, 332-8.

Castellan, N. J. 1979: The analysis of behaviour sequences. In R. B. Cairns, (ed.) *The Analysis of Social Interactions: Methods, Issues and Illustrations*, New Jersey: Lawrence Erlbaum Associates, 81-116.

Chatfield, C. and Lemon, R. E. 1970: Analysing sequences of behavioural events. *Journal of Theoretical Biology*, 29, 427-45.

Clarke-Stewart, K. 1973: Interactions between mothers and their young children: characteristics and consequences. *Monographs of the Society for Research in Child Development*, 38, 6 and 7.

Cohn, J. F. and Tronick, E. Z. 1982: Communicate rule and the sequential structure of infant behaviour during normal and depressed interaction. In E. Z. Tronick, (ed.) *Social Interchange in Infancy: Affect, Cognition and Communication*, Baltimore: University Park Press, 59-79.

Conger, A. J. 1985: Kappa reliabilities for continuous behaviours and events. *Educational and Psychological Measurement*, 45, 861-8.

Crittenden, P. M. 1985: Maltreated infants: vulnerability and resilience. *Journal of Child Psychology and Psychiatry*, 26, 1, 85-96.

Dawkins, R. 1976a: *The Selfish Gene*. London: Palladin.
—— 1976b: Hierarchical organisation: a candidate principle for ethology. In P. P. G. Bateson and R. A. Hinde (eds), *Growing Points in Ethology*, Cambridge: Cambridge University Press, 7–54.
Douglas, J. M. and Tweed, R. L. 1979: Analysing and patterning of a sequence of discrete behavioural events. *Animal Behaviour*, 27, 1236–52.
Dowdney, L., Mrazek, D., Quinton, D. and Rutter, M. 1984: Observation of parent–child interaction with two- to three-year olds. *Journal of Child Psychology and Psychiatry*, 25, 3, 379–407.
Drummond, H. 1981: The nature and description of behaviour patterns. In P. P. G. Bateson and P. H. Klopfer (eds), *Perspectives in Ethology*, vol. 4, New York: Plenum Press, 1–33.
Dunbar, R. I. M. 1976: Some aspects of research design and their implications in the observational study of behaviour. *Behaviour*, 58, 78–98.
Eibl-Eibesfeldt, I. 1975: *Ethology, the Biology of Behaviour*, (2nd edn). New York: Holt, Rinehart and Winston.
Everitt, B. 1974: *Cluster Analysis*, London: Heinemann.
Fagen, R. M. and Goldman, R. N. 1977: Behavioural catalogue analysis methods. *Animal Behaviour*, 25, 261–74.
Fagen, R. M. and Young, D. Y. 1978: Temporal patterns of behaviour: durations, intervals, latencies and sequences. In P. W. Colgan, (ed.), *Quantitative Ethology*, Toronto: John Wiley, 79–114.
Fienberg, S. 1972: On the use of Hansen frequencies for estimating rates of behaviour. *Primates*, 13, 323–6.
Flowers, J. H. 1982: Some simple Apple II software for the collection and analysis of observational data. *Behaviour Research Methods and Instruments*, 14, 2, 241–9.
Gordon, R. L. 1969: *Interviewing: Strategy, Techniques and Tactics*, Homewood, Ill.: Dorsey Press.
Gottman, J., Rose, F. T. and Mettelal, G. 1982: Time series analysis of social interaction data. In T. Field and A. Fogel (eds), *Emotion and Early Interaction*, New Jersey, Lawrence Erlbaum Associates.
Hansen, E. W. 1966: The development of maternal and infant behaviour in the rhesus monkey. *Animal Behaviour*, 27, 104–49.
Harper, V. 1977: Effects of the young on bouts of interaction. In R. Bell and V. Harper, *Child Effects on Adults*, New Jersey: Lawrence Erlbaum Associates.
Hartmann, C. P. 1977: Considerations in the choice of inter-observer reliability estimates. *Journal of Applied Behaviour Analysis*, 10, 103–16.
Hargrove, D. S. and Martin, T. A. 1982: Development of a microcomputer system for verbal interaction analysis. *Behaviour Research Methods and Instruments*, 14, 2, 236–9.
Harris, F. C. and Lahey, B. B. 1982a: Subject reactivity in direct observational assessment: a review and critical analysis. *Clinical Psychology Review*, 2, 523–38.
—— 1982b: Recording system bias in direct observational methodology. A review and critical analysis of factors causing inaccurate coding behaviour. *Clinical Psychology Review*, 2, 539–56.
Hinde, R. A. 1974: *The Biological Bases of Human Social Behaviour*. New York: McGraw-Hill Book Co.

—— 1975: The concept of function. In G. P. Baerends, C. Beer, and A. Manning, (eds), *Essays on Function and Evolution in Behaviour*, Oxford: Clarendon Press.

—— 1979: *Towards Understanding Relationships*. London: Academic Press.

—— 1982: Ethology and child development. In P. H. Mussen (ed.), *Handbook of Child Psychology*, vol 2, New York: John Wiley, 27-93.

Hollenbeck, A. R. 1978: Problems of reliability in observational research. In G. P. Sackett (ed.), *Observing Behaviour Vol. II: Data Collection and Analysis Methods*, Baltimore: University Park Press, 79-98.

Hutt, S. J. and Hutt, C. 1970: *Direct Observation and Measurement of Behaviour*. Springfield Illinois: C. C. Thomas.

Hyman, C. A., Parr, R. and Browne, K. D. 1979: An observational Study of mother-infant interaction in abusing families. *Child Abuse and Neglect*, 3, 1, 241-6.

Kahn, R. and Cannell, C. F. 1963: *The Dynamics of Interviewing*. New York: John Wiley.

Kaye, K. 1982: *The Mental and Social Life of Babies*. London: Harvester Press.

Kelly, M. B. 1977: A review of observational and data-collection and reliability procedures reported in this journal. *Journal of Applied Behaviour Analysis*, 10, 97-101.

Kempe, T. S. and Kempe, C. H. 1978: *Child Abuse*. London: Fontana/Open Books.

Kraemer, H. C. 1979: One-zero sampling in the study of primate behaviour. *Primates*, 20, 237-44.

Lewis, M. and Rosenblum, L. A. (eds) 1974: *The Effect of the Infant on its Caregiver*. New York: Interscience, John Wiley.

Lorenz, K. 1965: *Evolution and the Modification of Behaviour*. Chicago: University of Chicago Press.

Lytton, H. 1973: Three approaches to the study of parent-child interaction: ethological, interview and experimental. *Journal of Child Psychology and Psychiatry*, 14, 1-17.

—— 1980: *Parent-Child Interaction*. New York: Plenum Press.

Maccoby, E. 1980: *Social Development: Psychological Growth and the Parent-Child Relationship*. New York: Harcourt, Brace and Jovanovich, 87-9.

Martin, J. A, Maccoby, E., Baron, K. and Jacklin, C. N. 1981: Sequential analysis of mother-child interaction at 18 months. A comparison of macroanalytic techniques. *Developmental Psychology*, 17, 2, 146-57.

Moltz, H. 1965: Contemporary instinct theory and the fixed action pattern. *Psychological Review*, 72, 1, 1.

Moss, H. A. 1964: Methodological issues studying 'mother-infant interaction'. *American Journal of Orthopsychiatry*, 35, 482-6.

Nachmias, D. 1976: *Research Methods in Social Sciences*. London: Edward Arnold.

Newson, J. and Newson, E. 1980: Parental punishment strategies with eleven-year old children. In N. Frude, (ed.) *Psychological Approaches to Child Abuse*, London: Batsford, 62-80.

Papousek, H. and Papousek, M. 1983: Biological basis of social interactions: implications of research for understanding behaviour deviance. *Journal of Child Psychology and Psychiatry*, 24, 1, 117-29.

Phelps, R. E. and Slater, M. A. 1985: Sequential interactions that discriminate high and low problem single mother–son dyads. *Journal of Consulting and Clinical Psychology*, 53, 5, 684–92.

Poole, T. B. 1978: An analysis of social play in polecats (*mustelidae*) with comments on the form and evolutionary history of the open mouth play face. *Animal Behaviour*, 6, 36–78.

Powell, J., Martindale, B., Kulp, S., Martindale, A. and Bauman, R. 1977: Taking a closer look: time-sampling and measurement error. *Journal of Applied Behavioural Analysis*, 10, 325–32.

Rapoport, R. and Rapoport, R. 1980: *Growing Through Life*. London: Harper and Row.

Raphael-Leff, J. 1986: Facilitators and regulators: conscious and unconscious processes in pregnancy and early motherhood. *British Journal of Medical Psychology*, 59, 43–55.

Rosenblum, L. A. and Moltz, H., (eds) 1983: *Symbiosis in Parent-Offspring Interactions*. New York: John Wiley.

Sackett, G. P. 1978: Measurement in observational research. In G. P. Sackett, G. P. (ed.), *Observing Behaviour, Vol. II: Data Collection and Analysis Methods*, Baltimore: University Park Press, chapter 2.

Schaffer, H. R. (ed.), 1977: *Studies in Mother–Infant Interaction*. London: Academic Press.

Siegel, S. 1956: *Nonparametric Statistics for the Behavioural Sciences*. New York: McGraw-Hill.

Simpson, M. J. and Simpson, A. E. 1977: One–zero and scan methods of sampling behaviour. *Animal Behaviour*, 25, 726–31.

Skinner, B. F. 1953: *Science and Human Behaviour*. New York: Macmillan.

Slater, P. J. B. 1973: Describing sequences of behaviour. In P. P. G. Bateson and P. H. Klopfer (eds), *Perspectives in Ethology*, New York: Plenum Press, chapter 5.

—— 1978: Data collection. In P. W. Colgan, (ed.) *Quantitative Ethology*, Toronto: John Wiley, chapter 1.

Slater, P. J. B. and Lester, N. P. 1982: Minimising errors in splitting behaviour into bouts, *Behaviour*, 79, 153–61.

Smith, W. J. 1977: *The Behaviour of Communicating*. Cambridge, Mass.: Harvard University Press.

Stevenson, M. A. and Poole, T. B. 1976: An ethogram of the common marmoset (*Calithrix jacchus*) general behavioural repertoire. *Animal Behaviour*, 24, 2, 428–51.

Suomi, S. J. 1976: Neglect and abuse of infants by rhesus monkey mothers. *Journal of American Academy of Psychotherapy*, 12, 5–8.

Trivers, R. L. 1972: Parental investment and sexual selection. In B. G. Campbell, (ed.), *Sexual Selection and the Descent of Man 1871–1971*, Chicago: Aldine, 136–79.

—— 1974: Parent–offspring conflict. *American Zoologist*, 14, 249–64.

Tyler, S. 1979: Time-sampling: a matter of convention. *Animal Behaviour*, 27, 801–10.

Wilson, E. O. 1978: *On Human Nature*. Cambridge, Mass.: Harvard University Press.

Notes on Contributors

J. C. Berryman is Lecturer in Psychology at the Department of Adult Education, University of Leicester.

K. Browne is Lecturer in Psychology at the Medical School and Department of Psychology, University of Leicester.

K. Carson, Researcher, was formerly at the Edinburgh School of Agriculture, University of Edinburgh.

G. M. Collis is Lecturer in Psychology at the Department of Psychology, University of Warwick.

M. Herbert is Professor of Clinical Psychology and Head of the Department of Psychology at the University of Leicester, and Director of the Centre for Behavioural Work with Families.

J. D. Higley is a Research Fellow at the Laboratory of Comparative Ethology, National Institute for Child Health and Development, Bethesda, Maryland.

K. Immelmann is Professor of Biology and Head of the University of Bielefeld Research Zoo at the University of Bielefeld, West Germany. He is Chairman of the International Ornithological Committee, 1982-6, and President of the XIX International Ornithological Congress, Ottawa, 1986.

A. B. Lawrence is Specialist Adviser in Agricultural Ethology and Farm Animal Welfare at the Edinburgh School of Agriculture, University of Edinburgh.

C. Lewis is Lecturer in Developmental Psychology in the Department of Psychology, University of Reading.

H. A. Moser is a Researcher and was formerly at the Edinburgh School of Agriculture, University of Edinburgh.

O. A. E. Rasa is Associate Professor of Ethology at the Department of Zoology, University of Pretoria, South Africa.

H. R. Schaffer is Professor of Psychology at the Department of Psychology, University of Strathclyde.

M. Shaw is Senior Lecturer at the School of Social Work, University of Leicester.

Notes on Contributors

W. Sluckin was formerly Professor of Psychology and Head of Department at the Department of Psychology, University of Leicester.

R. Sossinka is Professor of Zoology in the Faculty of Biology, University of Bielefeld, West Germany.

S. J. Suomi is Head of the Laboratory of Comparative Ethology at the National Institute for Child Health and Development, Bethesda, Maryland.

D. G. M. Wood-Gush is Honorary Professor at the Edinburgh School of Agriculture, University of Edinburgh.

Index

acceptance/rejection, in parental behaviour, 369
accessibility/ignoring, in parental behaviour, 369
acknowledgement-of-difference strategies, 271–2
ad libitum sampling, 350
adoption, 269–78; and bonding, 223; in cows, 96
adoption agencies, 274–5
adoptive parents, 271–74
affectionate behaviour, 211
affectionate relationships, characteristics of, 366
affiliative interactions, 171
age: of adopted children, 276–7; of 'aunts', 187; of child, and maternal speech, 297; and paternal involvement in child care, 239
aggression: in children, and parental behaviour, 326, 332, 335; inter-sibling, in carnivores, 143
agonistic buffering, 183
Ainsworth, M. D. S., 291–2; on sensitivity, 286–7, 289, 299, 309; strange situation procedure, 230, 366–8, 369
alloparental care, 32, 128, 183–8, 190; *see also* aunting behaviour
altricial species, 22–3, 46, 152–3; *see also* precocity
altruism: and communal care, 31, 32, 74; and parental behaviour, 4–5; and preferential adopters, 275; *see also* helpers
ambivalence, in infants, 360, 361, 362, 368

ambivalence, period of, in primates, 162
anti-predator mechanism, 99, 100
apes, parental care in, 157, 159, 170–3, 178–9, 212–13
appropriateness of response, 287, 288–9, 292–3
arousal modulation, 161
assimilation and accommodation, concepts of, 328–9
attachment theory, 218, 229, 361, 366
aunting behaviour, 1, 155, 170, 184–8
authoritative mother, ideal of, 329
autism and bonding, 222

badgers, parental care in, 120–1
Barash, D. P., 69
bathing of children, by fathers, 242, 248–9
bears, parental care in, 123, 145
begging behaviour, *see* food begging
behaviour, description and analysis of, 2–5, 348–53
behavioural catalogue, 349–50
behavioural flexibility, 153, 171
behavioural timer, 352
Bell, S. M., 291–2, 309
Belsky, J., 370
'Benefactress of Children' foster parents, 266, 275
binomial test, in sequential analysis, 356–8
birds, parental behaviour in, 3, 6, 8–38; compared to mammals, 8–10, 15, 20–1, 34, 35, 37–8
birth, *see* parturition

Index 379

birth families, and substitute parents, 262, 276–7
birth parents, 261, 263, 264, 272
birth site, preparation of, 119, 121
bobcats, parental care in, 130
bonding doctrine, 210–17, 221–4, 229, 334, 340
bottle feeding, 209
bout analysis, 354
Bowlby, J., 218, 229, 232, 366
breast feeding, 209, 249
Brew, M. F., 337
brood-parasites, 3, 20
brood patch, 9, 17, 18
brooding, 23, 28
Browne, K. D., 353, 360
bunting, 101
burrows, carnivore, 119

caesarean sections, and bonding, 221
canids, parental behaviour of, 131–2
cannibalism, 48, 124–5, 127, 128, 145
care-giving behaviour, defined, 2
Carlsson, S. G., 212
carnivores: evolution of, 116–17; parental behaviour of, 118, 147; see also maternal behaviour, in carnivores
case history, form of, 341
Cater, J. I., 215, 222
cattle, maternal behaviour of, 90–6, 107
Cautley, P. W., 267, 268
chained responses, 355–60 passim
chemical signals, see communication, olfactory
Chess, Stella, 216, 224, 322
child abuse, 324–5, 330–41, 333, 365, 371; and bonding, 215, 222–3, 340; and parent-offspring interactions, 354, 359–63
child characteristics, effect of, on parental behaviour, 284, 299
childlessness, of adoptive parents, 271, 278
children as parental property, 260, 278
chimpanzees, parental behaviour of, 163, 170, 171–2, 178, 179
chronism, 124–5
class membership, and paternal role, 239, 244

clinging, infant, 158, 161, 168
cluster analysis, 358–9
Collis, G. M., 366, 367
colonial breeding, 14, 29, 69, 134–5
coloration, as maternal elicitor, 155–6, 185
Colvin, R., 266
commitment, parental, 366
communal nesting, 18
communal rearing, 3, 60, 74; see also alloparental care; den-helper systems; familial care
communal sucking, 105
communication: acoustic, 21–2, 26, 64–6, 73; mother–infant, 25–7, 60–6, 73, 88–90, 100; olfactory, 62–4; tactile, 26, 61, 62, 95; through thermal signals, 61–2, 86; visual, 26, 66, 294; see also crying; cues; parent–offspring interactions; vocalizations
compensation by parents, for child's deficiencies, 306–7
complete sample coverage, 349–50
complimentary care, 173, 174, 190
concaveation, 47–8, 72
conditioning, and maternal attachment, 218–20, 285
confidentiality in interviews, 364
consistency of response, 287, 288
contact comfort, 7
control–autonomy dimension, 320
control function of fathering, 335
co-operation/interference, in parental care, 369
co-operative hunting, 129–30, 131, 137, 141–2, 143, 146
core teaching situations, 304
court orders, and substitute parenting, 260–1
courtship behaviour, 11
coverage indicator, 349–50
crèches, formation of, 29, 129
critical periods in development, 211–12, 216, 305; see also sensitive periods
cross-lagged correlations, 301
crying, 290–3
cues, 5: auditory, 89, 92, 160; facial, 160, 294; and feeding, 25–7; olfactory, 88, 89, 92, 98, 100,

cues (cont.)
104, 135, 167; recognition by, 66, 88, 89, 92, 98, 100; visual, 89, 92; see also communication
cultural environment: and parent–offspring interactions, 289, 292, 296; and paternal role, 180, 220, 245; and sex differences, 305
culture, transmission of, through family, 316–19
custodianship, 270

Danzinger, K., 339
data collection, methods of, 349–53, 363–5
Dawkins, R., 358, 359
deficiencies, child, 306–7, 320; see also temperamental characteristics
delinquents, 326, 327, 331
delivery attendance, and paternal role, 238–9, 242–4
dendrogram, 359
den-helper systems, 117, 118, 124–6; see also communal rearing
dens, 120, 126–7
dependent behaviours, 324, 327, 332, 335
depression, parental, 337
deprivation of needs, 334, 340, 371 and emotional development, human
developmental stage, parenthood as, 328–9
dialogues, parent–offspring interactions as, 293
Dick-Read, Dr G., *Childbirth without Fear*, 209
dihydrotestosterone, 19
directiveness, maternal, 298, 304
disadvantage, cycles of, 284, 334
dispersal of young, *see* social contact
distraction display, 32–3
distress signals, 360, 362
domesticated species, behaviour of, 48, 106–8
domestic cats, parental behaviour of, 128–9
dominance hierarchies, 132, 133, 165, 186–7
double-bind situations, 340
double-blind experiments, 215

Drummond, H., 348
Dudley, D., 71
Dunn, J. B., 211
duration of group dependency, 146
duration of maternal dependency, 118, 123, 131, 142–7
duration of parental care, 10, 33, 152–3, 186–7
dysfunctional behaviour, learning of, 318; *see also* social and emotional development, human

Easton, P. M., 215, 222
ecological factors: and mode of social organization, 3–4, 141–2; and period of maternal dependency, 147; *see also* habitats
ecological perspective, on parental behaviour, 370
Edwards, S. A., 91–2
egg arrangement, exchange of, 16
egg-shell removal, 22
egg-tooth, 21
ego, development of, 339
Eibl-Eibesfeldt, I., 357
Elwood, R. W., 46, 54, 69
Emerson, P., 230–1
embryo, avian, 9–10, 21–2; *see also* foetus, equine
emotional abuse of children, 220
employment patterns, and paternal role, 240–1, 244, 247, 250
engrossment, paternal, 229–30, 249
entitlement, for adoptive parents, 273
enurination over young, 64
environment, interest of, 163
environmental conditions, 308, 347; *see also* cultural environment; social factors
Equidae, parental care in, 99–102
'Erbkoodination', 17
Erikson, E. H., 328, 329
ermine, parental behaviour of, 120
Espmark, Y., 97
ethological methods, 346–63, 370–1
evolution: and aunting, 187–8; of carnivores, 116–17; convergent, 38; and parental care, 38, 317; *see also* phylogeny; selection pressures
expectations, in foster care, 266, 269

experience, parental: and parental behaviour in animals, 25, 88, 107, 176, 181; and responsiveness in human interaction, 295-6
experience of childhood, and parental care, 238, 305, 325, 328, 334
exposure learning, and maternal attachment, 217-18, 219

face-to-face interactions, 293-6
facial stimulation, 160, 294
facilitator, parent as, 369
faeces, consumption of: by mother, 63-4; by offspring, 120, 128
familial care, 131-4, 137-41, 147, 176-7; see also alloparental care; communal rearing; den-helper systems
family, 250, 341-42; cohesion in, 337-8; life-cycle of, 239, 331; as transmitter of culture, 316-19
family-based societies, 121, 125, 131-4, 136
family structures, human: diversity of, 233-4, 259, 317, 330; in foster care, 260-2; pathological, 338-40; role of child in, 330; role of father in, 228, 233-4, 242-9, 253
Fanshel, D., 265, 266, 267, 273
farrowing, see parturition
fat reserves, infant, 144, 145
father, human: and child's psychological development, 229, 232, 252, 335; involvement of, in child care, 228, 234-49; psychological state of, 249-50; role of, 228-9, 251-3
feedback, use of, 301
feeding: in birds, 6, 24-7, 29; in humans, 223, 242, 248, 283
female groups, 127-8, 134-5, 137
feminism, view of paternal role, 246-7
Field, T., 233
financial pressures, and paternal role, 247
fine tuning, parental, 297-8, 300-1, 307
fledging, 28-31
focal individual sampling, 351

foetus, equine, 99; see also embryo, avian
following reactions, 37, 97-8, 99, 127
food begging, 25-7, 125-6, 132, 171
food caching, 132
food protection, 142
food provision: in birds, 10; in carnivores, 119-32 *passim*, 141-3; see also prey-capture techniques
food-sharing, 133, 142, 171, 177
foraging, 167
foster care, 261-71, 277-8
foster children, 261, 263, 264, 266, 269
foster parent attitude test, 267
foster parents, 3, 223, 263-5, 265-70; father and mother compared, 265, 267-8
frequency analysis, 353-4
Freud, Sigmund, 341
Fullerton, C., 57

gatherers, 117, 119, 123, 126, 141, 143
genetic advantage of communal care, 4-5, 74, 137-41; see also evolution; selection pressures
genetic determination: of behaviour, 346-7; in child development, 320; of maternal behaviour, 107, 215; of paternal bond, 218; see also selection pressures
gland development, and self-licking, in rats, 47
goats, maternal behaviour of, 85-6, 88
gonadotrophic hormones, 19, 53
gorillas, parental care in, 170, 171-2
green acouchi, parturition of, 49
Greenberg, M., 229-30
grooming, 47, 48, 164
Grota, L. J., 54
group care situations, 307-8
group-living, in monkeys, 169-70
group size, in hunting, 142
growth rates, 106, 166-7; see also maturation
'guarding', learning of, 126; see also protection
guiding behaviour, 28
guinea pigs, parental behaviour of, 49-50, 57, 65-6, 70

habitats, adaptations to, 45–6, 99; *see also* ecological factors
'hammer and anvil' behaviour, 121, 144
hamsters, parental behaviour of, 48, 49, 52, 53, 55, 72
Hansen frequencies, 351–3
hard-to-place children, 274–7
harem system, in pinnepeds, 134–6
Harlow, H. F., 154, 158, 161, 215
harriers, 117, 126–7, 131, 141–2, 143
hatching, 16, 18–19, 21–2
heat, provision of, 9, 10–11, 17, 65, 71–2, 73, 104; *see also* temperature, regulation of
helpers: in birds, 31–2; in canids, 137–41; in felids, 128, 129, 130–1; human fathers as, 249; *see also* aunting behaviour; den-helper systems; familial care
hierarchy of behavioural acts, 358–9
Higley, J. D., 155
Hinde, R. A., 6, 7, 347, 365, 366
homiothermic environment, *see* heat, provision of
hormonal influences on parental behaviour: in birds, 9, 11, 19; in humans, 305; and placentaphagia, 157; in primates, 154, 176; in rodents, 48–9, 50–3, 57, 72, 73; in ungulates, 86–8, 93, 102, 107
hospitalization, effect of, on child care, 244, 245
howler monkeys, parental care in, 170
humidity, regulation of, in incubation, 15–16
hunting, *see* co-operative hunting; prey-capture techniques; solitary hunter type parental care
hunting, filial participation in: in cats and lions, 128–9, 130, 146; in other carnivores, 120, 123, 125–6, 127, 133
hunting dogs, parental behaviour of, 133–4, 142, 145–6
hyenas, parental behaviour of, 126–7, 145
hysterectomy, and maternal behaviour, 50–1

hystricomorphs, parental behaviour of, 45–6, 48, 55–7, 64, 65–6, 69–70, 74

imprinting: filial, 2, 10, 30, 35, 91, 100, 217; maternal, 87–8, 92–3, 98, 212, 213
inclusive and exclusive foster care, 264
inclusive fitness of helpers, 32, 74, 137–41
incubation, 6, 15–20, 35–7
indirect reports, 363, 365
infancy, period of, in primates, 170–1, 189
infant behaviour, as maternal elicitor: in primates, 154–8, 212–13; in rodents, 52–66, 72–3; in ungulates, 88, 89; *see also* communication
infant carrying, 164, 167–8, 170
infant dress, 209
infanticide: in carnivores, 117, 118, 124; human, 324, 337; parental, in rodents, 48, 68, 69–70
infant independence, 162–3
infantilization, by mother, 332–3
information, transfer of, in parent-offspring interaction, 283, 290–1
'injury feigning', 33
inspections, maternal, 160
instantaneous and scan sampling, 352
intellectual development, father's role in, 232
intensity level of parental feeling, 322–3
interaction frequencies, 353–5
interactionist perspective, 287, 319, 348
interference, parental, 323, 369
inter-species comparisons, 2, 6–7, 160, 186–7, 189–90
interview method, 234–6, 363–5
intra-sexual tolerance, 127–8, 129, 130–1, 147
isolation-seeking, in parturient ungulates, 86, 91, 97, 98
Itani, J., 180

Jakubowski, M., 48
Japanese macaques, parental behaviour of, 163, 179, 180
joint enterprise contexts, 301–4
juvenile mortality, 118–19, 130

Kempe, C. H., 325
Kempe, R. S., 325
Kennell, J. H., 210, 211, 213–15, 216–17, 229, 305
kinesthetic stimulation, 161
kin selection, and helping, 137–41, 188
Kirk, H. D., 271–2
Klaus, M. H., 210, 211, 213, 214, 216–17, 229, 305
Kleiman, D. G., 56
Klopfer, P. H., 212
Konner, M. J., 292

labelling, 88, 157–8, 212
laboratory conditions, and study of behaviour, 67–8, 69, 348
labour: and motherly love, 208–9; prolongation of, in ungulates, 107; *see also* parturition
lactation, 95, 101, 105, 144
Lamb, M. E., 216, 224, 233, 300
language development, 296–8, 300–1, 303–4, 339–40
langurs, parental behaviour of, 165, 179
law on parenthood and parental rights, 257–9
learning: in birds, 10, 33–5; and establishment of maternal bond, 215–17; of hunting techniques, 129, 146; in mongooses, 126; of parental behaviour, 2, 5, 107, 245, 346; parental influence on, 163–4, 288, 318; and periods of dependency, 144–7
Leboyer, F., 210
legal parents, 261
Lehrman, D. S., 7, 53, 54
Leiderman, P. H., 216, 223
Leifer, A. D., 213
lemurs, male care in, 175
LeNeindre, P., 86–8
Leon, M., 62, 63

lesion studies, on control of parental behaviour, 154
Levy, D. M., 332
licking: and gland development, 47; as labelling, 157–8; postpartum, in ungulates, 91, 97, 98, 100; and sanitary needs of young, 57, 62, 120
Lidz, T., 338
life-span approach, 370
lions, parental behaviour of, 129–30, 141–2, 143, 145–6
lipsmacking, 159, 160
litter size, 55, 57, 118, 124, 131, 144
local authorities, as legal parents, 260–1, 263
locomotion, infant, 160–1, 169, 171–2

Maccoby, E., 369
Madaley, R., 353
male care, in primates, 173–83, 190, 220
male paternal certainty, 182–3
mammals, parental care in, 3, 8–10, 15, 20–1, 32, 33, 37–8
manic-depressive parents, 337
Margison, F. R., 336, 337
'marital skew', 338–9
Markov chain analysis, 356
mate desertion, 46, 73
maternal attachment, human, 208, 210–19, 221, 336–7; compared to paternal, 217, 229–31; security of, 299–301, 368; *see also* bonding doctrine; over-protective attachment
maternal behaviour: in birds, 14–15, 18, 21, 22; in carnivores, 118–19, 120–2, 128, 132, 134–6, 147; 'male', *see* substitutive care; in primates, 154–73, 188, 189, 212–13; psychological advantages of, 247–8; in rodents, 44–5, 47–66, 71–3; stimulation of, by offspring, *see* infant behaviour; and survival of mother, 118–19; in ungulates, 85–108, 212–13
maternal bonds: in apes, 212–13; in humans, *see* bonding doctrine

maternal bonds (*cont.*)
and maternal attachment; in ungulates, 90, 102, 106, 108
maternal dependency, period of, *see* duration of maternal dependency; infancy, period of; social contact, maintenance of
maternal influence on behaviour, 90
maternal rejection and punishment: in humans, 324, 333–5, 337, 368, 369; in primates, 162–3; *see also* parental rejection
mating systems, 9, 14, 120, 121, 134
maturation: and presence of adult, 61–2, 71–2; rate of, and parental care, 10, 35, 99, 100, 123, 144, 146, 152–3, 166–7; sexual, 118, 122, 166–7
mechanical treatment of eggs, 15, 16
megapodes, parental behaviour of, 3, 21, 35–7
methodology, 211, 215, 232–3, 319, 347–71
mice, maternal behaviour of, 48, 49, 51–2, 53, 54
microcomputers, use of, in data collection, 352–3
Mischel, W., 319
mismatching of responses, 296
Mitchell, G., 173
mock hunts, 146
modelling of behaviour, 334, 335
mole rats, parental behaviour of, 60, 71
Moltz, H., 62, 63, 346–7
mongooses, parental behaviour of, 124–6
monkeys, New World, parental care in, 168–70, 175–9, 182
monkeys, Old World, parental care in, 159–66, 171, 178–9, 186
monogamy, 1, 9, 14, 120; *see also* pair bonding
monotropy, 229, 233
morphological adaptations, 27
Morris, N., 229–30
mother–child contact: human, 209–10, 213–17, 305, 332; in monkeys, 160; in rodents, 61; *see also* clinging, infant; skin-to-skin contact; social contact, maintenance of
mother–daughter bonds, 129
motherese, 297–8, 300–1, 303–4
motherly love, 217, 218, 219, 324
mounds, constructed by megapodes, 35–7
mustelids, parental care in, 120–2
mutual replaceability cluster analysis, 358–9
myomorph rodents, parental behaviour of, 45, 47–8, 53, 68–9, 72–3, 74; and communication, 62, 64–5

nappy changing, by father, 242, 246–7
naso-nasal contact, 104
needs, children's, 323, 334
nest, functions of, 10–11
nest-building: in birds, 6, 9, 10–15; in nocturnal prosimians, 167; in pigs, 103; in rodents, 48, 49, 51–2, 53–4, 55–6; *see also* birth site, preparation of
nesting behaviour, 61
nest sanitation, 27–8
neurological controls for maternal care, 154–5
nipple attachment, 64
nipple location: in mice, 64; in ungulates, 86, 91–2, 97, 100–1, 104, 107
Noirot, E., 54
nomadism, 124, 125, 133, 141
non-organic failure-to-thrive syndrome, 221–2
'normal' families, 259, 273–4
nosing, 105
nursing behaviour: and maternal bonding in humans, 214–15; in monkeys, 160; in ungulates, 85, 89, 95–6, 97, 98, 101, 105; *see also* breast feeding; suckling behaviour
nurturant functions, 283; *see also* survival functions

Oakley, A., 247, 250
observational data, collection and analysis of, 347–64

oestradiol, 50, 51, 52
oestrogen: in birds, 19; in rodents, 50–1, 52, 72; in ungulates, 86–7, 88, 93
oestrus: after infanticide, in lions, 130; lactational, 102; postpartum, 53, 57, 93; 'silent', 107
one-zero sampling, 351–2
ontogeny of parental behaviour, 2–3, 5–6, 346–7
orang-utans, parental care in, 170, 171–2
Orford, J., 337–8
origins of adopted children, 272–3
otters, parental behaviour of, 121
over-protective attachment, 220, 322–3, 332–3
own-body substances, feeding of, 24–5
oxygen acquisition, of embryo, 20–1
oxytocin, 95, 101

packs, as social system, 131–4, 144
pain-killing drugs, use of, during labour, 208–9
pair bonding, 131, 137, 147, 182; see also monogamy
pandas, parental care in, 122
parent, definition of, 261
parental agreement, substitutive care by, 258, 259
parental attachment, 2, 208, 331, 366–9; see also maternal attachment
parental behaviour, human: defined, 2, 208, 320; determinants of, 304–8, 328–9; effects of, on development of child, 283–6, 299–304, 319, 331–42; historical variation in, 223; interactionist perspective on, 286, 319–20, 327–8; pathology in, see pathology; responsiveness in, 286–98, 304–8; and socialization, 283, 318–20, 325–8; study of, 346, 349, 370–1
parental care: and developmental taxonomy, 22–3; duration of, 10, 118, 123, 131, 142–7, 152–3, 186–7; in primates, 152–3, 186–7, 189–90; in rodents, 73–4; stimulation of, 5
parental coalition, 338
parental education, 317, 341
parental investment, concept of, 347
parental rejection, 220, 323–6, 331, 335; see also maternal rejection
parental rights, 260, 270
parental satisfaction, in adoption, 273–4
parenting parents, 261
parenting stress index, 370
parent-offspring interactions: in birds, 35, 37; in primates, 153, 158–9; and social development of primates, 161–2, 171–3; see also communication; mother-child contact
parent-offspring interactions, human: and bonding, 213–14; and deficiencies in child, 306–7, 320; paternal, 228, 229, 231–4, 249, 250–2; and responsiveness, 290–8, 308–9; and social development of child, 283, 285–9, 299–304, 309, 319–20, 330–1, 339, 361–2; study of, 1–2, 347–8, 351, 353–63, 365–6, 368–9
'parking', 167
participant observation, 364
parturition: assistance with, 6, 49; in birds, 20–1; in carnivores, 120, 121, 134; in rodents, 48–50; in ungulates, 86, 91, 97, 98, 99, 103, 107; see also hatching; labour
paternal attachment, human, 219–20, 249–53; compared to maternal attachment, 229–31, 252–3
paternal behaviour: in birds, 34–5; in carnivores, 118, 124; in humans, see father; in New World monkeys, 168; in primates, see male care; in rodents, 44–5, 46, 67–72, 73–4; and social organization, 46, 68, 102; in ungulates, 85; see also sexual division of care-giving
paternity leave, 244
'pathological mothers', 322, 327

386 Index

pathology: in interactions, 354, 359–63; in parental behaviour, 4, 220, 320–8, 329–41, 370, 371
Patterns of Patterns algorithm method, 358
peer interactions, child, 302–3
permissiveness, parental, 326–7
personality: adult, and parental functioning, 180, 185, 333, 340, 370; development of, 223, 328–9, and interactive process, 296; sensitivity as trait of, 287, 304
'pheromonal bond', 62–4, 73
phylogeny: and explanation of behaviour, 2–3, 4–5; of nest and mound building, 11, 36–7; see also evolution; selection pressures
physically handicapped children, 277
Piaget, Jean, 326
pigs, parental behaviour of, 102–6
pigtail macaques, parental care in, 154, 162, 165, 171
pinnepeds, parental behaviour of, 134–6, 144–5
placentaphagia: in cats, 128, 129; in primates, as maternal elivitor, 156–8; in rats, 47–8; in ungulates, 97, 99–100, 103
play: in carnivores, 120; maternal, in humans, 211–12; maternal, in primates, 168, 171, 172–3; paternal, in humans, 251–3; paternal and maternal compared, 231–2, 252
Poindron, P., 86–8
polygyny and polyandry, 9, 14; see also harem system
Poole, T. B., 356, 357
population control, 118–19
power assertion, paternal, 333
precocity, 22–3; in birds, 28, 37; in carnivores, 131, 135; of primate neonates, 152; in rodents, 46; in ungulates, 99, 100, 104
predation, avoidance of, 120, 123, 125
preferential adopters, 275
premature births, 221, 306
prey-capture techniques, 3–4, 116–17, 136–7; see alos co-operative hunting; solitary hunter type parental care
Priestnall, R., 52, 53, 54
privacy, in the family, 317
probability in behavioural sequences, 355–8
procyonids, parental behaviour of, 122
professionalization of substitute parenting, 269, 278
progesterone, 19, 50–2, 53, 72, 86–7, 93
prolactin: and birds, 19, 24; and cows, 93, 95; and primates, 154, 176; and rodents, 50, 51, 52
promiscuous nursing, 56, 74
promptness of response, 287, 288
proportional frequency scores, 353
prosimians, maternal care in, 166–8, 174–5
prosocial behaviour, fostering of, 328–9
protection of young: in birds, 32–3; in carnivores, 117, 123, 125, 127, 129, 132, 135, 136; and length of maternal dependency, 145, 147; maternal, 117, 123, 125, 127, 129, 132, 135; paternal, 67, 70, 136, 174; in primates, 174; in rodents, 67, 70, 73
psychological development of child, 223, 264; see also social and emotional development, human
psychological disorders, parental, 333, 335–9
psychological factors, in paternal care, 238–9
psycho-social functions of parenting, 318, 323, 325–8
puberty, 102
punishment, use of, 326; see also maternal rejection and punishment

questionnaires, construction of, 363–5

racial identity, in adopted children, 276
rank acquisition, 166
Raphael-Leff, J., 369

rate of emission, of parental behaviours, 323
rats, parental behaviour of, 47-8, 49, 50-1, 53-4, 61-2, 72-3
Raynor, L., 274
reciprocity: in parent-offspring interactions, 285, 354, 360-2, 369; in personality development, 328-9
recognition of mother, 29-30, 100
recognition of offspring: in birds, 29, 30-1; in carnivores, 135; in primates, 156, 160, 167, 168; in rodents, 66; in ungulates, 88, 89, 92-3, 98, 100, 104
red deer, maternal behaviour of, 98-9, 107
Redican, W., 181
Redshaw, M., 214, 224
regulator, parent as, 369
regurgitation, feeding by, 24, 129-30, 132, 133, 141
reindeer, maternal behaviour of, 96-8
reinforcement, socialization by, 319-20, 334, 341
rejection-of-difference strategies, 271-2
relationships, description of, 365-6
relatives, and maternal restrictiveness, 165
reliability of observational data, 349
repetitiveness, in parental style, 294
responsiveness, parental, 230, 283, 285, 339, 354; see also sensitive responsiveness
restrictiveness, maternal, 165, 169-70, 186-7; and male care, 175, 177, 180-2
retrieval, 47, 48, 49, 53-5, 56, 62, 156
retrospective data, 340, 341
rhesus monkeys, parental behaviour of, 154, 155-6, 157, 158, 165, 180, 371
Richards, M. P., 211
'rodent run', 33
role handicap: for adoptive parents, 271, 275; for children, 330, 339
rooming-in, 209, 221
Rosenblatt, D. B., 216, 224
Rosenblatt, J. S., 49, 52, 53, 54, 154
routines, children's, 248, 362

Index 387

sampling all occasions of some behaviour, 351
sampling behaviour, techniques of, 350-2
sanitary needs of young, 57, 62
Schaffer, H. R., 230-1, 366, 367
schismatic conflict in family, 338
schizophrenia: in children, 338, 339-40; in parents, 337
sciuromorph rodents, parental care in, 45, 48, 55, 64, 65, 69, 74
sea otters, parental care in, 121-2, 144
secure base for exploration, in primates, 160-1, 169, 177, 179
security of attachment, 299-301, 368
Seidenberg, R., 337
selection pressures, 346; on birds, 8, 14; in domesticated species, 90-1, 107-8; and maternal bonding period, 211; on parental behaviour, 188, 190; for recognition mechanisms, 29; see also evolution; genetic advantage
'selfish gene' notion, 74, 345
self reports, 363, 365
sense of effectance, in child, 288
sensitive periods: for bonding, 214, 216; for learning, 285; see also critical periods
sensitive responsiveness, 286-309, 368, 369; determinants of, 304-8; lack of, and pathologies, 360-2; see also responsiveness, parental
separation of parent and child, 263, 334
sequence sampling, 351
sequential analysis, 354-8, 359-63
serum cortisol, 154
Seward, G. H., 57
Seward, J. P., 57
sex differences: and alloparental care in primates, 184; in child abuse, 324-5; in foster parents, 267-8; in parental attachments, 217, 229-31, 252-3, 324; in parent-child interactions, 231-3, 252, 295-6, 304-5
sex roles, 232, 238
sexual division of care-giving, 1, 6; in

sexual division (*cont.*)
  birds, 9, 14–15, 17–18, 23, 24; in carnivores, 130, 131–2, 136; in humans, 228, 234–49; in monkeys, 168; *see also* paternal behaviour
shared parenting, 262, 264
sheep, maternal behaviour of, 85–90, 107
shelter, provision of, 9, 17
shelter-seeking, 86
Siegel, H. I., 49, 52, 154
sifakas, male care in, 175
simulation experiments, 296
single people as adopters, 274
skin-to-skin contact, 210, 214, 215, 221, 223; *see also* mother–child contact
skunks, parental behaviour of, 121
social and emotional development, human: father's role in, 229, 232, 252, 335; and overprotective attachment, 220; and parental behaviour, 283–9; and parent–offspring interactions, 286–9, 299–304, 309, 319–20, 330–1, 339, 361–2; *see also* psychological development
social and emotional development, primate, 153, 161–3, 171–3, 188–9
social companions, 166
social contact, maintenance of: in group hunting carnivores, 129, 130, 143; in monkeys, 160–1, 162–3; in rodents, 120; in solitary hunting carnivores, 120–4 *passim*, 135, 142–3; in ungulates, 89–90, 96, 98–9, 100, 102, 104
social factors: in paternal role, 228, 239–41; in pathological parenting, 370; *see also* cultural environment
social interactions: between adults, and paternal care, 69, 70; description of, 365–6; between parent and offspring, *see* parent–offspring interactions
socialization, 283, 318–20, 325–8
social organization, forms of: analysis of, 357; and avoidance of predation, 120, 124, 125, 147; co-operative and group, 131–4, 137–47, 169–70; in *Equidae*, 102; and intra-sexual tolerance, 127–31; and prey-capture techniques, 116–17, 127–47; in rodents, 45–6, 68
social status, 183
social stimulation, provision of, 161–2, 294–9
social workers, 263–4, 267–8, 323
sociomatrices, analysis of, 359
sociometric matrix completion, 350
software packages, for research use, 353
solid food, ingestion of, 106
solitariness, and male care, 174
solitary hunter type parental care, 120–4, 126–7, 134–5, 136–7, 147
song learning, 34–5
Spock, Dr Benjamin, 209, 217
squirrel monkeys, parental care in, 154, 156, 169–70, 178, 179, 187
stalking hunters, 116–17, 119, 120, 127, 146
state-dependent responsiveness, 282
state of security and trust, 160–1
Stevenson, M. A., 354
strange situation procedure, 230–1, 299–300, 367–8
subject reactivity, 364
substitutive care: in carnivores, 132; in primates, 173–4, 175, 177, 179, 181, 190; in rodents, 68, 73–4
suckling behaviour, 3; in carnivores, 119, 120, 122, 127, 135, 141–2, 145; failure of, and undernourishment, 92, 97; in rodents, 56, 57; in ungulates, 85, 92, 95, 100, 101, 104–6; *see also* nursing behaviour
Suomi, S. J., 371
support systems for parents, 318, 338, 370
survey method, 364
survival functions, of parental behaviour, 318, 323–5
survival value: of family as insti-

tution, 316; and function of behaviour, 3–4, 346; of maternal failure, 118–19; of parent-offspring interactions, 347–8
Svejda, M. J., 215
synchronous breeding, 102, 122

tactile signals, 26, 61, 62, 95
taxonomy of developmental stages, 22–3
teaching, maternal: of guarding, in mongooses, 126; of hunting, in lions, 146; of language, see motherese; of locomotion, in apes, 171–2; and sensitivity, 301
teat order, 104, 105–6
teat ownership, 56
temperamental characteristics of child, 330–1
temperature, regulation of, 15, 16, 23, 36, 61; see also heat, provision of
Terkel, J., 48
territoriality, 118, 120
testosterone, 19, 68, 176
thermal signals, 61–2, 86
Thomas, A., 216–24
Tizard, B., 223
tolerance, male, of infants, 175, 177, 178–9
topic sharing, 301, 302, 361–2
'Tragestarre', 120, 125
transition matrix, 355–9
transport of young, 28–9
transracial adoption, 275–6
Trasler, G., 266
treatment, of abusive families, 362–3
treatment-oriented foster home, 262
tree shrews, maternal care of, 166
'tree-sitting', 167

ultransonic calling, 65
unavailability of fathers, 248–9

variation: in paternal roles, 238–40; in ungulate parental behaviour, 106–7
ventilation, regulation of, 15–16
verbal reference, in parent–child interactions, 302
violence towards offspring, human, 324–5, 326, 331, 333, 337
visual co-ordination, in parent–child interactions, 302
viverrids, parental care in, 123–6
viviparity, lack of, in birds, 8–10
vocal exchanges, human, 294–8, 307, 339,
vocalizations, 26, 64–6, 92; and danger of predation; learning of, 34–5; parental, 231; recognition by, 89, 100, 104, 135, 156
voice types, 89
Vygotsky, L., 303

walruses, parental care in, 136, 144
warmth–hostility dimension of parental behaviour, 320, 327
Wasz-Hockert. O., 290, 291
watchfulness, parental, 293–4, 309
water support, in birds, 27
weaning: in primates, 162–3, 171; and psychological development of children, 223; in rodents, 56; in ungulates, 85, 90, 98, 101–2, 106
Wilson, E. O., 346
winter lethargy, 123, 145, 147
Wolff, P. H., 290, 291
wolves, parental behaviour of, 132–3, 145–6

Yarrow, Marian Radke, *Child Rearing*, 342
Yogman, M., 231

zone of proximal development, 303

Index by Ken Hirschkop

HQ755.8 .P377 1986

A87009211401481C

WITHDRAWN
From Bertrand Library